Herman Strecker

Lepidoptera, rhopaloceres and heteroceres, indigenous and exotic

with descriptions and colored illustrations

Herman Strecker

Lepidoptera, rhopaloceres and heteroceres, indigenous and exotic
with descriptions and colored illustrations

ISBN/EAN: 9783744740531

Printed in Europe, USA, Canada, Australia, Japan

Cover: Foto ©berggeist007 / pixelio.de

More available books at **www.hansebooks.com**

LEPIDOPTERA,

RHOPALOCERES AND HETEROCERES,

INDIGENOUS AND EXOTIC;

WITH

Descriptions and Colored Illustrations,

BY

HERMAN STRECKER.

Reading, Pa., January 1, 1872.

Reading, Pa.:
OWEN'S STEAM BOOK AND JOB PRINTING OFFICE, 515 COURT STREET,
1872.

ADVERTISEMENT.

Not only is our own, and the Lepidopterous fauna of other countries, constantly receiving numerous additions, but innumerable species of already described Lepidoptera, both native and exotic, have never been figured, and as a consequent result are in very many instances unknown to the student. Having, during the course of over twenty years' study and collecting, amassed a great quantity of material, I have concluded to endeavor to carry out my long-cherished idea of publishing illustrations and descriptions of such undescribed species as I may possess or have access to. The number is immense, embracing many uniques, types, and other great rarities of the North American fauna; also to figure such species as have been heretofore described but not figured. I do not intend to confine myself strictly to North American species, but will illustrate new or unfigured species from any other part of the world—as the value of exotics for purposes of comparison, I think no one will dispute. Inasmuch as nature, or nature's God, did not divide the earth into kingdoms, counties, or townships, I don't see why we should do so in scientific matters. For my part, I consider an example from Europe, Africa, or elsewhere, as interesting an object of study as a North American one; but, of course, as it happened to be my luck to be born on that portion of the earth's mud, stones, and water, called America, I will give precedence to the species of this country. It is my purpose to issue one number every month—each number to have one plate; but where the size of the species will admit, I will put as many figures on one plate as possible. Where they are of small size, I will give sometimes as high as a dozen species on the same plate, but only less than two when the size is very large—as in the first illustration, where one necessarily occupied the whole page. Of course upper and under surface will always be given, and both sexes with larvæ and stages of transformation whenever possible. The figures will all be drawn and colored by myself from nature. My wish is to secure a sufficient number of subscribers, at as early a date as possible, to pay the expenses of printing and paper. The hope of being even in a small way useful in this my favorite science will be more than sufficient reward for my own trouble and labor. As soon as I can secure an adequate number of subscribers, I will add one other plate to each number, without increasing the price—which is fifty cents per number. I will always give as complete technical description and history of species as possible, also such observations or facts as I may deem of general interest to the Entomologist.

Trusting that the Entomological world, and friends of Science and Art generally, will feel enough interest in my undertaking to send in their names and subscription price for A. D. 1872, viz: six dollars, at as early a date as convenient.

I am yours, truly,
HERMAN STRECKER,
Box 111 Reading P. O., Berks Co., Pa.

Drawn by Herman Strecker

PLATYSAMIA GLOVERI. NOV. SP.

MALE. Expands 4½ inches.

Antennæ pectinated. Head and thorax dark brownish red; collar white; abdomen alternately banded with dark brown and white.

Upper surface, primaries, inner half dark reddish brown, with a dirty white band near the base, and a small white oval discal spot bordered on the outer and inner sides with black; the dark red colored space is bounded outwardly by a white transverse band extending from the costa to the interior margin, whitening at the latter termination; the space immediately beyond the white band, occupying fully two-thirds of the exterior half of the surface, is dark grey, composed of black and white scales, the exterior portion being the darkest; the remaining space between this and the exterior margin is dirty grey, traversed by a serpentine black line; a short space in from the apex on the costa is a small black spot, from which a zigzag white line extends a short distance to another larger black spot of an oval shape, which has within it a pale blue crescent; the space from this latter black spot to the costa, and from the zigzag line almost to the white transverse band is violet.

Secondaries have a white basal patch, also a white transverse band, as in Cecropia, the color between this band and the base is dark brownish red; in the centre is a moderately large lunate discal spot, white edged with black; of the remaining space, between the transverse band and the exterior margin, the inner two-thirds is composed of black and white scales, as in the primaries, and the outer third is light, dirty grey, which latter is divided by a black line, on the inside of which line, running parallel with it, is a row of irregularly shaped black spots.

Under surface same as above, with the exception of the ground color of the secondaries between the transverse bands and base, which is composed of black and white scales instead of being dark red as on the upper surface.

FEMALE. Expands 5½ inches.

Antennæ not as broadly pectinated as in the male. Head and thorax brownish or brick red, much the same as in ordinary forms of Cecropia; abdomen alternately banded with the same color and white.*

On the primaries the transverse white band is somewhat irregular, and rounded outwardly; the basal band also rounded; the discal spot larger than in male, and slightly lunate; the ground color between the transverse band and base is brownish red, of a slightly different tint than the thorax; the portion beyond the transverse band, light dirty grey as far as the serpentine line, the inner third of it powdered with a band of reddish scales, broader near the exterior margin, and narrowing towards the costa; margin outside of serpentine line dirty white, shaded slightly exteriorly; the black apical spots and violet patch same as in male.

* As the bodies of my examples were much rubbed when I received them, some allowance must be made for this portion of my description.

PLATYSAMIA GLOVERI.

Secondaries. Color same as in primaries; ornamentation same as in male, except the discal spot which is double the size.

Under surface marked same as the upper; coloration same as in under surface of male.

Habitat. Arizona.

This species I have named in honor of that most indefatigable of all hard working naturalists, Prof. Townend Glover, of Washington, D. C., who first showed me examples of it, which were said to have come from lower California, but as L. Weidemeyerii, Parn. Smintheus, and other northern Montane species were sent along in company with it, I expressed my doubts regarding that locality, which have since been confirmed by my receiving a female example from Arizona.

Of the distinctness of this species from P. Euryale, Boisd., (P. Californica Grote,) and P. Cecropia, there cannot be a particle of doubt, but what relation it may bear to P. Columbia, Smith, I am unable to say, (though the wide difference in locality convinces me they are distinct,) as I believe no figure has yet been published of the latter species, and Prof. S. J. Smith's types are all contained in some institution in Massachusetts, and the species must consequently remain a blank to the scientific world, until some one gives a figure of it. Apropos to this giving descriptions without figures, I may say that we Americans certainly occupy a most enviable pre-eminence; if we had more figures and fewer descriptions there would be, I have little doubt, more satisfaction and considerably less confusion among scientists. The idea of expecting anything short of the supernatural to identify a Lycaena, Hesperia, or any of the smaller noctuidae, by a mere description, is preposterous. Why, even larger species cannot thus be identified. I would like to see the Entomologist who could, by any description, identify or separate from each other Vanessa Polychloros, Californica, Xanthomelas, and Jchnusa—yet although probably sprung from the same root, they are different in appearance when placed side by side, and exist in localities widely remote from each other. I would say the same of Van. J. Album and V. Album, or of many of the Coliades. Many and many a time have I, when a whole evening was wasted, trying with aching head to find out whether some little butterfly was something or something else, consigned the discoverer of the species in question to all kinds of unspeakable torments. Here is a sample of the result of this state of affairs, as shown by the latest synopsis of North American Butterflies, by W. H. Edwards: Thecla Humuli, Harris, is Melinus, Hub., Favonius, Bois. & Lec., and Hyperici, Bois. & Lec., T. Edwardsii, Saunders is Falacer, Harris and Calanus, Grote & Robinson, while Calanus, Hub., is Falacer, Godt. T. Viridis, Edwards, is Dumetorum, Boisd. T. Henrici, Grote, is Arsace, Boise & Lec. Lycaena Anna, Edwards, is Cajona, Reakirt, Argyrotoxus, Behr, and Philemon, Boisd.

The female of Colias Eurydice has been in its time Gonepteryx Rhamni, Gonepteryx Lorquinii, and lastly Megonostoma Helena, Male! and if the brilliant colored male had not been at last coupled to his plain spouse, heaven knows what she would not have been. I might go on multiplying instances ad libitum, but until the descriptions of species are accompanied by correct figures, every new species described will but add to confusion confounded.

NOTES RELATIVE TO SOME VARIETIES OF LEPIDOPTERA.

HESPERIA POCHAHONTAS, Scudder, I am confident is a Melanotic Female variety of H. Hobomok, Harris. During the month of June, for a number of years, I have taken in some meadows, a mile or so from Reading, Pa., Hobomok male and female in large numbers; also Pochahontas at the same time and place, but in upwards of twenty years collecting, I have never yet captured or even seen a male Pochahontas, nor have I ever seen Pochahontas unless Hobomok was also in the same vicinity. A number of naturalists and collectors assure me of the same facts. Even a casual examination of the under surface will show that the markings in Hobomok and Pochahontas are the same in delineation, though not in color.

VANESSA LINTNERII, Fitch. I have of this variety of V. Antiopa, four examples, two of which were taken in the sexual act.

EUCHÆTES EGLE. If the white Euchætes Egle is an Albina var. of the slate colored one, why is it that where the white one is abundant, the grey one does not occur? On the line of the East Pennsylvania Railroad, near Reading, Pa., the white one can be taken by hundreds, but I have never met with a grey one in this county, (Berks), while Mr. Edw. Graef, of Brooklyn, N. Y., a most careful collector, says that he has taken the grey one in some localities in large numbers, but has never seen a white one in the same places.

ADVERTISEMENTS.

TO PUBLISHERS, &c., &c.

Advertisements of Publishers, Taxidermists, dealers in specimens of Natural History, Public Lecturers, Bird Fanciers, &c., inserted on the cover and fly leaves of each number, at very low rates, as my object is not to make money in this instance, but merely to cover the expenses of issuing this work. Parties choosing to avail themselves of this means of advertising will please address.

HERMAN STRECKER,
Box 111 Reading P. O., Berks Co., Pa.

The Butterflies of North America,

BY WM. H. EDWARDS.

Quarto. Published in quarterly parts, each containing 5 handsomely colored plates. Parts 1-8 now ready. Price per part, $2.50. Published by the American Entomological Society, at their Hall, 518 S. Thirteenth Street, Philadelphia, Pa.

☞ The finest work on Lepidoptera yet published in this country.

THE Canadian Entomologist,

PUBLISHED BY
The Entomological Society of Ontario,
INCORPORATED 1871.

General Editor :—The Rev. C. J. S. BETHUNE, M. A., Trinity College School, Port Hope, Ont.
Price in Canada, $1.00 per annum.
Price in United States, 1.25 "

Remittances and other business communications should be addressed to the Secretary-Treasurer of the Society, E. BAYNES REED, Esq., London, Ontario. All exchanges and articles for insertion, etc., to the General Editor, or to any member of the Editing Committee.

Remittances will be acknowledged in the ENTOMOLOGIST. Extra copies furnished when required—10 cents each; $1 per dozen.

JOHN AKHURST,
Naturalist & Taxidermist,

9 1-2 Prospect St., Brooklyn, N. Y.

Particular attention paid to preserving of all kinds of quadrupeds, birds, reptiles, &c. Native and foreign bird skins on hand.

N. B. Will always exchange Coleoptera and Lepidoptera from all parts of the world.

Trout-Dale Ponds,
NEAR
BLOOMSBURY, NEW JERSEY.

Trout of all sizes, and trout eggs, for sale. Also CONE'S HATCHING BOXES, and all necessary apparatus constantly on hand.

Fish Ponds laid out and full and complete practical INSTRUCTIONS GIVEN IN PISCICULTURE.
☞ Sites suitable for fish farms for sale.
Address, J. H. SLACK, M. D.,
Bloomsbury, New Jersey.

PUBLICATIONS
OF THE
PEABODY ACADEMY OF SCIENCE,
SALEM, MASS.

THE AMERICAN NATURALIST. A popular Illustrated Monthly Magazine of Natural History, 8 vo., 58 pages and illustrations in each number. Subscription $4.00 a year. Single numbers, 35 cents.
Vols. I. to IV., 1867-71. In numbers, $4 each; in cloth, $5.

GUIDE TO THE STUDY OF INSECTS, and a treatise on those injurious and beneficial to crops. For the use of Colleges, Farm Schools, Agriculturists, and Entomologists. By A. S. Packard, Jr. M. D., 8 vo., with eleven plates, and upwards of 650 engravings in the text. Cloth, 8 vo., $5.00.

RECORD OF AMERICAN ENTOMOLOGY. Year books of the Progress in American Entomology during 1868 and 1869, 2 vols. Price $1 each.

For any of the above publications address,
W. S. WEST, Office of the "American Naturalist,"
SALEM, MASS.

COLLECTING TOUR IN LABRADOR.

The undersigned intends to leave next spring, in the first vessel from Quebec, on a collecting tour in Labrador. Insects of all orders will be collected, so many species will be, no doubt, unique, undetermined, or new to science, those who are anxious to obtain specimens of LEPIDOPTERA and COLEOPTERA, will please communicate with me as early as possible. Terms in accordance with number and specialties.

WM. COUPER,
Montreal, Canada.

OWEN'S
Steam Book & Job Printing Office,
315 COURT STREET, READING, PA.

No. 2. PRICE 50 CENTS.

LEPIDOPTERA,

RHOPALOCERES AND HETEROCERES,

INDIGENOUS AND EXOTIC;

WITH

Descriptions and Colored Illustrations,

BY

HERMAN STRECKER.

Reading, Pa., April, 1873.

Reading, Pa.:
OWEN'S STEAM BOOK AND JOB PRINTING OFFICE, 515 COURT STREET,
1873.

NOTICE.

As none of us, unfortunately or otherwise, can always do as we please, I was unable, owing to adverse circumstances, to continue this work last year further than the first number, but I have now so arranged it as to be able to issue a part regularly each month.

Subscribers who prefer to do so can remit the money for each part as they receive it, which will be in United States, 55 cents per part, inclusive of postage. As soon as I have subscribers sufficient to pay the extra expense of printing, I will add another plate, so I trust Lepidopterists and Naturalists generally will exert themselves to increase my subscription list, which is not at present as large as that of the London Times.

TO PUBLISHERS, &c.

Advertisements of Publishers, Taxidermists, dealers in specimens of Natural History, Public Lecturers, Bird Fanciers, &c., inserted on the cover and fly leaves of each number, at very low rates, as my object is not to make money in this instance, but merely to cover the expense of issuing this work. Parties choosing to avail themselves of this means of advertising will please address,

HERMAN STRECKER,
Box 111 Reading P. O., Berks Co., Pa.

PAPILIO INDRA. Reakirt.

Proc. Ent. Soc. Phil., VI. p. 123, 1866.

MALE. Expands 3 inches.

Antennæ, head and thorax black, two small yellow spots behind the eyes, sides of collar and patagiæ dull yellow; abdomen black with a yellow dash on each side of the anal segment.

Upper surface black, primaries with a sub-marginal row of eight pale yellow lunate spots which become gradually smaller as they approach the posterior angle; also a band of nine larger spots of same color, extending from costa to inner margin, the second one having a black mark on the inner end; disco-cellular nervules defined by a yellow line.

Secondaries have the yellow band of the primaries continued; this band is divided by the black veins into seven parts, the three nearest to the costal and the two nearest to inner margin are of parallelogram form, and pretty much of one size; the two remaining parts, laying between the second sub-costal and third median veinlets are of irregular shape, extending in obtuse points beyond the line of the others; along the outer margin are five yellow spots, the one nearest the outer angle is a mere dot; the next, which is the largest, is oval, and the three remaining ones are lunate; the anal spot is large, fulvous, and encloses a black pupil; in the space between the inner band and sub-marginal spots is a series of clusters of blue scales, almost obsolete towards the costa, but becoming more distinct as they approach the interior margin, where the last and best defined one surmounts the anal ocellus; emarginations regular and pale yellow, the tail, if it deserve such a distinction, is but little more than a tooth.

Under surface, ground color, paler, that of markings much the same as on upper surface; on the primaries the sub-marginal lunules are larger than above; the inner band remains the same; on secondaries the markings of the upper side are also reproduced, with the addition of one more lunule placed between the first and second median nervules, also, the one nearest the outer angle, which is on the upper surface indicated by a mere yellow dot, is here advanced to the dignity of a respectable sized crescent, tinged in the middle with fulvous; some greyish yellow scales are in conjunction with the blue ones intermediate between the lunular and mesial bands.

Of the female I am not fortunate enough to be able to say anything, for the one reason, that, as far as I am aware of, no examples of the sex have yet turned up; all the specimens I know of being the males in the museums of the Am. Ent. Soc., W. H. Edwards and myself, which were taken by Mr. Ridings at Pike's Peak, Colorado, in 1864, nor has any collector since been lucky enough to obtain it.

Mr. Reakirt in his description of this species in the Proc. Ent. Soc. says: "I cannot reconcile this beautiful species with Dr. Boisduval's description of Pap. Aristor Godt." neither can I nor do I think it was much worth the while to say so considering that Pap. Aristor is described as a tailed species with a band of five spots, some red, some yellow on the under side of secondaries, not to consider the fact that it is a tropical species probably Central American or West Indian; he further adds in continuance " to which with Asterius it must be closely allied," it doubtless stands nearer the Asterius group than to any other, but it is distinct enough to stand on its own merits; we all know that Asterius has on the abdomen four rows of spots, two dorsal and two lateral, as also has P. Sadalus Luc., the Nov. Spec. described on the succeeding page, and some other tropical American species of whose identity I am not quite certain, whilst in P. Indra the abdomen were it not for the two small lateral dashes on the anal segment would be entirely black, moreover instead of a small round spot on each side of the collar as in Asterius, not only the sides of the collar but the whole surface of the patagiæ are yellow, of an obscure shade; in this species, on the under side there is not the slightest indication of fulvous on the inner bands as is invariably the case with Asterius and near allies; it may also be worth while to note that this is the only tailless American Papilio so far found north of Mexico.

Mr. Ridings captured at the same time with Indra, examples of Asterius, which differed in nowise from those found in other localities; in connection with this fact I would quote what Mr. W. C. Hewitson whose authority few would be inclined to dispute, says: in Proc. Zool. Soc. of London, 1859, " that two insects differing but slightly are most likely distinct species if they come from the same locality; but if they come from a distance they are most likely the same species changed by the difference of locality." As no illustration accompanies the original description, nor has any since been published, I thought it might not be amiss to head my second plate with a figure of this pretty insect. I believe I have now said all I at present know in regard to this species, which is one of the few out of the many described by Reakirt that will be able to hold its own.

PAPILIO ANTICOSTIENSIS, NOV. SP.

MALE. Expands 2¾ inches.

Antennæ, head and body black, a small yellow spot behind each eye, two larger spots of dull yellow on back of collar, patagiæ dusky yellow, four rows of yellow spots on abdomen, as in Asterius.

Upper surface black, primaries with two rows of yellow spots running parallel with the exterior margin, the outer one composed of eight spots, of which the three nearest the apex are round and nearly of a size; the next four are oval and a little larger; the last one is geminate; the spots composing the inner band are nine in number, and much larger; the one nearest the costa is oval, and not in line with the others, being nearer the discal cell; the second one is an oblong triangle, almost divided at the inner end by a black dash; the next five are also triangular, and increase in size as they near the interior margin; the eighth is the largest, and square in form; the ninth is narrow; a yellow discal bar; fringes alternately black and yellow.

Secondaries have six yellow sub-marginal lunules, the one nearest to anal angle much the smallest; also an inner band of seven yellow spots, the two nearest the costal margin almost square, the next four oblong, and the last triangular; between these two macular bands is a row of spots composed of blue atoms; anal spot, deep fulvous, edged below with yellow, and contains a black pupil; emarginations yellow; tails one-fourth of an inch in length.

Under surface dark brown, ornamentation much as above; outer row of spots on superiors larger than on upper side, paler in color, and more round in form; inner row pale fulvous, margined with light yellow; on secondaries the centres of the four outer spots between the costa and third median veinlet are fulvous; the spots comprising the inner band are also fulvous, edged on inner sides with yellow; a small yellow discal spot; space between outer and inner bands filled with greyish yellow scales, also a few blue ones nearest the inner band.

FEMALE. Expands 3½ inches.

The description of the male will apply almost equally well to the female, excepting that the ground color is not quite so dark, the inner bands are much broader and the black pupil in the anal eye, which is round in the former sex, is oblong in this; the foregoing with the figure in the accompanying plate will, I trust, be sufficient for purposes of identification, for, after all, one good figure will do more towards determining a species than any quantity of written description however careful.

Habitat. Fox Bay, Anticosti Island; Labrador.

For this species I am indebted to my valued friend Mr. Wm. Couper, of Montreal, who took several specimens of both sexes, whilst on a collecting tour last summer, (1872,) in the above localities. He says: " when I arrived at Fox Bay, Anticosti, last June it was extremely rare ; and I captured only four specimens in fifteen days, the specimens were fresh on the 20th of June, they generally flew low frequenting the flowers of a species of Wild Pea, which occurs abundantly on the banks of rivers in Anticosti and Labrador. I experienced great difficulty in approaching them with the net ; its flight is rapid and low, extending along the margin of rocky cliffs and in grassy places near the Bay, near tide mark ; I never noticed them in the woods, they appeared to keep entirely within the circuit of the Bay and I remarked the same fact on the Labrador coast, where I also found them hovering about the flowers of the Wild Pea ; towards the end of July their strength gives way and if the weather be cool, tattered specimens may be taken by hand, it is the only species of Papilio, so far noticed by me, either in Anticosti or Labrador."

When I received from friend Couper the box of Anticosti Lepid, my first impression as I glanced at its contents was that this species was Asterius and that both examples were males at that, but a closer examination soon convinced me to my surprise that the one with the most yellow was a female, I then thought it might be Saunder's P. Brevicauda described in a foot note in Packard's Guide to Entomology, page 246, but on consulting that publication, I found it did not agree with his description in several important particulars, in Brevicauda on upper side of primaries the spots composing the inner band, with the exception of the one nearest the costa are *fulvous*, in my species they are all yellow without the slightest indication of fulvous; on secondaries the spots of inner band are " fulvous from near the middle to the outer edge, " in Anticostiensis these spots are entirely yellow; the tails in Brevicauda, as its name would indicate are " very short, scarcely one-eighth of an inch long—not more than half the length of those of Asterius ;" in the species I have just described, they are the same length as those of specimens of Asterias, from the Eastern, Middle and Western States and Colorado, but shorter than those from Florida. Brevicauda was taken at St. John's Newfoundland.

I now again had the pleasant excitement incidental to endeavoring to study out bare descriptions, unaccompanied by figures, and in my misery I wrote to Mr. Couper, in Montreal, requesting him to try to see the types of Brevicauda, and compare his examples with them, or if that was impossible, to write to Mr. Saunders, of Ontario, Canada, who described it, and with whom he was acquainted, concerning the species; after some time Mr. Couper wrote: "I communicated with the Rev. Canon Innes, (in whose collection are specimens of Brevicauda,) and Mr. W. Saunders, asking for information regarding P. Brevicauda; up to this instant, no answer from either;" this certainly was not very satisfactory, but as I was not particularly anxious to make a fool of myself by re-christening old species, I importuned Mr. Couper to try the gentleman with another epistolary shot; in due time, under date March 17th, 1873, came another letter from Couper, thus: "I have purposely delayed a reply to your favor of 2d, because since its receipt I wrote again to Mr. W. Saunders for the desired information, and my letter was written in terms which could not deter him from answering; however, no answer has been received;" after receiving this letter, I, of course, concluded that Mr. Saunders' time was of too much value to be encroached upon, and requested Mr. Couper to by no means trouble him again, as his dignified silence at last brought me to a proper sense of my true position, and was a merited punishment to both Couper and myself for our temerity.

However, I believe this is distinct from Brevicauda, and if it be not, it is an absurdity to retain that name; the probability, after all, is, that Brevicauda and Anticostiensis, (if they be not the same,) are both varieties of Asterias; if such is the case, they have the merit, at least, of being very marked and interesting ones, which, I trust, will be considered sufficient reason for my having figured the latter; I did not figure the male, as I considered the female the most remarkable on account of the greater width of the macular inner bands. Mr. Couper took on Anticosti Island at the same time with this species, Colias Interior, Scud ; Pieris Frigida, Scud.; Chionobas Jutta, Hub.; Ly. Lucia, Kirby; Ly. Scudderii, Edw.; Ly. Nov. Sp.? Thyatira Pudens, Guen.; Sesia Uniformis, Grote; and many other Lep. as well as Coleoptera; he starts again for same locality next month on another collecting tour; he expects to be able to find the larvæ of Anticostiensis, and I hope he will meet with the success he so well deserves.

April, 1873.

PAPILIO PILUMNUS. Boisduval.

BOISD. SP. GEN. 1. p. 340, n. 154, (1836.)
MENETRIES, CAT. MUS. PETR. LEP. II. p. 110, t. 7. f. 2. (1857.)

MALE. Expands 3½ inches.

Body yellow; a broad black dorsal, narrow lateral and a broader ventral band; antennæ black.

Upper surface, chrome yellow, primaries, costa narrowly black, five transverse black bands; first basal, second extending from inner third of costal margin to same distance on inner; third, a mesial and convergent band extending from the costa to first median nervule, covering the disco-cellular veins; fourth, short situated midway between third and fifth bands, extending from costa to first radial vein; fifth and terminal one very broad, extending along whole outer margin, covering one-third of the whole area of the wing, containing two rows of imperfect yellow lunules, nearly confluent, outer ones large and distinct, the inner of segregated atoms.

Secondaries, three transverse black bands, continuations of the first, second and fifth of primaries, first and second converging to a point on the abdominal margin about three-fourths its length and separated from the terminal border by two fulvous crescents, preceeded by a narrow yellow line, the outer band with six yellow long straight or lunulate bars all of which are more or less tinged with fulvous, interior to these the band is irrorated with four shining blue crescent-shaped patches, a black discal mark; tri-tailed of which the outer is the longest and tapering, yellow ciliæ on inner side, other tails one-half and one-quarter the length of the outer one; emarginations yellow.

Under surface, paler than above, bands of upper surface repeated, but brown instead of black; the six lunulate bars near outer margin of secondaries, fulvous; on inner side these are joined by black, irregular shaped patches, which are in turn surmounted by shining blue crescents, edged above with black.

I have seen but one female, and as nearly as I can recollect, she resembles the male very closely, but was larger, probably expanding four inches, or over.

It is a matter of astonishment that so large and beautiful a butterfly of our own fauna should be so rare in N. American collections; in fact, I know of but two examples; the female above alluded to, which came from New Mexico, and is in the collection of Mr. W. H. Edwards, and the male, from which the accompanying figure was drawn, I received from Vera Cruz, Mexico. I have no better reason for giving an illustration of a butterfly that has been already both described and figured, than that I think it the finest of its genus found in North America, and, secondly, Menetries' Catalogue is not a work likely to be found at every book-stall and finally, must I confess it, I did not know that there was a plate of it, until, after I had drawn mine; so if confession is good and wholesome, I trust I am somewhat benefited thereby if no one else is.

PIERIS MENAPIA. Felder.

Wein. Ent. Monat. III. p. 271, n. 18. (1859.)
Reise Nov. Lep. II. p. 181, n. 172. t. 25, f. 7. (1865.)
Pieris Tau. Scudder. Proc. Boston Nat. Hist. Soc. VIII. p. 183, (1861.)
Pieris Ninonia, Boisd. Lep. Cal. p. 38, n. 5. (1869.)
Neophasia Menapia, Behr Trans. Am. Ent. Soc. (1869.)
Edwards Butterflies of N. Am. Part 8, (1871.)

FEMALE. Expanse 2 inches.

Antennæ black, throax black with white hairs; abdomen blackish above near the thorax, rest white.

Upper surface white, primaries, costa edged with black rather broadly till to the discal vein on which it is continued to its extremity; a black border with inner edge sinuous, broader at apex and diminishing to a mere line at inner angle, extends from costa downwards, within this border are six irregular white spots, the one nearest the inner angle being much the smallest.

Secondaries have an irregular sub-marginal black line, from which to the exterior margin the veins are edged with black, which widens at their tips, forming as it were six large white spots, from the space between the veins and sub-marginal line and exterior margin.

Under surface white, primaries nearly the same as above, spots in border a little larger. Secondaries, veins all margined with brownish black; sub-marginal band as above; costa and interior margin edged with rose or flesh color; a row of marginal lunulate spots and a line of same color running parallel with and adjoining the inside of the sub-median nervure.

This species, of which I am not aware any figure of the female has heretofore been published, has had almost as rough a time of it as had Colias Eurydice; it is evident, that until very lately, the female was entirely unknown to Lepidopterists in this country; Mr. W. H. Edwards, in his Butterflies of N. America, part 8, gives three excellent representations, all males, although one of them he supposed was a female, and described it as such in the accompanying text. Scudder, in the Proc. Boston Soc. Nat. History, 1861, where he redescribed it under the name of Pieris Tau, gives elaborate descriptions of both sexes, of which he says: "a large number of specimens are in the Museum of Comparative Zoology obtained by Mr. Agassiz at Gulf of Georgia;" now it is evident from the description that either all of the specimens above alluded to were males, or else Mr. Scudder gave them a very careless examination, indeed, for the only difference he mentions between the sexes is that on the upper surface of secondaries the female " repeats slightly at the outer angle, the markings of the lower surface;" then, after stating that " it represents in Washington Territory the *P. Sisimbrii* Boisd. of California," (which it does not resemble a bit, either in size, shape or markings,) he goes on to say that the way he distinguishes between males and females in his foregoing description was founded on the " cut of the hind margin of secondaries;" he then goes on and gives some minute differences in the shape of the wings, but does not make any mention whatever of the black mesial line on upper surface, or the very remarkable flesh-colored edging of costa, and six marginal lunules of same color on under surface of secondaries; so we must come to the conclusion that Scudder fell into the same error as W. H. Edwards, and described both sexes from male specimens only.

I would suggest whilst on this subject that it might be better perhaps, instead of heeding imaginary differences in the cut of the wings, which only lead to error : to bear in mind that in none of the Pieridæ are the male and female marked alike.

Dr. Felder in the Weiner Monatschrift is more accurate, and appears to be the only one who has heretofore really known the female; he notices that while the male has but five white spots, in the black apical patch, the female has six and also mentions that on secondaries of female, below, the costa, the basal and other spots are livid.

For the example from which I made my drawing I am indebted to my valued friend, Mr. Henry Edwards, of California, who received it from Vancouver's Island.

April, 1873.

NOTES ON SOME SPECIES.

Papilio Burtonii, Reak., Proc. Acad. Nat. Sc., Phil., p. 89, 1868, is a synonym of P. Columbus, Hew., Trans. Ent. Soc., Ser. II, Vol. 1, p. 98; t. 10, f. 1, 1850.

Papilio Caleti, Reak., Proc. Ent. Soc., Phil., II, p. 138, 1863, is Papilio Polymetus, Godt, Enc. Meth., 1819.

Eresia Yorita, Reak., Proc. Ent. Soc., Phil., V., p. 224, 1865, is Eresia Ezra, Hew., Ex. Butt., III, Eres. t. 4, f. 29, 1864. In Kirby's Catalogue, this is set down as variety of Eresia Theona, Men.; on what grounds I do not know; there certainly is little or no resemblance.

*Colias Semperi, Reak.————is C. Dimera, Doubl., Hew., Gen. D. L., t. 9, f. 3, 1847. Although Reakirt's name might be retained for the white female variety, which I believe he was the first to notice and describe.

Lycæna Helloides, Boisd., L. Castro, Reak., L. Ianthe, Edw., concerning these there is some confusion, either Helloides is unknown in N. American collections or two of the above are synonyms, which latter I am inclined to believe is the real state of affairs, for I do not believe any one American collection can produce examples of all three; Mr. Edwards has in his I believe Helloides and Ianthe; Mr. Mead has Ianthe and Helloides, and I have helloides and Castro, but my Helloides is the same as the Ianthe of the others and my Castro is like their Helloides, a specimen of the typical Helloides from Dr. Boisduval would be of some use here I trow.

April, 1873.

*I have in my possession the types of Reakirt's Semperi male and female, but where he described it I cannot recollect or at the moment ascertain.

Of the following species I am anxious to obtain examples, either by exchange or purchase, any Naturalists having duplicates of any of them will confer a great favor by communicating with

HERMAN STRECKER,
Box 111 Reading, P. O.,
Berks Co., Pennsylvania, U. S. of N. America.

Ornithoptera Hippolytus, Cram.
Ornithoptera Amphrysus, Cram.
Ornithoptera Helena, Linn.
Ornithoptera Croesus, Wall.
Ornithoptera Brookiana, Wall.
Papilio Evan, Doubl.
Papilio Pericles, Wall.
Papilio Blumei, Boisd.
Papilio Macedon, Wall.
Papilio Philippus, Wall.
Papilio Arcturus, West.
Papilio Phorbanta, Linn.
Papilio Homerus, Fabr.

Papilio Gundlachianus, Feld.
Dynastor Napoleon, Doubl, Hew.
Argynnis Sagana, Doubl, Hew.
Zeuxidia Aurelias, Cram.
Urania Rhipheus, Cram.
Urania Sloanus, Cram.
Nyctalemon Orontes, Linn.
Nyctalemon Cydnus, Feld.
Erasmia Pulchella, Hope.
Actias Maenas
Saturnia Dercetó, Mén.
Saturnia Argus, Drury.
Any Asiatic species of Parnassius.

These are a few of the very many of the rarer species that I am eager to procure, of course there are numberless others from all parts of the world, equally desirable and coveted by me.

North Atlantic Express Co.,

NEW YORK,
OFFICE, 71 BROADWAY.

Chartered by Special Act of Incorporation.

CENTRAL EUROPEAN OFFICE:
5 RUE SCRIBE, PARIS.

PRINCIPAL OFFICE IN GREAT BRITAIN:
4 Moorgate St., London, E. C.

B. W. & H. HORNE, AGENTS.

BRANCH OFFICES: Golden Cross, Charing Cross; George & Blue Boar, High Holborn; 108 New Bond Street; 474½ New Oxford Street.
OFFICE IN LIVERPOOL: 5 Knowsley Buildings Tithebarn Street.
CONTINENTIAL OFFICES: 5 Rue Scribe, Paris; 51 Kleine Reichen-strasse, Hamburg; 116 Lagen-strasse, Bremen.

Merchandise, specie, bullion, stocks, bonds, or other valuables and packages and parcels of every description, personal effects, baggage, etc., forwarded to and from Europe and all parts of the United States, the States and Territories of the Pacific Coast, British Columbia and the Canadas included, at *fixed Tariff rates, with no extra charges whatever* for Custom-House brokerage, commissions, delivery, etc., etc., the shipper or receiver being *under no other care or expense than the stipulated freight from the point of shipment to place of delivery*, and the amount of duties and government fees actually paid at the Custom-houses of the United States or Europe.

For the convenience of shippers, where agencies of the Company are not established, packages or heavy goods may be forwarded to either of the offices or agencies of the Company, by either of the express or transportation companies in the United States, or by post, by railway, through the parcel delivery companies, or forwarding houses in any part of Great Britain or the Continent of Europe.

All packages, trunks, or parcels forwarded by this Company will be landed on arrival simultaneously with the mails, or immediately thereafter and will be entered at the Custom-house, duties paid and delivered to the parties to whom addressed in any part of Europe, the United States, the Canadas, or British Columbia, with the greatest possible dispatch. Transportation charges and duties collected on delivery, or may be prepaid, at the option of shipper.

Insurance against marine risk taken by the Company, when desired by the shipper, at the lowest current rates; premium payable in all cases in advance.

Shippers to or from any part of America, and Americans traveling in Europe, will find this the quickest, cheapest and most reliable medium of transportation, the business of this Company being conducted upon the well known prompt American express system, which has become so great a commercial necessity and convenience throughout the United States.

Purchases made, and collections and communications in every part of Europe and the United States promptly executed.

☞ Circulars sent on application to

S. D. JONES,
Manager,
71 BROADWAY, N. Y.

No. 3. PRICE 50 CENTS.

LEPIDOPTERA,

RHOPALOCERES AND HETEROCERES,

INDIGENOUS AND EXOTIC;

WITH

Descriptions and Colored Illustrations,

BY

HERMAN STRECKER.

Reading, Pa., May, 1873.

Reading, Pa.:
Owen's Steam Book and Job Printing Office, 515 Court Street,
1873.

NOTICE.

Subscribers who prefer to do so can remit the money for each part as they receive it, which will be in the United States, 55 cents per part, inclusive of postage. As soon as I have subscribers sufficient to pay the extra expense of printing, I will add another plate, so I trust Lepidopterists and Naturalists generally will exert themselves to increase my subscription list, which is not at present as large as that of the London Times.

TO PUBLISHERS, &c.

Advertisements of Publishers, Taxidermists, dealers in specimens of Natural History, Public Lecturers, Bird Fanciers, &c., inserted on the cover and fly leaves of each number, at very low rates, as my object is not to make money in this instance, but merely to cover the expense of issuing this work. Parties choosing to avail themselves of this means of advertising will please address,

HERMAN STRECKER,
Box 111 Reading P. O., Berks Co., Pa.

CATOCALA TRISTIS. Edwards.
PROC. ENT. SOC. PHIL. VOL. II, p. 511. (1864.)

(PLATE III, FIG. 1 ♂)

Expands 1½ inches.

Head and thorax light grey; abdomen above dark brown; beneath white.

Primaries, from exterior margin to the undulate band, greyish with darker shades; from thence to transverse posterior line white; the space interior of this to the base, with the exception of the white open sub-reniform, is light grey; reniform dark brown; joining the sub-reniform on inside is a very distinct black spot; another is on edge of costa at termination of the transverse anterior line; whole inner edge of wings shaded with black to about one-fourth their width.

Secondaries black, outer angle tipped with white.

Under surface, primaries black, outer angle white, a broad white sub-marginal band. Secondaries black with white edge at outer angle as on upper side.

Habitat. New York, New Jersey, Pennsylvania, Rhode Island.

For the original of my figure I am indebted to Mrs. Bridgham, who collected several examples near Providence, Rhode Island. The smallest and rarest of all our known black winged species.

CATOCALA VIDUATA, Guenée in Appendix.
CATOCALA VIDUA. Guenée, SPEC. GEN. VOL. VIII, p. 94.

(PLATE III, FIG. 2 ♀)

Expands 3½ inches.

Head and thorax grey; collar banded with chestnut; abdomen greyish brown; beneath white.

Upper surface, primaries ashen with brown shadings and powdered with black atoms; a conspicuous black arc sweeps from the sub-apical dash, which forms a part of it, downwards to the reniform, thence obliquely upwards to the costa; transverse posterior line black, accompanied outwardly by a brown band which is in turn succeeded by the grey undulate sub-marginal band; reniform distinct and brown; fringes grey.

Secondaries black; white fringes middle of which are penciled with black; basal hairs, heavy and greyish.

Under surface white, on primaries the transverse bands are confluent along the interior margin.

Secondaries have marginal band very broad, mesial moderate and strongly angulate, white space between the two bands, very narrow.

A southern species although taken in a few rare instances in Pennsylvania. Examples are in the Mus. of the Am. Ent. Soc. and my own. The most robust and with the exception of the Californian C. Marmorata, the largest American Catocala. With this species has frequently been confounded C. Desperata Guen. a smaller and slighter built insect, common throughout the Middle and Southern States and which is figured in Abbot & Smith, under the name of Phalaena Vidua.

CATOCALA LACHRYMOSA. Guenée.
SPEC. GEN. VOL. VII p. 93.

(PLATE III, FIG. 3 ♂)

Expands 3 inches.

Upper surface, primaries very dark and dusted with minute pale grey scales, transverse lines black, sub-terminal distinct and sometimes shaded interiorly with grey; the grey shadings of the transverse lines are broader and brighter between the sub-median vein and interior margin; whole surface of wings frosted and powdered in such a way as to make the markings very indistinct.

Secondaries black, fringes white, divided by black at terminations of nervules.

Under surface much like C. Viduata.

Habitat. Pennsylvania.

I have not seen examples from any other state; it appears to be exceedingly local; two years since a dozen or so were taken in a small piece of woods, four miles from Reading, but in none of the neighboring localities have I ever met with it. It is subject to much variation; of six examples now before me, none agree in the depth or quantity of the dark color of primaries; the one figured on Plate III has the black, sub-terminal line, margined with grey of unusual brightness, whilst in another there is no accompanying grey at all; yet another has the third of the wing along the interior margin deep black, like in C. Tristis, and the most notable var. is one in which the whole space between the transverse anterior and sub-terminal lines is black, whilst the space from sub-marginal line to exterior margin is remarkably light and even colored, exactly after the manner of C. Scintillans; these were all taken the same day in one place.

I must confess I can see in this species none of the resemblance to C. Epione alluded to by Mr. Grote,[*] more than that they both have black inferiors; under side of Lachrymosa is white, with usual black bands; that of Epione is black, with, on primaries, a narrow white sub-terminal band, midway between which and the base is a small white patch commencing on costa and running diagonally to middle of wing; secondaries have the merest trace of a very narrow, almost obsolete white band running from costa a short way in.

[*] Trans. Am. Ent. Soc., Vol. IV, pp. 2 & 19.

CATOCALA OBSCURA, NOV. SP.
(PLATE III, FIG. 4 ♂)

Expands 2¼ to 2½ inches.

Thorax above, dark grey; abdomen black; beneath dirty white.

Upper surface, primaries dark smoky grey, pretty evenly colored; transverse anterior and posterior lines black, varying somewhat in width and distinctness in different examples; a sub-terminal grey band between two brownish ones; in some examples an almost obsolete reniform, but in the majority this appears to be replaced by a small round deep brown or black discal spot; sub-reniform open.

Secondaries black; fringes white, and in some cases black and white.

Under surface white, with black bands.

Habitat. New York.

I received a number of this species from Mr. Angus who took them in the vicinity of the village of West Farms, N. Y.; he says they are difficult to discover, as they secrete themselves in the crevices beneath the bark of trees, and the rustling of leaves, &c., will not start them from their hiding places; nothing short of hard raps against the tree trunks will do it. It does not appear to be very rare; I have seen examples in various collections for a number of years past under the name of Lachrymosa, Vidua, and Insolabilis, so to get it out of its obscurity, and as its appearance is more obscure than any of its allies, I have christened it accordingly.

May 1st, 1873.

CATOCALA RELICTA. Walker.
? CATOCALA FRAXINI, GUENEE, SPEC. GEN. VOL. VII, p. 83.

(PLATE III, FIG. 5 ♂, 6 ♀)

MALE. Expands about 3 inches.

Collar white; thorax above white, mottled with black; abdominal segments blackish, edged exteriorly with white; beneath whole body is white; anal tuft white.

Upper surface, primaries white, with a distinct row of black terminal spots; basal patch black, transverse bands and lines almost lost in centre of wing, but become more distinct as they near the costal and interior margins; reniform spot tolerably distinct, sub-reniform almost obsolete.

Secondaries black, with a regular narrow white median band; the discal spot of under surface is visible, the basal part of the wing not being as dark as the balance; fringes white.

Under surface, primaries white, with broad dark brown marginal and median and paler basal bands; secondaries dark brown, basal patch large, white, contains a black discal spot which connects exteriorly with the black of the remaining portion of the wing, the white median band of the upper side repeated.

FEMALE. Expands 3¼ inches.

Colors and markings as in male, but bands of primaries much intensified, heavier and better defined, and nearly whole surface more or less powdered with dark grey atoms.

Under surface less black on all wings than the male, discal spot of inferiors smaller, lunate, and disconnected from the median black part.

This is one of the rarest, as it certainly is the most beautiful of the N. American Catocalidae; it is found occasionally in various parts of New York, seldom in Pennsylvania, but occurs in some plenty near Providence, Rhode Island.

I have little doubt, that when Guenée in his Species Général (Vol. VII, p. 83,) credited N. America with C. Fraxini it was from examples of C. Relicta that he drew his conclusion, although there are and have been rumors of a blue banded Catocala like the former occurring on the Pacific Coast, and time may resolve the rumors to a certainty, for we all know what a wonderful resemblance bears the Lepidoptera of our Western Slope to those of Europe, and it would almost seem that eventually every European Species is to find its analogue with us.

CATOCALA BRISEIS. EDWARDS
PROC. ENT. SOC. PHIL. II, p. 508. (1864.)

(PLATE III, FIG. 7. ♀)

FEMALE. Expands 2⅞ inches.

Head and thorax above blackish grey, abdomen dark brown; beneath dirty white.

Upper surface, primaries blackish grey; a sub-terminal white zig-zag band joined interiorly by a much broken space of mixed yellow and white; reniform obscure, sub-reniform white, a white spot also joins the reniform on the inner side.

Secondaries deep scarlet; a broad marginal band with two indentations on the inner edge towards the anal angle; median band broad and a little elbowed at centre; some black mixed with the red hairs of the basal portion.

Fringes on all wings have the outer larger part white and the inner part adjoining the wings black or dark grey.

Under surface, primaries white with the usual three black bands; secondaries have inner three-fourths scarlet, remaining fourth white; marginal and mesial bands as above; a discal lune which connects with inner edge of median band.

Habitat. New York, Rhode Island.

Mus. Am. Ent. Soc., Mrs. Bridgham, Strecker.

The type is in the museum of the Am. Ent. Soc.

Briseis, which is the rarest of its genus found in the Atlantic States, belongs to the same group as Unijuga, Walk., Irene, Behr, Californica, Edw., and the species described below, but the grey of the upper and red of the lower wings is much darker than in any of its congeners. With much regret that I can say so little of this beautiful moth, I will proceed to

CATOCALA FAUSTINA, NOV. SP.
(PLATE III, FIG. 8 ♂)

MALE. Expands 2½ inches.
Body above grey, beneath white.
Upper surface, primaries blueish grey, powdered with brown atoms, marginal spots, transverse lines and bands well defined, reniform distinct, and surrounded by an outer circle which is produced in two points on exterior; sub-reniform white: above this and interior to the reniform is a white space; fringe light grey.
Secondaries scarlet, median band moderately wide, angulated at centre outwardly, and terminates somewhat abruptly about two lines from abdominal margin; marginal band with a deep indentation between the first and second median nervules, apical spot and emarginations rosy; fringe on exterior margin white, on interior margin grey.
Under surface, primaries white; secondaries, interior two-thirds rosy, towards costa this color becomes lost in white; almost imperceptible indications of a discal lune.
The single type from which the accompanying figure was drawn I received from Mr. W. H. Edwards, who stated that it had been taken in Arizona by Lt. Wheeler's Expedition in 1871. It is a pretty, medium sized species, and, like most of those from the western side of the Rocky Mountains, resembles wonderfully some species of Europe.

May 24, 1873.

CATOCALA COCCINATA. Grote.
TRANS. AM. ENT. SOC., VOL. IV., p. 6. (1872.)

(PLATE III, FIG. 9, ♂)

MALE. Expands 2½ inches.
Upper surface, primaries pale grey, variegated with brown; reniform small and pale, space between this and sub-basal transverse line white, with a pink tinge caused by the red of under side being reflected through; sub-reniform large and same color, transverse median line acutely dentate, a white sub-marginal spot formed of the space caused by the central inflection of this line.
Secondaries bright scarlet, marginal band regular, widest towards outer angle, median

band not reaching interior margin, apical spot narrow and red; basal hair greyish, fringe white and grey.

Under surface, superiors red, inferiors, inner two-thirds same color, outer third white; a black discal lunule from which extends along centre of discal cell almost to base of wing, a black streak.

Habitat. Pennsylvania.

This species is of rare occurrence; the only one I ever saw on the wing was the male from which I made my illustration on Plate III, and which I captured a number of years since in some oak woods then near Reading, but the grounds are now in that city, and the noble old trees are replaced by varied and execrable examples of domestic architecture. Mr. Grote's types are in the Mus. of the Am. Ent. Society; several examples also in fine collection of Mr. Wilt, of Phila.

CATOCALA CEROGAMA. Guenée.
SPEC. GEN., VOL. VII, p. 96.

(PLATE III, FIG. 10. ♀)

Expands 3¼ inches.

Thorax above light grey, with brown markings; abdomen brown, beneath yellowish white.

Upper surface primaries, pale grey of various shades and mottled with brown, transverse lines distinct and dark brown, reniform moderate size, sub-reniform white; exterior to these in the median space, a dark patch; two diagonal white spaces, one interior to the reniform and above the sub-reniform, the other interior to the outer half of the transverse posterior line; fringes brown.

Secondaries black with a narrow yellow median band of equal width throughout; basal part covered with long brown hair beneath which it is yellow; apical spot and emarginations yellow; fringes white cut with black at the terminations of the nervules. Beneath all wings are yellow; secondaries with black marginal and median bands.

Habitat. New York, Pennsylvania, Maryland, &c.

By no means a common species although at times occurring in some plenty in particular localities.

CATOCALA SERENA. Edwards.
PROC. ENT. SOC. PHIL., VOL. II, p. 510.

(PLATE III, FIG. 11. ♀)

Expands 2¼ inches.

Head and collar chestnut brown; thorax smoky grey; abdomen brown.

Upper surface, primaries same grey as thorax with transverse brownish shades, transverse lines narrow but distinct.

Secondaries deep yellow; marginal band indented in middle; median narrow, irregular and prominently elbowed at centre; basal hairs brown.

Under surface yellow, all the bands pretty heavy; bases of wings obscured with brown which color more or less shades the whole under surface.

Habitat. New York, Pennsylvania.

Rare. Mr. Edwards' type is in collection of Mr. Wilt, Philadelphia. I took a single female examples about a mile from Reading, I also received specimens from my good friend Mr. Angus of West-Farms, N. Y., in which locality he captured them.

I trust this plate of Catocalidæ may prove acceptable; the confusion existing in regard to the nomenclature of the various species is truly wonderful; I do not think I would be going too far in asserting that there are not over three collections in America in which they are all correctly determined; this is particularly the case with our typical black species.

As I either possess or have access to examples of all our known species, I will in future numbers of this work illustrate the whole of them, probably every third plate I issue will be of this genus until they or I am exhausted.

NOTES ON SOME SPECIES.

In List of Lepidoptera of N. America by Grote and Robinson, Part 1, (1868,) on page VIII, Arctia Speciosissima, Mosch., is incorrect, it should be Arctia Speciosa, Mosch.

On same page, right below, A. Quenselii, Geyer, and A. Gelida, Mosch., are given as distinct, whereas they are one and the same, Geyer's name having priority.

The authors fell into the same error with regard to Aretia Parthenos, Harris, and Aretia Borealis, Moschler, which are likewise synonyms, Harris' name being much the older; this can easily be seen by comparing examples of Parthenos with fig. 3 on plate 9 of Vol. IV., Wein. Ent. Monat.

Eresia Sydra, Reak. Proc. Acad. Nat. Sc., Phil., 1866, p. 335, n. 36, is Eresia Acesas, Hew. l. c. f. 48, 49. 1864.

Aretia Americana, Aretia Caja. I never had much faith in the genuineness of Harris' Aretia Americana, but before I had given the matter much attention I thought the white collar would seem to be entitled to some value as a specific distinction. Dr. Harris who described A. Americana says: "This moth closely resembles the European *Caja*, and especially some of its varieties, from all of which however it is essentially distinguished by the white edging of the collar and shoulder covers and the absence of black lines on the sides of the body." As far as the examples of Americana and Caja in my possession go, the above amounts to nothing; four examples of Caja are before me; the first, from Saxony, expands 2¼ inches, has the collar edged with red, patagiæ *narrowly edged with white* on outer edge; five brown spots on side of abdomen and five on back.

The second, from Osterode-am-Harz, expands 2¼ inches, has collar edged with red, no white whatever on patagiæ; *no marks of any kind on sides of abdomen*, four black spots on back; the blue spots on secondaries of this example are very small, with the exception of two near the exterior margin; they are little more than dots.

No. 3, from England, expands 2⅜ inches, has *front edge of collar and outer edge of patagiæ white*; five very small black spots on side of abdomen and four large ones on back. Primaries of this example have the brown markings very narrow; there is as much white as brown.

No. 4, from S. France, expands 2⅜ inches, front edge of collar red, outer edge of patagiæ *narrowly white*; abdomen, except the segment nearest the thorax, black above, on sides and below, a little red on sides, hair on sides of anal segment red; in this specimen the primaries are very dark brown with but little white; secondaries very dark orange with spots of unusually large size, those nearest the base confluent forming a band.

Of my examples from British America, the one expands 2¼ inches; collar and outer edge of patagiæ white; no spots on sides of abdomen; *six* black spots on back; primaries with but little white; secondaries have four very large spots, three sub-marginal and the other half way between these and the base.

The other example expands 2¾ inches, has collar white; outer edge of patagiæ narrowly white; faint indications of five spots on sides and five black spots on back of abdomen; brown and white of primaries in same proportion as in ordinary forms of Caja; spots on secondaries likewise.

I wrote to Mr. Moschler in Germany, concerning Caja and Americana, he says in reply "In my collection are 16 Aret. Caja from here* *2 examples have distinct white collars* exactly like my examples from N. America, Caja and Americana are surely one species." And I must say that I agree with his conclusion.

*Germany.

Of the following species I am anxious to obtain examples, either by exchange or purchase, any Naturalists having duplicates of any of them will confer a great favor by communicating with

HERMAN STRECKER,
Box 111 Reading, P. O.,
Berks Co., Pennsylvania, U. S. of N. America.

Ornithoptera Hippolytus, Cram.
Ornithoptera Amphrysus, Cram.
Ornithoptera Helena, Linn.
Ornithoptera Crocsus, Wall.
Ornithoptera Brookiana, Wall.
Papilio Evan, Doubl.
Papilio Pericles, Wall.
Papilio Blumei, Boisd.
Papilio Maceleou, Wall.
Papilio Philippus, Wall.
Papilio Arcturus, West.
Papilio Phorbanta, Linn.
Papilio Homerus, Fabr.
Papilio Garamas, Hub.
Papilio Caiguanabus, Pocy.
Argynnis Radra, Moore.
Argynnis Oscarus, Evers.
Argynnis Cnidia, Feld.
Argynnis Jerdoni, Lang.
Argynnis Dexamene, Boisd.
Argynnis Jainedeva, Moore.
Argynnis Ruslana, Motsch.
Argynnis Anna, Blanch.
Argynnis Childreni, Gray.
Argynnis Aruna, Moore.
Acherontia Satanus.

Papilio Gundlachianus, Feld.
Dynastor Napoleon, Doubl. Hew.
Argynnis Sagana, Doubl. Hew.
Zeuxidia Aurelius, Cram.
Urania Rhipheus, Cram.
Urania Sloanus, Cram.
Nyctalemon Orontes, Linn.
Nyctalemon Cydnus, Feld.
Erasmia Pulchella, Hope.
Actias Maenas
Saturnia Dercoto, Moon.
Saturnia Argus, Drury.
Any Asiatic species of Parnassius.
Citheronia Phronima.
Castnia Daedalus, Cram.
Pyrameis Gonerilla, Fabr.
Pyrameis Abyssinica, Fehl.
Pyrameis Dejeanii, Godt.
Pyrameis Tameamea, Esch.
Vanessa v. Hygiaea, Hdrch.
Vanessa v. Elymi, Rbr.
Dasyopthalmia Rusina, Godt.
Morpho Phanodemus, Hew.
Any species of Agrias.
Any species of Callithea.

These are a few of the very many of the rarer species that I am eager to procure; of course there are numberless others from all parts of the world, equally desirable and coveted by me.

North Atlantic Express Co.,

NEW YORK,
OFFICE, 71 BROADWAY.

Chartered by Special Act of Incorporation.

CENTRAL EUROPEAN OFFICE:
5 RUE SCRIBE, PARIS.

PRINCIPAL OFFICE IN GREAT BRITAIN:
4 Moorgate St., London, E. C.
B. W. & H. HORNE, AGENTS.

BRANCH OFFICES: Golden Cross, Charing Cross, George & Blue Boar, High Holborn; 1st New Bond Street; 474½ New Oxford Street.

OFFICE IN LIVERPOOL: 5 Knowsley Buildings Tithebarn Street.

CONTINENTAL OFFICES: 5 Rue Scribe, Paris; 81 Kleine Reichenstrasse, Hamburg, 11d Lagestrasse, Bremen. Merchandise, specie, bullion, stocks, bonds, or other valuables and packages and parcels of every description, personal effects, baggage, etc., forwarded to and from Europe and all parts of the United States, the States and Territories of the Pacific Coast, British Columbia and the Canadas included, at *fixed Tariff rates*, *with no extra charges whatever for Custom-House brokerage, commissions, delivery, etc., etc., the shipper or receiver being under no other care or expense than the stipulated freight from the point of shipment to place of delivery, and the amount of duties and government fees actually paid at the Custom-houses of the United States or Europe.*

For the convenience of shippers, where agencies of the Company are not established, packages or heavy goods may be forwarded to either of the offices or agencies of the Company, by either of the express or transportation companies in the United States, or by post, by railway, through the parcel delivery companies, or forwarding houses in any part of Great Britain or the Continent of Europe.

All packages, trunks, or parcels forwarded by this Company will be landed on arrival simultaneously with the mails, or immediately thereafter and will be entered at the Custom-house, duties paid and delivered to the parties to whom addressed in any part of Europe, the United States, the Canadas, or British Columbia, with the greatest possible dispatch. Transportation charges and duties collected on delivery, or may be prepaid, at the option of shipper.

Insurance against marine risk taken by the Company, when desired by the shipper, at the lowest current rates; premium payable in all cases in advance.

Shippers to or from any part of America, and Americans traveling in Europe, will find this the quickest, cheapest and most reliable medium of transportation, the business of this Company being conducted upon the well known prompt American express system, which has become so great a commercial necessity and convenience throughout the United States.

Purchases made, and collections and communications in every part of Europe and the United States promptly executed.

☞ Circulars sent on application to

S. D. JONES,
Manager,
71 BROADWAY, N. Y.

No. 4. PRICE 50 CENTS.

LEPIDOPTERA,

RHOPALOCERES AND HETEROCERES,

INDIGENOUS AND EXOTIC;

WITH

Descriptions and Colored Illustrations,

BY

HERMAN STRECKER.

Reading, Pa., June, 1873.

Reading, Pa.:
OWEN'S STEAM BOOK AND JOB PRINTING OFFICE, 515 COURT STREET,
1873.

NOTICE.

Subscribers who prefer to do so can remit the money for each part as they receive it, which will be in the United States, 55 cents per part, inclusive of postage. As soon as I have subscribers sufficient to pay the extra expense of printing, I will add another plate, so I trust Lepidopterists and Naturalists generally will exert themselves to increase my subscription list, which is not at present as large as that of the London Times.

TO PUBLISHERS, &c.

Advertisements of Publishers, Taxidermists, dealers in specimens of Natural History, Public Lecturers, Bird Fanciers, &c., inserted on the cover and fly leaves of each number, at very low rates, as my object is not to make money in this instance, but merely to cover the expense of issuing this work. Parties choosing to avail themselves of this means of advertising will please address,

HERMAN STRECKER,
Box 111 Reading P. O., Berks Co., Pa.

1 PAPILIO EURYMEDON ♂. 2 P. MARCHANDII ♂. 3 COLIAS DIMERA ♀.
4 C. ab. SEMPERI ♀. 5 CHIONOBAS UHLERI ♂. 6 SATYRUS RIDINGSII ♀.
7 S. STHENELE ♂. 8 S. var. HOFFMANI ♀.

PAPILIO EURYMEDON. Boisduval.

BOISD. ANN. SOC. ENT. FR. 1852, p. 280.
LUCAS, REV. ZOOL. 1852, p. 140.

MALE. Expands 3½ inches.

Antennæ, head and thorax black; a yellowish white line runs from behind the eyes on the neck and the thorax to the termination of the latter; abdomen black above, white on the sides with black lateral lines.

Upper surface, yellowish white, with black bands disposed in same manner as on Turnus, but these are much broader, and the third one on primaries extends to the sub-median nervure; costa and veins black; discal mark of secondaries black; anal spot fulvous surmounted by a blue crescent, the two or three sub-marginal lunules nearest the anal angle fulvous; tail ½ inch long and a little spatulate; emarginations white.

Under surface same as above, except that the sub-marginal spots on border of primaries are confluent, forming a continuous line; also another narrower and more obscure band towards the inner edge; on border of secondaries, above all the crescents, are shining blue bars or lunules.

FEMALE. Expands 4 inches and is the same in color and markings as the male.

Habitat. California, Washington Ty., Vancouver's Island.

In the examples from Vancouver's Island and Oregon the ground color is almost white, whilst in the Californian specimens it is a decided yellow tint, on those from the latter locality there is also much less black than on the more northern types.

In Cat. Lep. Ins. Brit. M. I. p. 24, Gray erroneously considered this species as a variety of P. Rutulus.

It is one of the common species of California and is found in most parts of that State and adjoining Territories.

PAPILIO MARCHANDII. Boisduval.

BOISD. SP. GEN. I., p. 350, n. 192. (1836.)
LUCAS, ANN. SOC. ENT. FR., p. 532. (1852.)

MALE. Expands 3½ inches.

Antennæ black; head and body black above, with yellow lines behind the eyes and on thorax above the wings; beneath yellow, a black line on each side of abdomen.

Upper surface orange colored; primaries black at base, also three short black bands, first extending from costal margin downwards a short distance beyond the median nervure; the second extends from costa to median nervure, from thence it narrows and runs along the third median nervule till it joins the broad marginal band; the third covers the disco-cellular nervules and is broad at costa, but diminishes to a point at the junction of the median nervure and second discoidal nervule, where it also connects with the marginal band; within the latter is a row of eight orange colored spots, the five nearest to costa oval, the others slightly lunulate.

Secondaries have a broad black marginal band, covering half the area of the wing, and containing a sub-marginal row of four orange spots between the outer angle and third median nervule, and three white crescents between third median nervule and anal angle; inner margin of wing black; a small orange anal spot; tails slender, ⅞ inch long, with outer half black and inner half orange.

Under surface yellowish, paler than above; primaries have the markings of upper surface reproduced, but pale brown instead of black; the sub-marginal spots are connected, forming a band.

Secondaries with black marginal band, within which the yellow and white spots are disposed as on superior surface, with the addition of a narrow yellow bar interior to the sub-marginal spots, and extending from costa to radial nervure; a black basal stripe which extends along near the inner margin of wing; this portion and that interior to and joining the marginal band is tinged with fulvous.

Habitat. Costa Rica, Panama, Honduras.

Dr. Boisduval first described this insect from examples in Mus. of M. Marchand, to whom he dedicated the species.

The specimen from which I made the drawing was given me by my old entomological friend, Mr. H. Sachs, of New York, in whose collection are several fine examples; it belongs to the same group as P. Calliste, and is as graceful and beautiful as an emanation from some Fairy Isle.

I have never looked at this lovely thing, with its delicate form and brilliant hue, without my thoughts reverting to the long past builders of the temples and altars of Palenque and Copan, the butterfly flitted through the tropical groves in their day, as now, but the inhabitants of the old dead cities have passed away, their names, their history unknown! birds, reptiles and insects now alone tenant the forest where once stood the populous cities, the kings and priests of which, with their slaves and sycophants, long ages ago have gone to rest; naught remains of their past greatness but the moss-coated and time-worn ruins of altar and idol, and the frail, golden butterfly hovers, suspended in mid air, over the monster face of some fallen Dagon, which far back beyond even "the night of time," received its meed of human sacrifice; in imagination, we can see the temples restored, the long trains of devotees, all the paraphernalia of pagan worship, we can hear the sound of music, the shrieks of the agonized prisoner about to be offered as a propitiation to some monstrous conception of barbaric superstition; but all now is hushed; priest, cacique and victim, alike, are gone, fallen are the idols, giant trees grasp with their roots the ruins of the temples, and creeping vines and gorgeous flowers mingle with the sculpture of the marvelous shrines; scarce a sound is heard save the rustling of some snake gliding stealthily to his hole, or shimmering lizard running over leaf or twig; from these thoughts we turn to others more sad; it seems almost incredible that a great country like the United States would allow these monuments of Central America to decay away, day by day, and make no effort to secure and save them from ruin. Whilst hundreds of thousands of dollars are annually squandered, publicly and privately, for follies and worse than follies, not a movement is made or a dollar appropriated to obtain these relics of a past race; each day time passes his hand with a sharper sweep over the graven record, and long ere some Gliddon may appear to find the key thereto, the characters in which that record is sculptured will have been entirely effaced; what a sad commentary on a mighty nation—on a great government; thousands are criminally wasted in schemes for self-aggrandizement by those in high positions of honor and trust; all is "rottenness and foul corruption," from the meanest to the highest serf, all preying on the resources of the nation, with no thought for aught that is noble or good.

That our government does not make some move towards purchasing and removing to some place of safety at least a portion of these memorials of a grand past is matter of equal astonishment and regret; the idols or columns are all of a size that is manageable, and if not too many incorruptible officials are to be feed by the bidders for the contract of removing and transporting them, the expense would not materially affect the financial stability of the country; that they could be purchased from the resident claimants, if their be any such, I do not think there is much doubt, as a dollar doubtless attains to a most respectable magnitude in Central America. But I suppose it is folly to ever hope to see anything of the sort accomplished, as ere government could possibly arrange all preliminaries and the countless men in place who took an interest in the matter, receive each the perquisites for selling out the various contracts for transportation, pins, sailors, sugar, laborers, assistant engineers, envelopes, crowbars, assistants for laborers, milk, sealing-wax, ice, rollers, chief engineers, servant's assistants, derricks, red-tape, &c., &c., &c., &c., either some other country will have secured and removed these wonderful remains, or time will have accomplished the work of destruction.

COLIAS DIMERA. Doubleday, Hewitson.

Doubl. Hew. Gen. D. L. 9. Fig. 3 ♂. (1847.)!
C. Erythrogramma, Koll. Denkschr. Akad. Wiss. Wien, Math. Nat,. Cl. I. p. 363. n. 31. t. 45, Fig. 13, 14. (1850.)
C. Semperi, Reakirt.

FEMALE. Expands 1½ inches.

Antennæ and head rose colored; thorax black above with yellow hairs; abdomen black above; yellow below.

Upper surface, primaries orange, tinged with black at base; a broad black marginal band prolonged far inwards on veins; two yellow spots in this band near the apex, discal dash sagittate and joined to the marginal band by a black streak; costa and ciliæ rosy.

Secondaries, lemon yellow with some black scales at base and narrowly bordered with black from outer angle to middle of margin, the black extending some distance up the veins; discal spot orange and elongated outwardly in a line almost to the marginal band; ciliæ rosy.

Under surface, primaries yellow, powdered with black along the costa; a small black discal point. Inferiors yellowish green, a small silver discal spot surrounded with pink which extends in a long dash towards the margin, another dash of same color runs from base outside along side of the median nervure; a sub-marginal row of pink spots, the one nearest the inner margin being the largest; costa and fringes of all wings rosy.

Habitat. Honduras, Panama, New Granada.

Ab. ♀ SEMPERI same size as the preceding.

Antennæ and head rosy; body black above, yellowish white beneath.

Upper surface, ground color greenish white; ornamentation of primaries same as in normal form. On secondaries the marginal border is little more than a line; no discal spot; costa and ciliæ of all wings rose colored.

Under surface, primaries yellowish with a white central patch; discal point small. Secondaries greenish yellow marked as in the ordinary ♀ form; ciliæ and costa as above.

Habitat. Insagasuga.

Reakirt described a ♂ C. Dimera and this form of the ♀ under the name of C. Semperi, and inasmuch as the name of Dimera has precedence I propose to retain that of Semperi for this white ♀ form, but for the life of me, although I have read the description and possess Reakirt's types, I cannot tell where he described it, probably it is the Proc. Acad. Nat. Sciences of Philadelphia, which work I unfortunately do not possess, and can only have occasional access to by going to Philadelphia, as our town here has no library containing works on the natural sciences, which is one of the many disadvantages incidental to residing in a provincial city, the expense precludes too frequent visits to consult the libraries of Philadelphia or New York, and the Philadelphia & Reading Railroad Company would as soon think of running their locomotives Juggernaut style over the body of a naturalist as to be guilty of the folly of giving him a free pass over their lines, although to give them their due, I believe, as an extraordinary favor, under all kinds of restrictions, they do something of the sort for Clergymen.

June 8th, 1873.

CHIONOBAS UHLERI. Reakirt.
PROC. ENT. SOC., PHIL., VOL. VI, p. 143. (1866.)

Male. Expands 1¾ inches.

Head and body brown.

Upper surface ochraceous, costa of primaries greyish, exterior margin bordered narrowly with greyish; in fifth cell towards outer margin is a small black oval spot, and in the types are two additional small spots, one in the second, the other in the third cell.

Secondaries bordered outwardly with grey same as primaries, the reticulated markings of under surface shows through the scantily scaled wings; a round black spot in the space between the second sub-costal veinlet and radial vein, and another between the first and second median veinlets; in one of the types is an additional spot situated between the radial vein and third median veinlet, and in the other type are two more spots, making four in all, which occupy the spaces between the second sub-costal and first median; ciliæ alternately grey and white.

Under surface greyish white; primaries pale ochraceous towards inner margin and penciled with brown lines, the most decided of which are those nearest the costa; the black spot or spots of upper surface repeated and pupiled with white.

Secondaries, whole surface striated with brown of various depths of color; the black spots of upper surface reproduced.

FEMALE. Expands 2 inches.

Superiors broader in proportion than in male.

Upper surface same color and ornamentation as male.

Under surface likewise as in male but with a tendency in the linear markings of inferiors to form a mesial band.

Habitat. Colorado.

From the above description it will be seen that the two types have more sub-marginal spots on all wings than the one from which the accompanying figures were made; this latter, along with many more of the same species, was taken in Colorado by Mr. Th. L. Mead in 1871, the types were captured by Mr. Ridings, also in Colorado, in 1864. The fact of these latter having three black spots on superiors and three and four on inferiors, is probably owing to local variation, as I believe all of Mr. Mead's have but one spot on the superiors and two on inferiors, although some of the females have a minute or slight rudimentary third spot on the secondaries.

Reakirt's description of male and female was taken from two males, both of which are in my possession.

This species bears on the upper surface a considerable resemblance to the Caucasian Ch. Tarpeia, Pall., but on the lower surface the resemblance almost altogether ceases.

SATYRUS RIDINGSII. EDWARDS.

Proc. Ent. Soc., Phil., Vol. IV, p. 201, (1865.)
Reakirt Proc. Ent. Soc., Phil., Vol. VI, p. 145. (1866.)
Chionobas Stretchii, Edwards, Trans. Am. Ent. Soc., Vol. III, p. 192. (1870.)

MALE. Expands 1½ inches.

Head and body grey.

Upper surface ashy brown, primaries with a broad sub-marginal yellowish white band, separated where it is crossed by the second radial vein into two parts; the one nearest the costa is palmated, being produced in four points outwardly, and has in the middle an oval black spot pupiled with white; the other is divided into three oblong portions by the crossing of the first and second median veins, in the space between these latter is another black spot, smaller than the one towards the costa, but, like it, pupiled with white. The basal third of wing is of a lighter color than the remainder, and in some examples yellow white, almost as pale as in the sub-marginal bands.

Secondaries also have a sub-marginal pale band which is toothed on outer edge.

Edges of all wings margined with a narrow brown line; fringes grey and white.

Under surface yellowish and pale brown; the bands and spots same as on upper side; on primaries all the space not occupied by the broad band is marked with little wavy brown lines as is usual on under surface of most species of Satyrus. Secondaries have a sub-basal line, the space between which and the inner edge of sub-marginal band is darker than rest of wing, after the manner of the Chionobas; whole surface of wing covered with little lines; veins white.

FEMALE. Expands 1¾ inches and resembles the male.

The types were taken by Mr. Ridings at Burlington, Colorado; the two which Mr. Edwards afterwards redescribed as Ch. Stretchii, were captured by Mr. Stretch in the vicinity of Virginia City, Nevada, on the hill-sides, early in Spring ere the snow had melted away.

Mr. Mead took a number of this species in Colorado in 1871, his examples as a general thing are I believe lighter colored than the original types.

S. Ridingsii belongs to the same group and strikingly resembles the Caucasian S. Hippolyte, Esp.* and still more so the Turkish S. Beroe, Frr.,† it also bears much similarity in appearance to Coenonympha Thyrsis Frr.‡ found in Crete.

June 9th, 1873.

SATYRUS STHENELE. BOISDUVAL.
ANN. SOC. ENT. FR. 1852. p. 308, n. 60.

MALE. Expands 1½ inches.

Head and body brown.

Upper surface brown of much the shade and depth as in Alope and Boopis; on primaries are two small black round sub-marginal spots with minute white centres; below the median vein is the oblong patch of short dense dark grey down peculiar to this sex. On secondaries is a dark sub-marginal line; Fringes of all the wings pale with dark alternations at terminations of nervures and nervules.

Under surface, primaries brown, in sub-marginal space are two large round black spots with white pupils, and surrounded by yellow rings; a dark brown sub-marginal line; the space interior to the two ocelli is marked with the lines so peculiar to the Satyrides. Costa and apex have some whiteish grey.

*Esper. Schmett. I. 2. p. 164. (1784.)
†Herrich-Schaeffer Schmett. Eur. I. f. 108-111. (1843.)
‡Freyer, Neuere Beiträge, V. t. 475. f. 3. (Jan. 1846.) Herr-Schaeff. Schmett. Eur. I. f. 297-300. (1846.)

Secondaries have a very irregular shaped dark brown mesial band, joining which on either side is a white space which is in turn much broken and shaded off into brown, which latter extends to the base and outer margin; inner margin whitish, as also is inner half of outer margin; two white points surrounded by dark shading towards the anal angle; base of wing very dark; whole surface is mottled with little lines. Fringes much as on upper side.

Female, a little larger than male, and not quite so dark in color on either surface; the two black spots on upper surface primaries are much larger, and are surrounded with ochraceous rings, beneath they are the same size as above; the under surface, throughout, is not so strongly marked as in the male.

Tolerably common in most parts of California and adjoining territory, where it frequents, in common with most of its genus, sunny spots in woods, flying low in and out under the bushes.

SATYRUS HOFFMANI, NOV. ? VAR.

MALE. Expands 2 inches.

Upper surface dark brown; primaries have two round black spots with small white pupils, the one nearest the costa is geminate, being joined on its lower edge by a smaller spot, those spots are surrounded by a narrow circle of paler color. On the secondaries, towards the outer margin between the first and second median nervules, is another ocellus about one-fourth the size of those on the superiors; between the second and third median nervules is a small black point. On all wings is an indistinct sub-marginal line; fringes brown.

Under surface pale greyish and brown; primaries have ocelli as on upper side, but surrounded by broad yellow rings, the one encircling the lower ocellus has a small yellow spot emanating from it on the upper edge; a narrow wavy sub-marginal and central band; the whole surface penciled with short transverse brown lines.

Secondaries with six ocelli arranged as in S. Pegala, in two rows of three each, the middle one of the row nearest the outer angle is oblong, and produced in a point outwardly, the others are round; all are encircled with yellow and pupiled with white. Whole surface of wing marked with short brown streaks which tend towards segregation near the outer margin; an irregular narrow band or line across the disc.

FEMALE. Expands 2½ inches.

Upper surface pale yellowish brown; on primaries the outer half is much the palest, and contains two large ocelli, the uppermost geminate, as in male; secondaries, also, have an ocellus, accompanied by a small black spot situated as in male. Marginal, sub-marginal and mesial lines brown.

Under surface white, ocelli and other markings as in male, but sharper and better defined throughout.

Taken in 1871, at Owen's Lake, Nevada, by Dr. W. J. Hoffman, of this city, whose name I have perpetuated through it.

SATYRUS HOFFMANI.

I was at first of the opinion that this might be S. Gabbii,* Edw., but Mr. Mead informs me that Mr. Edwards also has examples of it, taken by the Wheeler Expedition in 1871, and he says that it is a different thing from Gabbii; as I have no example of this latter, nor so far have I had any opportunity of comparing the two, but, on carefully reading Enwards' description of Gabbii, I find there is, at any rate, one constant difference, that is in the ocellus nearest the costa, on primaries, being geminate in Hoffmani; I have one ♂ and ten ♀ examples, and in every one of the eleven is that ocellus double, but I am still inclined to believe it a segregated form of Gabbii, at least until I have opportunity to compare it with male and female examples of the latter. It is, to say the least, difficult to draw lines between Alope, Pegala, Nephele, Boopis and this one; the ornamentation is about the same in all of them, in size they don't vary much, the principal difference is in the shades of the ground color; and, in suggesting this, I am but echoing the opinion of one of the greatest living Lepidopterists.

June 10, 1873.

I am afraid this No. (IV) will not be considered altogether orthodox by many of the Entomologists of the United States who persist in seeing nothing to please or interest them in an insect not found within the boundaries of our States or Territories, but in presenting the five figures numbered 2, 3 and 4 on plate IV, I am probably only anticipating, for some day or other Mexico will be annexed to, or swallowed up by this Government, and then what a trouble there will be among our patriotic entomologists when they will have to add so many more rare and costly species to their purely Federal collections.

However, to make amends, the forthcoming July No. (V) will contain figures and descriptions of N. American Catocalidæ, and will have as many species crowded into the plate as the size of the sheet will admit.

*Described in Trans. Am. Ent. Soc., Vol. III, p. 193. (1870.)

Of the following species I am anxious to obtain examples, either by exchange or purchase, any Naturalists having duplicates of any of them will confer a great favor by communicating with

HERMAN STRECKER,
Box 111 Reading, P. O.,
Berks Co., Pennsylvania, U. S. of N. America.

Ornithoptera Hippolytus, Cram.
Ornithoptera Amphrysus, Cram.
Ornithoptera Helena, Linn.
Ornithoptera Crocsus, Wall.
Ornithoptera Brookiana, Wall.
Papilio Evan, Doubl.
Papilio Pericles, Wall.
Papilio Blumei, Boisd.
Papilio Macedon, Wall.
Papilio Philippus, Wall.
Papilio Arcturus, West.
Papilio Phorbanta, Linn.
Papilio Homerus, Fabr.
Papilio Garamas, Hub.
Papilio Caiguanabus, Pocy.
Argynnis Rudra, Moore.
Argynnis Osarus, Evers.
Argynnis Cnidia, Feld.
Argynnis Jerdoni, Lang.
Argynnis Dexamene, Boisd.
Argynnis Jainadeva, Moore.
Argynnis Ruslana, Motsch.
Argynnis Anna, Blanch.
Argynnis Childreni, Gray.
Argynnis Aruna, Moore.
Acherontia Satanus.
Papilio Ascanius, Cram.
Cicrous Chorinaeus, Fabr.
Parthenos Tigrina, Voll.
Charaxes Epijasius Reiche.
Charaxes Kadenii, Feld.
Charaxes Jupiter, But.
Charaxes Etheocles, Cram.
Papilio Gundlachianus, Feld.
Dynastor Napoleon, Doubl. Hew.
Argynnis Sagana, Doubl. Hew.
Zeuxidia Aurelius, Cram.
Urania Rhipheus, Cram.
Urania Sloanus, Cram.
Nyctalemon Orontes, Linn.
Nyctalemon Cydnus, Feld.
Erasmia Pulchella, Hope.
Actias Maenas
Saturnia Derceto, Mssn.
Saturnia Argus, Drury.
Any Asiatic species of Parnassius.
Citheronia Phronima.
Castnia Daedalus, Cram.
Pyrameis Gonerilla, Fabr.
Pyrameis Abyssinica, Feld.
Pyrameis Dejeanii, Godt.
Pyrameis Tameamea, Esch.
Vanessa v. Hygiæa, Hdrch.
Vanessa v. Elymi, Rbr.
Dasyopthalmia Rusina, Godt.
Morpho Phanodemus, Hew.
Any species of Agrias.
Any species of Callithea.
Pandora Chalcothea, Bates.
Pandora Hypochlora, Cates.
Pandora Divalis, Bates.
Colias Viluiensis, Men.
Colias Ponteni, Wallengr.
Bunaea Deroyllei, Thom.
Bunaea Phaedusa, Dru.
Rinaca Zuleica, Hope.

These are a few of the very many of the rarer species that I am eager to procure; of course there are numberless others from all parts of the world, equally desirable and coveted by me.

North Atlantic Express Co.,

NEW YORK,
OFFICE, 71 BROADWAY.

Chartered by Special Act of Incorporation.

CENTRAL EUROPEAN OFFICE:
5 RUE SCRIBE, PARIS.

PRINCIPAL OFFICE IN GREAT BRITAIN:
4 Moorgate St., London, E. C.
B. W. & H. HORNE, AGENTS.

BRANCH OFFICES: Golden Cross, Charing Cross; George & Blue Boar, High Holborn; 108 New Bond Street; 474 New Oxford Street.

OFFICE IN LIVERPOOL: 5 Knowsley Buildings Tithebarn Street.

CONTINENTAL OFFICES: 5 Rue Scribe, Paris; 31 Kleine Reichenstrasse, Hamburg; 116 Lagerstrasse, Bremen.

Merchandise, specie, bullion, stocks, bonds, or other valuables and packages and parcels of every description, personal effects, baggage, etc., forwarded to and from Europe and all parts of the United States, the States and Territories of the Pacific Coast, British Columbia and the Canadas included, *at fixed Tariff rates, with no extra charges whatsoever for Custom-House brokerage, commission, delivery, etc., etc., the shipper or receiver being under no other care or expense than the stipulated freight from the point of shipment to place of delivery, and the amount of duties and government fees actually paid at the Custom-houses of the United States or Europe.*

For the convenience of shippers, where agencies of the Company are not established, packages or heavy goods may be forwarded to either of the offices or agencies of the Company, by either of the express or transportation companies in the United States, or by post, by railway, through the parcel delivery companies, or forwarding houses in any part of Great Britain or the Continent of Europe.

All packages, trunks, or parcels forwarded by this Company will be landed on arrival simultaneously with the mails, or immediately thereafter and will be entered at the Custom-house, duties paid and delivered to the parties to whom addressed in any part of Europe, the United States, the Canadas, or British Columbia, with the greatest possible dispatch. Transportation charges and duties collected on delivery, or may be prepaid, at the option of shipper.

Insurance against marine risk taken by the Company, when desired by the shipper, at the lowest current rates; premium payable in all cases in advance.

Shippers to or from any part of America, and Americans traveling in Europe, will find this the quickest, cheapest and most reliable medium of transportation, the business of this Company being conducted upon the well known prompt American express system, which has become so great a commercial necessity and convenience throughout the United States.

Purchases made, and collections and communications in every part of Europe and the United States promptly executed.

☞ Circulars sent on application to

S. D. JONES,
Manager,
71 BROADWAY, N. Y.

No. 5.   Price 50 Cents.

LEPIDOPTERA,

RHOPALOCERES AND HETEROCERES,

INDIGENOUS AND EXOTIC;

WITH

Descriptions and Colored Illustrations,

BY

HERMAN STRECKER.

Reading, Pa., July, 1873.

Reading, Pa.:
Owen's Steam Book and Job Printing Office, 515 Court Street,
1873.

NOTICE.

Subscribers who prefer to do so can remit the money for each part as they receive it, which will be in the United States, 55 cents per part, inclusive of postage. As soon as I have subscribers sufficient to pay the extra expense of printing, I will add another plate, so I trust Lepidopterists and Naturalists generally will exert themselves to increase my subscription list, which is not at present as large as that of the London Times.

TO PUBLISHERS, &c.

Advertisements of Publishers, Taxidermists, dealers in specimens of Natural History, Public Lecturers, Bird Fanciers, &c., inserted on the cover and fly leaves of each number, at very low rates, as my object is not to make money in this instance, but merely to cover the expense of issuing this work. Parties choosing to avail themselves of this means of advertising will please address,

HERMAN STRECKER,
Box 111 Reading P. O., Berks Co., Pa.

CATOCALA INSOLABILIS. Guenée.

Spec. Gen., Vol. VII, p. 94.

(PLATE V, FIG. 1, ♂.)

Expands 3 inches.
Thorax, above, dark grey; abdomen blackish; beneath white.
Upper surface, primaries greyish white, powdered with minute brown scales; the transverse lines are black; reniform small and surrounded by a brown annulus; a black apical dash; interior margin shaded with black; fringes dark grey.
Secondaries entirely black, with black fringes.
Under surface has bases of all wings white, rest black, with exception of slight indications of narrow white bands, most noticeable on the secondaries.
Habitat. New York, Pennsylvania, N. Jersey and Maryland.
Easy enough to distinguish from the other species by the dark shading of interior margin of upper surface of primaries, and the black fringes of secondaries; it is a slighter built insect than either Viduata, Lachrymosa or Desperata, to none of which does it bear any particular resemblance when placed side by side. This may rank among our rarer species as nowhere has it been found in any plenty.

CATOCALA DESPERATA. Guenée.

Spec. Gen., Vol. VII, p. 95.
Phalæna Vidua. Abbot & Smith Lepid., Georgia, Vol. II, p. 181, Pl. 91.

(PLATE V, FIG. 2, ♂.)

Expands 3 inches.
Head and thorax above, light grey, with distinct dark lines; abdomen blackish brown, beneath dirty white.
Upper surface, primaries light grey; transverse anterior line double, black, and, as well as the transverse posterior line, very distinct and well defined; reniform moderately large, oval, and surrounded by a double line; a black dash, broken in the middle, runs from base to sub-reniform; the usual black sub-apical dash, from which a dark shade passes to the reniform and from thence inwards and upwards to the costa; sub-terminal line joined inwardly with

very pale grey; the space from this latter to the transverse posterior line is brown of no very decided tint.

Secondaries, base covered with greyish hairs, rest of wing black, with broad pure white fringes.

Under surface, primaries white, with black marginal, median and sub-basal bands, which are confluent near interior margin; fringes white, with grey at the terminations of veins.

Secondaries white, with broad black marginal and narrower mesial bands; fringes white.

The caterpillar which is figured by Abbot feeds on various species of oak.

The commonest of all the black winged Catocalæ, and is found in most localities from New York to Florida.

There has been the most interminable confusion in regard to the identity of this species; for years it has been confounded with, and represented in American collections the C. Viduata or Vidua of Guenée, a larger and entirely distinct species peculiar to the Southern States; by comparing the figure of the latter on plate III of this work with that of the present species on plate V, the many obvious points of difference will be readily perceived without inflicting on me the misery of pointing them out piecemeal.

CATOCALA SUBNATA. Grote.

Proc. Ent. Soc. Phila. Vol. III, p. 326. (1864.)
Trans. Am. Ent. Soc. Vol. IV, p. 9. (1872.)

(PLATE V, FIG. 3 ♂.)

Expands 3¾ inches.

Head and thorax, above, pale grey with dark brown lines; abdomen bright ochre yellow; beneath yellowish white.

Upper surface, primaries greyish white with pale blueish and brown shades; transverse lines and other markings dark brown and very distinct; reniform medium size, sub-reniform large and open; fringes brown.

Secondaries bright yellow; marginal and mesial bands irregular and not extending to the interior margin; fringes yellow.

Under surface yellow, with all the black bands narrow.

Habitat. Middle and Southern States, of rare occurrence.

This has the appearance of being an improved edition of and is closely allied to C. Neogama, but can be easily distinguished from that species by its much greater size, the more brilliant yellow of abdomen and secondaries, and by the open sub-reniform, also the ground color of primaries is much lighter and the markings generally more prominent.

The ♂ figure on plate IV, Vol. III, Proc. Ent. Soc. Phil., which accompanied Mr. Grote's original description of C. Subnata, resembles it in size and shape, but the markings mainly, and the colors precisely are those of Neogama, it has even the closed sub-reniform which is one of Grotes' great points of distinction between the two species, as he says in his description* of Subnata the "Sub-reniform large, open, formed by a deep sinus of the t. p. line. †"

The type is in the Museum of the American Ent. Society.

CATOCALA NEOGAMA. Abbot & Smith.

Lepid. Georgia, Vol. II, p. 175, Pl. 88.
Guenee, Spec. Gen., Vol. VII, p. 96.
Duncan's Naturalists' Library, Vol. VII, p. 202, Pl. 26, fig. 1.

(PLATE V, FIG. 4 ♂, 5 ♀.)

Expands 3 inches.

Thorax above, grey; abdomen brownish yellow; beneath pale yellow.

Upper surface, primaries grey, with brown shades, markings dark brown, varying in distinctness in different examples; reniform, which is rather small and inconspicuous, is surrounded by a brown double line; sub-reniform small and not connected with the transverse posterior line.

Secondaries dark yellow, with irregular marginal and median bands which do not extend to the abdominal margin; apical spot and fringes yellow.

Under surface yellow, the black bands narrow and irregular.

The larva is figured by Abbot, who states that it feeds on the black American Walnut (Juglans Nigra); it is brown in color, with dark spots on the sides and two dark lines near the back, and "resembles the color of the bark so much as not to be discernable from it."

One of our commonest species found throughout the Atlantic States.

CATOCALA CLINTONII. Grote.

Proc. Am. Ent. Soc. Phil. Vol. III, p. 89, Pl. III, Fig. 4, ♀ (1864.)

(PLATE V. FIG. 6. ♂.)

Expands 2 inches.

Thorax whitish grey; abdomen yellow.

Upper surface, primaries very pale grey, tinged a trifle in the centre and on the exterior and interior margins with blueish; basal and other transverse lines fine but tolerably distinct,

*Trans. Am. Ent. Soc. Vol. IV, p. 10.
†Transverse posterior line.

a black longitudinal line runs from the base to the transvers anterior line ; reniform and sub-reniform pale and indistinct, the former surrounded with white.

Secondaries yellow ; median band which does not extend to the abdominal margin is narrow in the middle and broadest near the costa; marginal band of moderate width and broken between the first and second median veinlets, forming an oval spot near the anal angle ; apical spot yellow, fringes white.

Habitat. New York, Massachusetts, Rhode Island.

The species is evidently rare, the only examples I have yet seen are the types in museum of Am. Ent. Soc. and the example above described which was taken near Providence, Rhode Island.

It will be seen by referring to Mr. Grote's original description that there are a few points of difference between his type and my example, the principal of which is in the marginal band on upper surface of secondaries, which in the type is "continued" to within a short distance of secondaries, whilst in mine it is broken as above described, Grote's specimen also is a little larger, expanding 2 2-10 inches.

CATOCALA ANTINYMPHA, Hubner.

Ephesis Antinympha, Samml. Ex. Schmett.
Catocala Affinis, West, Drury, Vol. I, Pl. 24, fig. 6.
Catocala Melanympha, Guenee, Spec. Gen., Vol. VII, p. 98.

(PLATE V, FIG. 7 ♀.)

Expands 2½ inches.

Head and thorax above black ; abdomen brown ; beneath smoky grey.

Upper surface, primaries black, with the markings of a deeper and more lustrous shade.

Secondaries yellow, with black basal hairs, and rather regular marginal and mesial bands ; apical spot yellow ; fringes black, except at apex, where they are white.

Under surface yellow, with usual black bands, and otherwise much obscured with black.

Habitat. New York, Massachusetts, Pennsylvania, Maryland, &c.

A rare species, and easily distinguished from all others by the black primaries. In some examples the sub-reniform is white, which color is continued from thence upwards on inner side of reniform towards the costa, thus forming a diagonal white bar or patch across the middle of the wing.

CATOCALA FRATERCULA. Grote & Robinson.

Proc. Ent. Soc., Phila., Vol. VI, p. 21. (1866.)
Trans. Am. Ent. Soc., Phila. Vol. IV, p. 17. (1872.)

(PLATE V, FIG. 8 ♂.)

Expands 1½ inches.

Thorax, greyish white, abdomen grey.

Upper surface, primaries white, tinged and powdered with brown; sub-costal and transverse anterior lines fine and distinct, transverse posterior line produced outwardly in a single tooth opposite the reniform; a brown median shade starts from the inside of transverse posterior line at the second median veinlet, from thence passing inwards and upwards to the costa; the reniform, which is embraced within this brown space, is edged exteriorly and on side towards base of wing with a dark heavy line, but the sides towards the costal and interior margins have no defined boundary; sub-reniform small, white and closed; sub-terminal band or line white, from which to exterior margin the intervening space is brown; sub-apical dash distinct and dark brown; fringes brown.

Secondaries yellow, marginal band unusually broad and abruptly terminated near the second median veinlet, but is replaced near the anal angle by an oval spot; mesial band narrow and curved upwards where it terminates at the abdominal margin; some grey hairs mixed with the yellow at base of wings; fringes blackish, except at apex, where they are white.

Under surface pale yellow, darkest near base; very broad uninterrupted marginal and narrow median bands; no indication of sub-basal of primaries.

The description and figure were taken from an example captured at Providence, R. I., and for which I am indebted to Mrs. Bridgham of New York.

The types are in the Museum of the American Ent. Society.

CATOCALA UNIJUGA. Walker.

(PLATE V, FIG. 9 ♂.)

Expands 3 inches.

Thorax above, dark grey, abdomen light brown; beneath white.

Upper surface, primaries pale grey, densely powdered with black atoms; all the transverse lines geminate; reniform large and doubly ringed with black; joining the reniform interiorly is a white space; sub-reniform pale; sub-marginal crescents large and distinct; fringes white.

Secondaries red; marginal and mesial bands taper towards the abdominal margin which the latter does not quite reach; apical spot white; fringes of exterior margin white, of interior grey.

Under surface, primaries white, with broad black marginal and median bands, sub-basal band of paler color, and does not extend to interior margin.

Secondaries have inner two-thirds of same red as on upper surface, the costal third is white; black bands same as above, excepting that the mesial reaches in some examples quite to the margin; a discal lune connects with mesial band on its inner edge.

Habitat. Canada, New England, Middle and Western States.

This fine species, which is rather rare, belongs to the group of which C. Nupta is the type.

CATOCALA PARTA. Guenée.
Spec. Gen. Vol. VII, p. 84.

(PLATE V, FIG. 10 ♀.)

Expands 3 inches.

Thorax, above, grey; abdomen ochraceous; beneath, greyish white.

Upper surface, primaries grey shaded with yellowish brown, the whole surface has a smooth silky look, quite different from the squamose powdery appearance of C. Unijuga; the transverse lines, which are arranged much as in that species, are geminate, and as well as all the other markings are very distinct and clear. The sub-reniform is large and entirely disconnected from the transverse posterior line; between the sub-median vein and first median veinlet is a dark longitudinal line or shade, extending from the transverse posterior line to exterior margin; interior to the reniform is a pale patch; fringes grey.

Secondaries bright red; median band narrow and terminates in a point, two lines from the abdominal margin.

Under surface same as in C. Unijuga, except that the sub-basal band of primaries is as dark in color as the others.

Habitat. Canada, New England and Middle States.

In many localities quite common; the larva feeds on various species of Salix and the imago appears middle or end of July.

CATOCALA PERPLEXA. NOV. ? VAR.

(PLATE V, FIG. 11 ♂.)

Expands 3¼ inches.

Thorax, above, blueish grey, abdomen brownish yellow; beneath, white.

Upper surface, primaries dark blueish grey; transverse lines black and distinct as in C. Parta. Two white bars cross the wing, one formed by the intervening space between the transverse posterior and sub-terminal lines, the other by the large sub-reniform and the space adjoining it interior to the reniform; sub-apical dash black; fringes concolorous with the ground of wing.

CATOCALA PERPLEXA.

Secondaries red, of a somewhat deeper shade than Parta; median band extends almost to the abdominal margin; fringes white.

Under surface resembles that of Parta in a great measure, the principal difference being that the mesial band extends almost to the interior margin, in which respect it is nearer to some examples of Unijuga.

Two examples from which the above description was taken were captured in the vicinity of Brooklyn, N. Y., by Mr. Julian Hooper of that city, one of which, the original of the figure on plate V, he generously added to my cabinet.

I am not partial or addicted to the divertisement of hunting Lepidopterological mare's nests, but I must confess that this insect has perplexed me considerably; I showed it in company with Parta to a valued entomological friend, asking him if he thought it might be the latter, "I would not like to figure it as the typical form" was the answer, so, without arriving at any definite conclusion, I have offered the figure for the inspection of lepidopterists, and with much doubt provisionally cite it as a variety of Parta; the first and principal differences are the dark blueish color, and two conspicuous white bars of primaries, neither is there that soft smooth appearance so noticeable in Parta, there being more of a tendency to squamosencss as in Unijuga; then again the sub-reniform is connected with the transverse posterior line whilst in Parta it is entirely isolated, there are besides many minor points of difference and altogether, after frequent examinations, I am completely at a loss what to think about it, especially as both Parta and Unijuga, the two species to which it is the nearest, (if it be not identical with one or the other,) have less tendency to variation than any others I wot of, and it would be perhaps venturing too far to hazard the conjecture that it be the result of a love affair between those two.

July, 1873.

CATOCALA CONCUMBENS. WALKER.
Cat. B. M.

(PLATE V, FIG. 12 ♂)

Expands 2¾ inches.

Head and collar chestnut brown; thorax ashen grey; abdomen light brown; beneath white.

Upper surface, primaries almost unicolorous, pale silvery grey with slight shades of light brown; sub-basal, transverse anterior and posterior lines black, very fine and broken, being in many places obsolete; reniform indistinct and margined with white or light grey; sub-reniform open; fringes same color as wing.

Secondaries rose color with both bands broad and even, neither of them extend to the abdominal margin; fringes yellowish white.

Under surface, primaries white with usual dark bands; inner base of secondaries rosy, outer half white; mesial band contracted at both ends.

In Vol. II, Proc. Ent. Soc. of Phila., Mr. W. Saunders, of Canada, thus describes the larva: "length, two to two and-a-half inches, onisciform. Head flat, dark greyish, intermixed with red. Upper surface dirty brown, with a lightish chain-like dorsal stripe and a very small fleshy protuberance on each side of this stripe on each segment. On ninth segment is a small protuberance of a brownish color, and on the eleventh a mark resembling an oblique incision. A thick lateral fringe of short hair close to the under surface. Under surface pinkish, with a central row of round black spots which are larger about the middle of the body and much smaller towards the extremities. Food-plant, willow."

Habitat. Canada, Eastern and Middle States; rare in Pennsylvania, but more plentiful in Massachusetts and other New England States.

This lovely insect is nearer allied to the European C. Pacta than to any American species; it is a little larger than Pacta and the color of the abdomen is different, (being rosy in that species,) otherwise it resembles it very closely in most respects.

I hope this second plate of Catocalidæ will meet with the same hearty approval as did the first, (plate III of this work,) and, as I promised in that number, I will, if I live, in due time give figures of every known North American species.

Anarta Cordigera, Thnbg.—Anarta Luteola, Grote and Robinson.

I have compared examples of Anarta Cordigera with the types of Anarta Luteola, in the Museum of the Am. Ent. Soc., and can find not the slightest difference between the two, although Grote and Robinson say, in their description[*] of Luteola, "the differences between the species are perhaps sufficiently great to render a detailed comparison unnecessary," perhaps like the large lettered names on maps and placards, they are so great that no one ever notices them; in the above instance, after the closest examination, I cannot find a single point that would in the slightest degree indicate a specific distinction.

[*] Proc. Ent. Soc., Phila., Vol. IV, p. 493-4, (1865,)—the fig. of Luteola is on Plate III.

From the Canadian Entomologist Vol. 5, No. 6, pp. 117, 118, 119. June, 1873.

PERSONAL.—In part No. 2, "Lepidoptera, Rhopaloceres and Heteroceres," the author, Mr. Herman Strecker, makes a most uncalled-for and ungentlemanly attack on me, which in justice to myself, much as I dislike introducing matters of this sort into a scientific periodical, I can scarcely allow to pass unnoticed.

It appears that Mr. Strecker received last summer, from Mr. Couper, specimens of a Papilio which he had taken on the Island of Anticosti while on a collecting tour there. At first Mr. S. says he thought it might be my *P. brevicauda*, described in a foot note in "Packard's Guide," but on comparing the description there given with his specimens, he found them to differ in some important particulars. He then proceeds to say (I copy verb. et lit.) "I now again had the pleasant excitement incidental to endeavoring to study out bare descriptions, unaccompanied by figures, and in my misery I wrote to Mr. Couper, in Montreal, requesting him to try to see the types of Brevicauda, and compare his examples with them, or if that was impossible, to write to Mr. Saunders, of Ontario, Canada, who described it, and with whom he was acquainted, concerning the species; after some time Mr. Couper wrote 'I communicated with the Rev. Canon Innes (in whose collection are specimens of Brevicauda) and Mr. W. Saunders, asking for information regarding P. Brevicauda; up to this instant no answer from either.' This certainly was not very satisfactory, but as I was not particularly anxious to make a fool of myself by re-christening old species, I importuned Mr. Couper to try the gentleman with another epistolary shot; in due time, under date March 17, 1873, came another letter from Couper thus: 'I have purposely delayed a reply to your favor of 2nd, because since its receipt I wrote again to Mr. W. Saunders for the desired information, and my letter was written in terms which could not deter him from answering; however, no answer has been received.' After receiving this letter, I, of course, concluded that Mr. Saunders' time was of too much value to be encroached upon, and requested Mr. Couper to by no means trouble him again, as his dignified silence at last brought me to a proper sense of my true position, and was a merited punishment to both Couper and myself for our temerity."

I did receive the two letters referred to from Mr. Couper. In the first, dated Jan. 21, Mr. C. asks me where I obtained the Papilio described as *brevicauda*, and whether I would loan him a specimen, as he wished to compare it with some Anticosti Papilios which had been named for him by his U. S. correspondents as *P. polyxenes*. There were other matters referred to in the letter which I wished to attend to before replying to Mr. Couper, and as I was then extremely busy, and was obliged to leave home for a while, not knowing either that there was any pressing need of an immediate answer, I deferred writing for a time. In the second letter, dated March 3rd, Mr. C. refers again among other matters to *P. brevicauda*, expresses no disappointment at my not answering his first, does not even now ask for a prompt reply, or hint that any of the information he desires was for anyone but himself. Indeed, after referring to some differences which he thought existed between his Anticosti specimens and my *brevicauda* from Newfoundland, he says: "It is my intention to investigate this matter further," and referred to the opportunities he hoped to have on revisiting the Island. To this second letter I replied as promptly as possible, within a few days, and gave Mr. C. all the information in my power in reference to *brevicauda*, as well as satisfactory reasons why I had not written sooner.

It was scarcely kind of Mr. Couper to give me no hint of the terrible state of excitement under which his friend, poor Mr. Strecker, was at that time laboring, boiling over, as he evidently was, with indignation towards one who was perfectly innocent of all knowledge of his wants. Had I known the state of his mind my sympathies would at once have been aroused and I should have written promptly, when I suppose this formidable bull of his would never have been fulminated against me, and I should have been spared from being impaled on the sharp end of Mr. Strecker's irony, where, like a beetle on a pin, I am now supposed to be wriggling and writhing in great discomfort.

I do not know Mr. Strecker and have never had any correspondence with him, but I do feel sorry for him, that he should in his anger have allowed himself to use language so discourteous in reference to one who was a perfect stranger to him, without taking pains to enquire whether it was deserved or not. I can scarcely designate such a proceeding under such circumstances, as anything less than contemptible, and quite unworthy of a naturalist or a gentleman.

Mr. Strecker further remarks in the paragraph following that last quoted: "However, I believe this is distinct from Brevicauda, and if it be not, *it is an absurdity to retain that name*; the probability after all is that Brevicauda and Anticostiensis (if they be not the same) are both varieties of *Asterius*." Why Mr. Strecker considers it *absurd* to call a species *brevicauda* he does not deign to inform us; can it be that he has a conscientious objection to any further references to the tails of insects under any circumstances, or is it the *evident superiority* in length and grandiloquence of sound which *Anticostiensis* has over *brevicauda* which makes the use of the latter to his mind so absurd? It does seem strange that with all Mr. Strecker's anxiety to avoid "re-christening old species," he should astonish the Entomological world with such a name as *Anticostiensis* nov. sp., when at the same time he states his belief in the probability of its being but a variety of *Asterius*. Such a proceeding seems at least contradictory, and, it will appear to some, as if he had thus placed himself, in his anxiety to have his name attached to a species, in the very position he professes a wish to avoid, and which he has designated in such choice ! language.—W. SAUNDERS, London, Ontario.

My thanks are due to Mr. W. Saunders of the editing committee of the Canadian Entomologist for the above splendid and entirely unexpected advertisement of this work, which I find in the June number of that publication just issued ; in looking over the ever-welcome pages of the Entomologist, I found commendatory notices of twenty-five or more serials, but the present work on Lepidoptera was not, from some accidental cause, I supposed, among the number, but what were my feelings on turning a few pages further to find, that whilst such standard works as Hardwicke's Science Gossip, Proc. Acad. Nat. Sc., Phil., Newman's Entomologist, &c., had received but a few lines of editorial notice, my own poor work, owing to the generosity of Mr. Saunders, one of the able corps of editors, had assigned to it two and-a-half pages ! the emotions of Mr. Saunders must have been intense ; he gives me his sympathy, his compassion, he feels sorry for me to a degree that evidently lacks language to express itself, and, finally, in his wonderful and rare self-abnegation, his extreme modesty, (that unerring index of all great minds,) he affects to ask so unworthy an individual as myself for information, not directly it is true, but delicately by intimation ; of course Mr. Saunders knows already anything I can tell him, and every one knows that he does too, which only further enhances the delicacy of the compliment ; but still, far be it from me to be so lacking in respect to the Ent. Autocrat of the Canadas as not to go through the form of a reply, at least ; the reason I did not, to quote Mr. Saunders, "deign to inform us," (the "us," I humbly presume, means Mr. Saunders,) why it was absurd to call a species Brevicauda was simply because the information was already given on the last few lines of page 11 and the first couple of page 12 of this work ; by reading the whole of the article on Papilio Anticostiensis, it will be seen that the expression, that it would be absurd to retain the name of Brevicauda, was provisional, and only in the event that Brevicauda and Anticostiensis were the same, for the reason that the tails of the latter are no shorter than in many other species, such for instance as P. Zolicaon, Machaon, Hospiton, Philenor, Sadalus, Agamemnon, Indra, Xuthulus, &c., so that if it should eventually be proven that Anticostiensis is identical with Brevicauda I certainly would, in my woeful ignorance, persist in considering the name Brevicauda absurd when applied to a species with tails no shorter than twenty others of the same genus.

Regarding my astonishing "the Entomological world with such a name as Anticostiensis," I can only say if the said world is astonished, it has experienced the same sensation before, at least that part of it whose studies extend beyond the narrow confines of some province or county, for such names as Neelgherriensis,[*] Ladakensis,[†] Mephistopheles,[‡] Goschkevitschii,[§]

[*] Lethe Neelgherriensis, Guerin.
[†] Colias Ladakensis, Feld.
[‡] Heterochroa Mephistopheles, Butl.
[§] Lasiommata Goschkevitschii, Men.

Hampstediensis,* Chimborazium,† Zamboanga,‡ Mahallakoena,§ Madagascariensis,|| &c., were bestowed on rare and lovely insects by the giants of Entomology, men from whose dictum there can scarcely be an appeal.

I have now I trust done my duty towards Mr. Saunders, I can but make my salaam, reiterate my thanks, and adjure him not again to let his emotions so far overpower him as to make him lose his temper in the futile attempt to prove I had lost mine, a thing which I scrupulously avoid, as it unfits one for business or rational pleasures and, worst of all, it spoils digestion. HERMAN STRECKER, Reading, Pa.

The following letter, which I received just as the above was going to press, needs no comment of mine, and as I have the writer's permission to use it as I please, I think the very best use I can make of it is to lay it before the "Entomological World," not with the impression, however, that it will be astonished thereat after reading the foregoing pages:

ROOM 4, No. 117 BROADWAY, NEW YORK, August 5, 1873.

My Dear Sir:—You have doubtless read Mr. Saunders' reply to your observations in the 2nd No. of your book. Not very satisfactory to you, I presume, and not very creditable to Saunders.

The reply, however, recalls to my memory a circumstance which occurred now about a year since, and which is strongly illustrative of Saunders' supercilious behavior, as I deem it.

Believing him to have access to better libraries and larger collections than were within my reach, and furthermore induced by the invitation extended to amateurs in columns of the Canadian Entomologist, to send their collections to the Society for determination, I sent by the hand of a personal friend a box of insects with the proper request.

Not being sure that either Mr. Baynes Reed or Mr. Saunders was in London, I requested my friend to deliver the box to either party.

I did not feel quite sure of my friend, so after waiting a couple of months, without receiving any notice of the receipt of the box, I wrote both to Reed and Saunders enquiring if they had received such box. No answer came from either. I then caused enquiries to be made of my friend as to whom the box had been delivered, and the answer came, "to Mr. Saunders, on the day after my arrival in London." I again waited, perhaps another month, but no box, or acknowledgement of its receipt, arrived.

I then wrote again to Saunders, stating all the circumstances, and requesting return of box, but up to this moment no reply or notice of any kind has been received.

As I am Agent here for the Entomologist, and have certainly done something toward extending its circulation, I call this rather cavalier treatment, while even towards a stranger I think Mr. Saunders' conduct, (to use his own words,) "to be unworthy of a naturalist or a gentleman." Yours, truly,
W. V. ANDREWS.

You are at liberty to make any use you please of this note.

It is reported that Commander Greer of the Tigress, the vessel which sailed a short time since in search of the crew of the Polaris, has said that there is to be no time wasted pickling fish, bottling bugs, &c.; that the expedition will attend only to the object of its mission—the finding of the Polaris. If he even thought so, it is a disgrace to give utterance to such expressions, for, if but one new fact in science were attained, what, in comparison, are whole hecatombs of paltry human lives, which, as one flickers out, legions arise to fill the place. When thousands have again and again been ignobly, ruthlessly, sacrificed in useless and foolish wars, the offspring of insane ambition, why should any one murmur at life endangered, or lost, in the noble cause of science.

*Cynthia Hampstediensis, Steph.
†Nymphidium Chimborazium, Bates.
‡Pieris Zamboanga, Feld.
§Lycaena Mahallakoena, Walker.
||Godartia Madagascariensis, Lucas, and Crenis Madagascariensis, Boisd.

Of the following species I am anxious to obtain examples, either by exchange or purchase; any Naturalists having duplicates of any of them will confer a great favor by communicating with

HERMAN STRECKER,
Box 111 Reading, P. O.,
Berks Co., Pennsylvania, U. S. of N. America.

Ornithoptera Hippolytus, Cram.
Ornithoptera Amphrysus, Cram.
Ornithoptera Helena, Linn.
Ornithoptera Croesus, Wall.
Ornithoptera Brookiana, Wall.
Papilio Evan, Doubl.
Papilio Pericles, Wall.
Papilio Blumei, Boisd.
Papilio Macedon, Wall.
Papilio Philippus, Wall.
Papilio Arcturus, West.
Papilio Phorbanta, Linn.
Papilio Homerus, Fabr.
Papilio Garamas, Hub.
Papilio Caiguanabus, Pocy.
Argynnis Rudra, Moore.
Argynnis Oscura, Evers.
Argynnis Cnidia, Feld.
Argynnis Jerdoni, Lang.
Argynnis Dexamene, Boisd.
Argynnis Jainedeva, Moore.
Argynnis Ruslana, Motsch.
Argynnis Anna, Blanch.
Argynnis Childreni, Gray.
Argynnis Aruna, Moore.
Acherontia Satanas.
Papilio Ascanius, Cram.
Coerous Chorimens, Fabr.
Parthenos Tigrina, Voll.
Charaxes Epijasius Reiche.
Charaxes Kadenii, Feld.
Charaxes Jupiter, Bat.
Charaxes Etheocles, Cram.
Catagramma Excelsior, Hew.
Papilio Wallacei, Hew.
Papilio Slateri, Hew.
Papilio Endochus Boisd.
Dyctis Bioculatis, Guerin.
Romaleosoma Sophron, Dbldy.
Romaleosoma Pratinos, Dbldy.
Romaleosoma Arcadius, Fabr.
Nymphalis Calydonia, Hew.
Saturnia Epimethea, Dru.

Papilio Gundlachianus, Feld.
Dynastor Napoleon, Doubl, Hew.
Argynnis Sagana, Doubl, Hew.
Zeuxidia Aurelius, Cram.
Urania Rhipheus, Cram.
Urania Sloanus, Cram.
Nyctalemon Oroates, Linn.
Nyctalemon Cydnus, Feld.
Erasmia Pulchella, Hope.
Actias Moenas
Saturnia Derceto, Mssn.
Saturnia Argus, Drury.
Any Asiatic species of Parnassius.
Citheronia Phronima.
Castnia Daedalus, Cram.
Pyrameis Gonerilla, Fabr.
Pyrameis Abyssinica, Feld.
Pyrameis Dejeanii, Godt.
Pyrameis Tameamea, Esch.
Vanessa v. Hygiaea, Hdrch.
Vanessa v. Elymi, Rbr.
Dasyophthalmia Rusina, Godt.
Morpho Phanodemus, Hew.
Any species of Agrias.
Any species of Callithea.
Pandora Chalcothea, Bates.
Pandora Hypochlora, Cates.
Pandora Divalis, Bates.
Colias Viluiensis, Men.
Colias Ponteni, Wallengr.
Bunaea Desvoyllei, Thom.
Bunaea Phaedusa, Dru.
Rinaea Zuleica, Hope.
Papilio Disparilis, Herr-Sch.
Acraea Perenna, Dbldy.
Opsiphanes Boisduvalii, Dbldy.
Clothilde Jegeri, Herr-Sch.
Pieris Celestina, Boisd.
Euplea Eurypon, Hew.
Papilio Panariste, Hew.
Limenitis Lymire, Hew.
Io Beckeri Herr-Sch.
Eacles Kadenii, Herr-Sch.

These are a few of the very many of the rarer species that I am eager to procure; of course there are numberless others from all parts of the world, equally desirable and coveted by me.

North Atlantic Express Co.,

NEW YORK,
OFFICE, 71 BROADWAY.

Chartered by Special Act of Incorporation.

CENTRAL EUROPEAN OFFICE:
5 RUE SCRIBE, PARIS.

PRINCIPAL OFFICE IN GREAT BRITAIN:
4 Moorgate St., London, E. C.
B. W. & H. HORNE, AGENTS.

BRANCH OFFICES: Golden Cross, Charing Cross; George & Blue Boar, High Holborn; 108 New Bond Street; 474 New Oxford Street.
OFFICE IN LIVERPOOL: 5 Knowsley Buildings Tithebarn Street.
CONTINENTAL OFFICES: 5 Rue Scribe, Paris; 31 Kleine Reichenstrasse, Hamburg; 116 Lagerstrasse, Bremen.

Merchandise, specie, bullion, stocks, bonds, or other valuables and packages and parcels of every description, personal effects, baggage, etc., forwarded to and from Europe and all parts of the United States, the States and Territories of the Pacific Coast, British Columbia and the Canadas included, *at fixed Tariff rates, with no extra charges whatever for Custom-House brokerage, commissions, delivery, etc., etc., the shipper or receiver being under no other care or expense than the stipulated freight from the point of shipment to place of delivery, and the amount of duties and government fees actually paid at the Custom-houses of the United States or Europe.*

For the convenience of shippers, where agencies of the Company are not established, **packages or heavy goods** may be forwarded to either of the offices or agencies of the Company, by either of the **express or transportation** companies in the United States, or by post, by railway, through the parcel delivery companies, or forwarding houses in any part of Great Britain or the Continent of Europe.

All packages, trunks, or parcels forwarded by this Company will be landed on arrival simultaneously with the mails, or immediately thereafter and will be entered at the Custom-house, duties paid and delivered to the parties to whom addressed in any part of Europe, the United States, the Canadas, or British Columbia, with the greatest possible dispatch. Transportation charges and duties collected on delivery, or may be prepaid, at the option of shipper.

Insurance against marine risk taken by the Company, when **desired by the shipper, at the lowest current rates**; premium payable in all cases in advance.

Shippers to or from any part of America, and Americans travelling in Europe, will find this the quickest, cheapest and most reliable medium of transportation, the business of this Company being conducted upon the well known prompt American express system, which has become so great a commercial necessity and convenience throughout the United States.

Purchases made, and collections and communications in every part of Europe and **the United States promptly** executed.

☞ Circulars sent on application to

S. D. JONES,
Manager,
71 BROADWAY, N. Y.

No. 6. PRICE 50 CENTS.

LEPIDOPTERA,

RHOPALOCERES AND HETEROCERES,

INDIGENOUS AND EXOTIC;

WITH

Descriptions and Colored Illustrations,

BY

HERMAN STRECKER.

Reading, Pa., August, 1873.

Reading, Pa.:
OWEN'S STEAM BOOK AND JOB PRINTING OFFICE, 515 COURT STREET,
1873.

PAPILIO DAUNUS. Boisduval.

Spec. Gen. I., p. 342, n. 182. (1836.)
Ridings, Proc. Ent. Soc., Phil., I, p. 278. (1862.)

(PLATE VI, FIG. 1 ♂, 2 ♀.)

MALE. Expands 3⅞ to 5½ inches.

Antennæ black, body yellow with a broad black dorsal, and narrow lateral and ventral bands.

Upper surface chrome yellow; primaries with six transverse black bands, all of which, with the exception of the marginal, are very narrow; first is basal; second extends from costa to interior margin; third from costa to first median veinlet; fourth along the disco-cellular veins, from third sub-costal veinlet to second radial vein; fifth terminates in some scattered atoms at the second radial veinlet; sixth, broad, extending along the whole of the exterior margin and divided transversely by a row of almost confluent yellow lunules.

Secondaries have the basal and second black bands of primaries continued to near the anal spot, where they are united; a broad marginal band, with six large crescents, the one nearest the anal angle fulvous, the others yellow, more or less tinged with fulvous; anal spot, which is also fulvous, is surmounted by a blue crescent, some patches of blue scales on the border, interior to the sub-marginal lunules; four tails, the outer and innermost of which are the shortest, that next the outermost is longest, and the remaining one is half the length of this latter; emarginations yellow.

Under surface much paler; primaries marked as on upper surface, except that the sub-marginal lunules are replaced by a broad yellow band.

Secondaries have the sub-marginal lunules, which are, with the exception of the two nearest the anal angle, larger than above, succeeded by yellowish grey atoms, edged interiorly with shining blue, which is surmounted with black, adjoining which, between the abdominal margin and second sub-costal veinlet, are five triangular rufous spots or flames; discal are black; second and third median veinlets edged with black scales.

Female same as male, but all the markings are much heavier, and, on upper surface, not so intensely black as in the male.

Habitat. Colorado, Kansas, Mexico, Guatamala.

The above description of this superb species applies more particularly to the tropical form, found in Mexico and Central America; the examples from Colorado present some few points of difference in the male, in that the third and fourth transverse bands of primaries do not extend beyond the median vein, and the fifth is almost obsolete; on secondaries the lunule nearest the anal angle is the only one that is fulvous, and there are no red flames on the under surface.

Mr. Ridings, who, in 1864, took several of this species in the Rocky Mountains, says it is rare and difficult to capture, owing to its high flight and the almost inaccessible nature of its haunts.

Friend Sachs, of New York, added another to the numberless favors already conferred by loaning me from his collection the original of the ♂ figures for the purpose of illustrating the accompanying plate; the ♀ is from one of a number which I received from Vera Cruz, Mexico.

PAPILIO ZOLICAON. Boisduval.

Ann. Soc. Ent. Fr., p. 281. (1852.)
P. Zolicaon, Lucas, Rev. Zool., p. 136. (1852.)
P. Machaon, var. Californica, Menetries, Cat. Mus. Petr. Lep. I, p. 69. (1855.)

(PLATE VI, FIG. 3 ♂.)

MALE AND FEMALE. Expands 3 to 3½ inches.

Antennae black; head and thorax black with two yellow lines; abdomen black with a lateral yellow band.

Upper surface rich yellow, primaries with a large black basal patch, between which and the disco-cellular veins is a broad black band, extending from costa to median vein, another covers the disco-cellular veins and reaches to the fourth radial vein, beyond this, between the costa and fourth sub-costal veinlet, is a black dash, immediately below this and joining it is a round spot; a black marginal band, containing a row of yellow spots, round near the outer angle, and becoming lunate as they approach the inner, the one nearest to which is geminate; nervures defined with black.

Secondaries, abdominal margin black; discal arc, as well as the veins, black; a very broad black marginal band; sub-marginal lunules yellow, above these, within the marginal band, a row of shining blue crescents; anal eye large, red, edged below with yellow and pupilled with black; tails same as in P. Machaon and kindred species; emarginations yellow.

Under surface paler; primaries marked much as above. Secondaries, interior to the sub-marginal lunules, a band of greyish yellow edged with blue; adjoining the marginal band the wing is tinged with fulvous.

Habitat. California, Oregon, Vancouver's Island.

Although bearing a striking resemblance to P. Machaon, and particularly to its variety Sphyrus[*], I believe this to be a distinct species, especially as the true P. Machaon is found in the northern parts of our possessions and in British America; but even this Mr. Scudder considers distinct from the typical P. Machaon, and has named it P. Aliaska[†], but I do not think on sufficient grounds, as, after a rigid comparison, I do not find it to differ from the European types more than do examples from the Himalayas, China, Turkey, &c., which is very little, indeed.

[*]Hubner, Samml. Exot. Schmett., f. 775, 776. (1818-1827.)
[†]Scudder, Ent. Notes, II, p. 45. (1869.)

PAPILIO ASTERIOIDES. Reakirt.
Proc. Acad. Nat. Sc., Phil., p. 331, n. 27. (1866.)

(PLATE VI, FIG. 4 ♀.)

Male. Expands 3½ inches.
Head and body black; patagiæ yellow; abdomen with dorsal and lateral rows of yellow spots like Asterius.

Upper surface black; primaries, a sub-marginal row of eight yellow spots, the one nearest the inner angle is oblong and sometimes connected at lower end with the yellow emargination, the others are round, or nearly so; an inner regular band of eight triangular yellow spots, the one between the first median nervule and sub-median nervure is broadest, the next below it is the narrowest, the rest are pretty much of one size; further in, near the costa, is a small round yellow spot.

Secondaries, six yellow sub-marginal lunules; a yellow mesial band, divided by the veins into seven parts, between this and the sub-marginal lunules are clusters of shining blue atoms, the narrowest and brightest of which surmounts the anal eye, which is orange, margined below with yellow and pupilled with black; emarginations yellow.

Under surface brown, ornamentation same as above, with the addition of a small discal bar on primaries, but the sub-marginal spots and lunules are paler, and the triangular spots composing inner band of primaries, with the exception of the one nearest the costa, are fulvous; those of mesial band of secondaries also fulvous, of a richer shade and margined interiorly with yellow; the sub-marginal lunules, except the two nearest the anal angle, tinged with fulvous on the inner side, between these and the mesial band is a row of irregular crescents, composed of yellowish and blue scales, after the manner of Asterius and allied species; anal eye as above; tails like Asterius.

Female same expanse and color as male; all the wings broader; inner band of primaries a little broader and of same width throughout; pupil of anal ocellus small.

Mr. Reakirt's type (♂) has the spots forming the inner band of primaries much suffused with black, the suffusion increasing as they near the costa, where the last few become obsolete, or almost so; there is also a variation in the anal spot, the pupil of which, instead of being a round spot in the centre, extends across its whole breadth, cutting it into two parts, the upper of which is red, the lower yellow; this, I am convinced, however, has no specific value, as I have met with the same peculiarity in Asterius. This type in the Museum of the Am. Ent. Soc., Phil., is from Mexico; the ♂ and ♀ in my collection are from Costa Rica.

I received in the sending, along with it, examples of Sadalus taken at the same time, in the same locality.

This section of the genus Papilio bids fair to become involved in almost as hopeless a state of confusion as that at present enjoyed the by Coliades.

The geographical range of Asterius and congeners is as follows:—the ordinary form of Asterius, with but little variation, occurs from Canada to Florida, inclusive, and from Maine westward to the Rocky Mountains; in Newfoundland is Brevicauda; the southern coast of Labrador and Island of Anticosti produce Anticostiensis; Colorado, and probably other of the western territories, has Indra and Asterius, and, finally, in Mexico and Central America are Asterioides and Sadalus.

P. Bairdii, another variety or closely allied species, which I have not yet had opportunity to closely examine, was taken also in Mexico.

In view of the above premises, I must adhere to my first belief, that Indra and Sadalus are true species; Asterioides, I have not a particle of doubt, is the tropical form of Asterius, and Anticostiensis may be the sub-Arctic, whilst Brevicauda, if it be not identical with the latter, is a segregated type peculiar to Newfoundland.

It is curious to note, that whilst Asterius, occupying a vast extent of country lying intermediate between those in which Anticostiensis and Asterioides, &c., are found, should have the macular bands on wings of female almost obsolete, whilst the contrary is the case in the extreme northern and southern forms which unite in the peculiarity of the female having the bands of as great and greater width than in the male. Thus it is strange to see how extremes meet, and how wonderfully, like "a circle that ever returneth into the self-same spot," are the works of nature brought to harmonize under the unerring direction of the Great Master.

All my examples of Asterius ♂ ♀ from Peninsular Florida are without the round black spot situated in the division of macular band of primaries nearest the costa; in all the specimens of Asterius from other localities that I have ever seen, also of Indra, Sadalus and Anticostiensis this spot is prominent, in the majority of instances it is so large as to divide the yellow space into two parts.

The macular band of primaries in the ♀ Asterius above alluded to, from Florida, is of greater width than in any examples I have yet seen from other localities. The females present no other differences from examples found elsewhere than in the absence of the black spot alluded to.

Three other curious varieties of Asterius have come under my observation; the first, and perhaps most remarkable of which, is the one described by Mr. Grote under the name of Papilio Calverleyi,[c] from a male captured in Queens County, Long Island, August, 1863; a female was subsequently taken by Mr. T. L. Mead in Florida. It is a beautiful insect, the same size and form as Asterius, and with the same dorsal and lateral spots on the body; the wings have the basal half black and the outer half yellow, inclining to orange on a portion of the secondaries; a very narrow black marginal band; faint indications of sub-marginal lunules. Under surface nearly the same as above, but with more orange on secondaries; male and female resemble each other closely. Messrs. W. H. Edwards and Grote contend it is distinct from Asterius; for my part, I think it a most interesting variety of that species.

The second example in Mr. T. L. Mead's collection is an undoubted female, but is marked precisely as in the male.

^c Grote, Proc. Ent. Soc., Phil., II, p. 441, ♂. (1864.)
Mead, Am. Nat., III, p. 332, ♀. (1869.)

The third is a hermaphrodite, taken by Prof. J. E. Meyer in Brooklyn, in 1863, now in possession of Mr. W. H. Edwards, who described it in Proc. Ent. Soc., Phil., IV, p. 390, both the right wings are male and the left ones female, with no suffusion or mixing of color.

Since describing P. Anticostiensis on page 10, Mr. Couper has made another entomological tour to the Island of Anticosti, and, among other results of his most commendable enterprise, are some forty specimens of this species, taken at Ellis Bay, about 117 miles west of Fox Bay, where he took the types figured on plate II. I have examined twelve of these, male and female, and find they agree with the types and appear to be subject to scarcely any variation, except in the length of the tail, which varies in different examples from 3-16 to 5-16 of an inch; the size of this appendage is, however, valueless for specific purposes, as in P. Philenor it is found from ⅛ to ½ of an inch, in P. Agamemnon from a mere tooth to nearly ½ of an inch, in P. Pammon from ⅛ to ⅜ of an inch, and the same difference in length occurs in many others.

Mr. Couper also secured the egg and larva; the former, he states, "are laid singly on the leaves of Archangelica Atropurpurea which occurs common throughout the whole extent of the Island; the egg is spherical and pale yellow." The larva, which I will figure on my next plate of diurnals, is pale green with a transverse row of black spots or dashes on each segment, the lateral ones running obliquely; from these spots emanate little points; unfortunately, Mr. Couper could not sojourn on the island long enough to obtain the fully matured larva; the one just described is ¾ of an inch long. Mr. Couper also took at Anticosti this summer a dozen examples of Colias which will doubtless tend to increase the muddle into which that interesting genus has been thrown through the indefatigable labors of our lepidopterists.

ANTHOCHARIS LANCEOLATA. Boisduval.

Ann. Ent. Soc., Fr. (1852.)
Lucas, Rev. Zool., p. 338. (1852.)
Anthocharis Edwardsii, Behr, Trans. Am. Ent. Soc., Vol. II, p. 304. (1862.)

(PLATE VI, FIG. 5, ♂.)

Expands 1⅞ inches.
Antennæ white, club black, tipped with white at extremity; head and body black above, beneath white.

Upper surface white; primaries, some black scales at base; a broken black or dark brown apical patch, varying in extent in different examples, but in none that I have seen is it as heavy as in A. Ausonides; a black discal spot. In the secondaries the marbleing of under surface is partially visible; some black at base as in primaries.

Under surface white; primaries, some fine brown reticulations near the anterior angle; discal spot black.

Secondaries marbled with fine brown lines which become almost confluent on inner half, especially at costa, where there are several irregular shaped pure white spots, the one nearest the outer angle being the largest.

Female resembles the male.

Habitat. California.

This has the reputation of being one of the rarest of the Californian butterflies; it resembles in form and general appearance the female of our eastern A. Genutia, but is larger, and the ornamentation of under surface of secondaries is different from any other American, or, in fact, any species I am acquainted with.

ANTHOCHARIS JULIA. Edwards.

Trans. Am. Ent. Soc., Vol. IV, p. 61. (1872.)

(PLATE VI, FIG. 6, ♂, 7, ♀.)

MALE. Expands 1½ inches.

Head and body black above, yellowish beneath; antennæ blackish, club with yellow tip.

Upper surface white; primaries, black at base; an orange apical patch, margined by a black band outwardly, the inner edge of which is serrated, inner half of the orange patch is powdered with black atoms; costa, from disco-cellular veins to base, blackish; an S shaped discal mark extends to the costa.

Secondaries black at base; ciliæ black at termination of veins.

Under surface white; primaries, apical part greenish, the orange spot not near so vivid as above; discal mark almost divided in two at the centre; costal margin with some indistinct markings.

Secondaries variegated with greenish grey in a manner nearer approaching A. Genutia than any other species, that is, all the marks are connected with each other and have a foliated, not a spotted appearance, as in A. Sara and ? var. Reakirtii; veins yellow.

FEMALE. Expands 1½ inches.

Upper surface lemon yellow, marked much as in male, but the orange spot does not extend as far outward, being bounded exteriorly by an irregular blackish sub-marginal line, the space between which and the outer margin is yellow.

Under surface yellow, marked as in the other sex.

All the examples so far known are those taken by Mr. Mead in the pine forests near Fairplay, Colorado, in June, 1871, and now in Mus. Am. Ent. Soc., T. L. Mead, W. H. Edwards, H. Strecker.

This pretty species, I think, will hold its own, the variegation of under surface of secondaries is peculiar and constant and quite different from A. Sara and A. Reakirtii, notwithstanding the great tendency to variation in the latter.*

*Sara and Reakirtii, I believe, are identical.

Of the following species I am anxious to obtain examples, either by exchange or purchase; any Naturalists having duplicates of any of them will confer a great favor by communicating with

HERMAN STRECKER,
Box 111 Reading, P. O.,
Berks Co., Pennsylvania, U. S. of N. America.

Ornithoptera Hippolytus, Cram.
Ornithoptera Amphrysus, Cram.
Ornithoptera Helena, Linn.
Ornithoptera Croesus, Wall.
Ornithoptera Brookiana ?, Wall.
Papilio Evan, Doubl.
Papilio Pericles, Wall.
Papilio Blumei, Boisd.
Papilio Macedon, Wall.
Papilio Philippus, Wall.
Papilio Arcturus, West.
Papilio Phorbanta, Linn.
Papilio Homerus, Fabr.
Papilio Garamas, Hub.
Papilio Caiguanabus, Pocy.
Argynnis Rudra, Moore.
Argynnis Oscarus, Evers.
Argynnis Cnidia, Feld.
Argynnis Jerdoni, Lang.
Argynnis Dexamene, Boisd.
Argynnis Jainedeva, Moore.
Argynnis Ruslana, Motsch.
Argynnis Anna, Blanch.
Argynnis Childreni, Gray.
Argynnis Aruna, Moore.
Acherontia Satanus.
Papilio Ascanius, Cram.
Coerous Chorineus, Fabr.
Parthenos Tigrina, Voll.
Charaxes Epijasius Reiche.
Charaxes Kadenii, Feld.
Charaxes Jupiter, But.
Charaxes Etheocles, Cram.
Catagramma Excelsior, Hew.
Papilio Wallacei, Hew.
Papilio Slateri, Hew.
Papilio Endochus Boisd.
Dyotis Bioculatis, Guerin.
Romaleosoma Sophron, Dbldy.
Romaleosoma Pratinas, Dbldy.
Romaleosoma Arcadius, Fabr.
Nymphalis Calydonia, Hew.
Saturnia Epimethea, Dru.

Papilio Gundlachianus, Feld.
Dynastor Napoleon, Doubl, Hew.
Argynnis Sagana ♂, Doubl, Hew.
Zeuxidia Aurelias, Cram.
Urania Rhipheus, Cram.
Urania Sloanus, Cram.
Nyctalemon Orontes, Linn.
Nyctalemon Cydnus, Feld.
Erasmia Pulchella, Hope.
Actias Maenas
Saturnia Derecto, Mssn.
Saturnia Argus, Drury.
Any Asiatic species of Parnassius.
Citheronia Phronima.
Castnia Daedalus, Cram.
Pyrameis Gonerilla, Fabr.
Pyrameis Abyssinica, Feld.
Pyrameis Dejeanii, Godt.
Pyrameis Tameamea, Esch.
Vanessa v. Hygiaea, Hdrch.
Vanessa v. Elymi, Rbr.
Dasyopthalmia Rusina, Godt.
Morpho Phanodemus, Hew.
Any species of Agrias.
Any species of Callithea.
Pandora Chalcothea, Bates.
Pandora Hypochlora, Gates.
Pandora Divalis, Bates.
Colias Viluiensis, Men.
Colias Ponteni, Wallengr.
Bunaea Deroyllei, Thom.
Bunaea Phaedusa, Dru.
Rinaca Zuleica, Hope.
Papilio Disparilis, Herr-Sch.
Acrea Perenna, Dbldy.
Opsiphanes Boisduvalii, Dbldy.
Clothilde Jaegeri, Herr-Sch.
Pieris Celestina, Boisd.
Euphea Eurypon, Hew.
Papilio Panariste, Hew.
Limenitis Lymire, Hew.
Io Beckeri Herr-Sch.
Eacles Kadenii, Herr-Sch.

These are a few of the very many of the rarer species that I am eager to procure; of course there are numberless others from all parts of the world, equally desirable and coveted by me.

North Atlantic Express Co.,

NEW YORK,
OFFICE, 57 BROADWAY.

Chartered by Special Act of Incorporation.

CENTRAL EUROPEAN OFFICE:
5 RUE SCRIBE, PARIS.

PRINCIPAL OFFICE IN GREAT BRITAIN:
4 Moorgate St., London, E. C.
B. W. & H. HORNE, AGENTS.

BRANCH OFFICES: Golden Cross, Charing Cross; George & Blue Boar, High Holborn; 168 New Bond Street; 474½ New Oxford Street.
OFFICE IN LIVERPOOL: 5 Knowsley Buildings Tithebarn Street.
CONTINENTAL OFFICES: 5 Rue Scribe, Paris; 82 Rue d'Orleans, Havre; 88 Rodingsmarkt, Hamburg; 29 Eulenhofs Strasse, Bremen.

Merchandise, specie, bullion, stocks, bonds, or other valuables and packages and parcels of every description, personal effects, baggage, etc., forwarded to and from Europe and all parts of the United States, the States and Territories of the Pacific Coast, British Columbia and the Canadas included, at fixed Tariff rates, with no extra charges whatever for Custom-House brokerage, commissions, delivery, etc., etc., the shipper or receiver being under no other cost or expense than the stipulated freight from the point of shipment to place of delivery, and the amount of duties and government fees actually paid at the Custom-houses of the United States or Europe.

For the convenience of shippers, where agencies of the Company are not established, packages or heavy goods may be forwarded to either of the offices or agencies of the Company, by either of the express or transportation companies in the United States, or by post, by railway, through the parcel delivery companies, or forwarding houses in any part of Great Britain or the Continent of Europe.

All packages, trunks, or parcels forwarded by this Company will be landed on arrival simultaneously with the mails, or immediately thereafter and will be entered at the Custom-house, duties paid and delivered to the parties to whom addressed in any part of Europe, the United States, the Canadas, or British Columbia, with the greatest possible dispatch. Transportation charges and duties collected on delivery, or may be prepaid, at the option of shipper.

Insurance against marine risk taken by the Company, when desired by the shipper, at the lowest current rates; premium payable in all cases in advance.

Shippers to or from any part of America, and Americans traveling in Europe, will find this the quickest, cheapest and most reliable medium of transportation, the business of this Company being conducted upon the well known prompt American express system, which has become so great a commercial necessity and convenience throughout the United States.

Purchases made, and collections and communications in every part of Europe and the United States promptly executed.

☞ Circulars sent on application to

S. D. JONES,
Manager,
71 BROADWAY, N. Y.

No. 7.   Price 50 Cents.

LEPIDOPTERA,

RHOPALOCERES AND HETEROCERES,

INDIGENOUS AND EXOTIC;

WITH

Descriptions and Colored Illustrations,

BY

HERMAN STRECKER.

Reading, Pa., September, 1873.

Reading, Pa.:
Owen's Steam Book and Job Printing Office, 515 Court Street,
1873.

Monograph of the known Species of Smerinthus in N. America, of which the following is the Catalogue and Synonymy.

SMERINTHUS, Latreille.

JUGLANDIS, Abbot & Smith, (*Sphinx J.*) Insects of Georgia, Vol. I, p. 57, t. 29, (1797.)
    Amorpha dentata Juglandis, Hubner, Sam. Ex. Schmett., Vol. I, (1806.)
    Polyptychus Juglandis, Hubner, Verz. Bek. Schmett., p. 141, (1816.)
    Smerinthus Juglandis, Harris, Sill. Am. Jnl. Sc., Vol. 36, p. 292, (1839); Ins. Injurious to Vegetation p. 328, Flint's Ed., (1862.) *Walker*, Cat. Brit. Mus. Lep. 8, p. 247, (1856.) *Clemens*, Jnl. Acad. Nat. Sc., Phila., p. 185, (1859.) *Morris*, Syn., p. 213, n. 7, (1862.) *Lintner*, Proc. Ent. Soc., Phila., Vol. III, p. 668, (1864.)
    Cressonia Juglandis, Grote & Robinson, Proc. Ent. Soc., Phila., Vol. V, p. 161, 166, (1865); List Lep. N. Am., I. p. iv, (1868.) *Sanborn*, Can. Ent. Vol. I, p. 48, (1869.) *Grote*, Bull. Buff. Soc. Nat. Sc., Vol. I, p. 24, (1873.)

PALLENS, Nov. Sp., Strecker, (1873.)

EXCAECATA, Abbot & Smith, (*Sphinx, E.*) Insects of Georgia, Vol. I, p. 49, t. 25, (1797.) *Smerinthus E., Harris*, Sill. Am. Jnl. Sc., Vol. 36, p. 290, (1839); Ins. Injurious to Vegetation, p. 327, f. 155, Flint's Ed., (1862.)
    Paonias Excaecatus, Hubner, Verz. Bek. Schmett., p. 142, (1816.) *Grote*, Bull. Buff. Soc. Nat. Sc., Vol. I, p. 23, (1873.)
    Paonias Pavonina, Geyer, Zutr. Sam. Ex. Schmett., p. 12, f. 835, 836, (1837.)
    Paonias Pavoninus, Grote, Bull. Buff. Soc. Nat. Sc., Vol. I, p. 23, (1873.) *Smerinthus P., Grote & Robinson*, Proc. Ent. Soc. Phila., Vol. 5, p. 160, 185, (1865.)
    Smerinthus Excaecatus, Walker, Cat. Brit. Mus. Lep. 8, p. 246, (1856.) *Clemens*, Jnl. Acad. Nat. Sc., Phila., p. 182, (1859.) *Morris*, Syn. p. 208, n. 2, (1862.) *Grote & Robinson*, Proc. Ent. Soc., Phila., Vol. V, p. 160, (1865); List Lep. N. Am., I, p. iv, (1862.)
    Sanborn, Can. Ent., Vol. I, p. 48, (1869.) *Packard*, Guide to Ent., p. 275, (1869.)

MYOPS, Abbot & Smith, (*Sphinx M.*) Insects of Georgia. Vol. I, p. 51, t. 26, (1797.)
    Paonias Myops, Hubner, Verz. Bek. Schmett., p. 142, (1816.) *Grote*, Bull. Buff. Soc. Nat. Sc., Vol. I, p. 23, (1873.)
    Smerinthus Rosaecerum, Boisducal, Sp. Gen. Lep. t. 15, (1836.)
    Smerinthus Myops, Harris, Sill. Am. Jnl. Sc., Vol. 36, p. 291, (1839); Ins. Injurious to Vegetation, p. 328, Flint's Ed., (1862.) *Walker*, Cat. Brit. Mus. Lep. 8, p. 245, (1856.) *Clemens*, Jnl. Acad. Nat. Sc., Phila. p. 181, (1859.) *Morris*, Syn. p. 207, n. 1, (1862.) *Grote & Robinson*, Proc. Ent. Soc., Phila., Vol. V, p. 160, (1865); List Lep. N. Am. I, p. iv, (1868.)

ASTYLUS, Drury, (*Sphinx A.*) Ill. Exotic Ent., Vol. II, p. 45, t. 26, f. 2, (1773.)
    Sphinx Io, Boisducal, Guerin, Ic. du Reg. An. t. 84, (1829-1844.) *Smerinthus Io, Boisducal*, Griffith's Ed. Cuv., Vol. II, t. 83, (1835.) *Wilson*, Treatise, Ent. Brit. Enc., p. 246, t. 236, (1835.)
    Smerinthus Integerrinae, Harris Cat. Ins. Mass. in Hitchcock's Rep. Geo. Bot. & Zool. of Mass., (1835.)
    Smerinthus Astylus, Westwood, Drury Ill. Ex. Ent., Vol. II, p. 48, t. 26, (1837.) *Harris*, Sill. Am. Jnl. Sc., Vol. 36, p. 290, (1839.) *Walker*, Cat. Brit. Mus. Lep. 8, p. 245, (1856.) *Clemens*, Jnl. Acad. Nat. Sc., Phila., p. 184, (1859.) *Morris*, Syn. p. 211, n. 6, (1862.) *Grote & Robinson*, Proc. Ent. Soc., Phila., Vol. V, p. 160, (1865); List Lep. N. Am. I, p. iv, (1868.)
    Calasymbolus Astylus, Grote, Bull. Buff. Soc. Nat. Sc., Vol. I, p. 23, (1873.)

GEMINATUS, Say, Am. Ent., Vol. I, p. 25, t. 12, (1824.) *Walker*, Cat. Brit. Mus. Lep. 8, p. 246, (1856.) *Leconte*, Ed. Say, Am. Ent., p. 25, t. 12, (1859.) *Clemens*, Jnl. Acad. Nat. Sc., Phila., p. 183, (1859.) *Morris*, Syn., p. 210, n. 4, (1862.) *Grote & Robinson*, Proc.

SMERINTHUS

Ent. Soc., Phila., Vol. V, p. 160, 185, (1865); List Lep. N. Am. I, p. iv, (1868.) *Packard*, Guide, Ins., p. 275, (1869.) *Lintner*, 24th Report, N. Y. State, Mus. Nat. Hist., p. 119, (1870.) *Grote*, Bull. Buff. Soc. Nat. Sc., Vol. I, p. 23, (1873.)
Smerinthus *Geminata, Harris*, Cat. Ins. Mass. in Rep. Geo. Bot. & Zool. of Mass., (1835); Sill. Am. Jul. Sc., Vol. 36, p. 291, (1839.)
Var. JAMAICENSIS, *Drury*, (*Sphinx Ocellatus Jamaicensis*,) Ill. Ex. Ent., Vol. II, p. 43, t. 25, (1773.) *Smerinthus J., Westwood*, Drury Ill. Ex. Ent., Vol. II, p. 47, t. 25, (1837.) *Grote & Robinson*, Proc. Ent. Soc., Phila., Vol. V, p. 160, (1865); *Lintner*, 24th Report, N. Y. State, Mus. Nat. Hist., p. 123, (1870.)

OPTHALMICUS, *Boisduval*, Annales, Soc. Ent. de France, t. III, 3me ser. xxxii, (1855.) *Clemens*, Jnl. Acad. Nat. Sc., Phila., p. 184, (1859.) *Morris*, Syn., p. 211, n. 5, (1862.) *Grote & Robinson*, Proc. Ent. Soc., Phila., Vol. V, p. 160, (1865); List Lep. N. Am., I, p. iv, (1868.) *Lintner*, 24th Report, N. Y. State, Mus. Nat. Hist., p. 125, (1870.) *Grote*, Bull. Buff. Soc. Nat. Sc., I, p. 23, (1873.)

CERISYI, *Kirby*, Fauna Boreali Americana, Vol. IV, p. 302, t. 4, f. 4, (1837.) *Lintner*, 24th Report, N. Y. State, Mus. Nat. Hist., p. 124, (1870.)
Smerinthus *Cerisii, Grote*, Proc. Ent. Soc., Phila., Vol. V, p. 40, (1865); Bull. Buff. Soc. Nat Sc., I, p. 23, (1873.) *Grote & Robinson*, Proc. Ent. Soc., Phila., Vol. V, p. 160, (1865); List Lep. N. Am., I, p. iv, (1868.)

MODESTA, *Harris*, Sill. Am. Jul. Sc. & Art, Vol. 36, p. 292, (1839); *Agassiz'* Lake Superior, p. 388, t. 7, (1850.) *Clemens*, Jnl. Acad. Nat. Sc., Phila., p. 183, (1859.) *Morris*, Syn., p. 210, n. 3, (1862.)
Smerinthus *Modestus, Walker*, Cat. Brit. Mus. Lep. 8, p. 248, (1856.) *Grote & Robinson*, Proc. Ent. Soc., Phila., Vol. V, p. 161, 185, (1865); List Lep. N. Am. I, p. iv, (1868.)
Smerinthus *Princeps, Walker*, Cat. Brit. Mus. Lep. 8, p. 255, (1856.)
Laothoe *Modesta, Grote*, Bull. Buff. Soc. Nat. Sc. I, p. 24, (1873.)

In 1816 Hubner in his Verzeichniss Bekannter Schmetterlinge divided and sub-divided the Smerinthi into several groups and genera. His first sub-family, Dentati, comprised the genera Colaces and Polyptychus, and the second, Angulati, contained Paonias and Maimantes; Polyptychus was for the reception of the plain grey species such as Denatus Cr. and Juglandis Abb. & S., whilst Paonias was for the species having the inferior ornamented with eye-like spots, of which Ocellatus L., and Excaecatus Abb. & S. were types. But in this Hubner was no more felicitous than in the numberless other instances in which he pursued the same course; in fact so few admired the system that it fell into almost entire disuse, until within the present year, when, Mr. Grote very injudiciously attempted to revive it, with additions of his own, in his Catalogue of the Sph. of N. Am., published in the Bulletin of the Buffalo Society, in which the unfortunate Smerinthi are treated in this wise on p. 23, 24.

"Tribe, ANGULATI *Hubner*.

PAONIAS, Hubner (1816.)
Type: Sphinx *Excaecatus, Abbot & Smith*.

ESCAECATUS, *Hubner*.
Sphinx *Excaecatus, Abbot & Smith*.
Canada; Massachusetts; New York; Southern States.

PAVONINUS, *Geyer*. ——
Pennsylvania (Auth. Geyer.) *In quot. pace?*

MYOPS, *Hubner*.
Sphinx *Myops, Abbot & Smith*.
Smerinthus *Rosacearum, Boisduval*.
New York; Pennsylvania; Southern States.

CALASYMBOLUS, Grote (1873.)
Type: Sphinx *Astylus, Drury*.

ASTYLUS.
Sphinx *Astylus, Drury*.
Sphinx *Io, Boisduval*.
Smerinthus *Integerrima, Harris*.
Massachusetts; New York; Pennsylvania.

SMERINTHUS, Latreille (1809.)
Type: Sphinx *Ocellatus, Linnaeus*.

OPTHALMICUS, *Boisduval*.
California.

GEMINATUS, *Say*.
Canada; Massachusetts; New York; Pennsylvania.

CERISYI, *Kirby*.
Hudson's Bay Territory (Kennicott.)

Tribe: DENTATAE (*Hubner*.)
LAOTHOE, *Fabricius restr.* (1807.)
Type: Sphinx *Populi, Linnaeus*.

MODESTA.
Smerinthus *Modesta, Harris*.
Smerinthus *Princeps, Walker*.
Lake Superior; Canada; Massachusetts; New York.

CRESSONIA, *Grote & Robinson* (1865.)
Type: Sphinx *Juglandis, Abbot & Smith*.

JUGLANDIS, *Grote & Robinson*.
Sphinx *Juglandis, Abbot & Smith*.
Canada; Massachusetts; New York; Southern States."

From the above, which I have quoted in full, it will be seen that the compiler has not been more fortunate in this arrangement than was the great originator of it, in whose footsteps he endeavors to tread. The genus Calasymbolus, erected for Astylus, he says: "differs from Paonias in the shape of the secondaries, and from Smerinthus in antennal structure," in what the difference of antennal structure consists the student is not informed; I, for my part would immensely like to know, for after a patient examination of all the species of this country, as well as of various Exotics, I am forced to the conclusion that any real material difference in the structure exists not in the antennae of the Smerinthi but in the fervid imagination of the founder of the Calasymbolus and Cressonia Dynasties.

The difference in the shape of the wings is of no possible moment, and from the fact that Mr. Grote does not mention how much more his genus differs from Paonias in the shape of the primaries than in that of the secondaries, we are unpleasantly led to suspect that perhaps the genus Calasymbolus was reared without its architect having the proper material at hand for a solid foundation to build it on, i. e., the necessary examples of Astylus, Myops, &c., &c., for comparison: Astylus certainly differs from the others more in the shape of the outer margin of the primaries than in that of the secondaries; and should any one be inclined to follow the plan adopted by Mr. Grote and found genera on such trivial grounds as the shape of the wings, he would be constrained to separate Myops from Excaecata on account of the dissimilarity in shape of primaries, and to join it to Geminatus on account of their close resemblance in this respect, whilst Opthalmicus, Cerisyi and Geminatus ought each to be placed in separate genera on account of each having a different outline of wing, and vice versa Quercus should be placed with Cerisyi, and Tiliae with Geminatus. And in fact it will be seen by comparison that there are scarcely any two, except Juglandis and Pallens, which could be placed in one genus if uniformity of shape in the wings were taken as the basis thereof; and it would be indeed a new era in natural science when the forepart of an insect belonged to one genus and the hindpart to another, thus, supposing Astylus to have the "antennal structure" and primaries the same as Excaecata or Myops, the two composing Paonias, whilst the secondaries are different from those of that genus and like those of Smerinthus, as we are to infer from Mr. Grote's language,* especially when taken with reference to the position of the genus Calasymbolus in his Catalogue above referred to, then we would have the anomaly of an insect in which the antenna and primaries belonged to the genus Paonias and the secondaries to Smerinthus, a compound only equalled by Mad. Merian's lower figure, on t. xlix,† and for which we would propose the name of Paoni—S—merinthus Astylus, the dash meaning that the head and body are left out until some other aspirant for Entomological distinction shall place them in a third genus.

Mr. Grote has been, to judge from his productions, in an alarming state of indecision regarding the Smerinthi, ever since he commenced to massacre them in 1865. In the Cat. N. Am. Sph., then compiled by himself and his colleague Mr. Robinson, Jamaicensis, Geminatus, Cerisii, Opthalmicus, Pavoninus, Excaecatus, Myops, Astylus and Modestus, nine species, are placed in Smerinthus, and Cressonia was erected for Juglandis, the authors entirely ignoring the previous sonorous Amorpha dentata Juglandis, which had been conceived and bestowed on the unhappy insect by Hubner, whilst Mr. Grote was yet an impalpability disporting through space.

In the accompanying notes to the above mentioned catalogue it was hinted that Modestus might at some future time be also separated, a threat which Mr. Grote has since fulfilled.

In Sep. 1868 appeared List of the Lep. of N. Am. by the same authors, (containing the Sphingidae and Bombycidae,) a work so replete with errors and inaccuracies that to eliminate them all would leave it in much the same condition as the result of that arithmetical problem where "nothing from nothing and nothing remains." In this stupendous work Jamaicensis has silently stolen away without any apology for such a piece of impoliteness, and Pavoninus has been degraded to a synonym prefixed with an ?, which with regard to Pavoninus was perfectly correct, all except the ?. After the appearance of this work there was a pause of five years, when Mr. Grote's last great literary effort appears in his Cat. Sph. N. Am., in which the unfortunate eight species represent five genera! In this Pavoninus, after a modest retirement of eight years, is again allowed to occupy a position as a true species, with a doubt(———) behind it, which is the author's mode of informing his readers that he knows nothing about the insect, which seems very strange for we certainly should think that knocking a species in and out of place for a term of eight years ought to give opportunity of forming some acquaintance with its true status, and besides, how could the learned author thus define its generic position with so much certainty if he were entirely unacquainted with it; In such cases it is usual to place the doubtful species at the end of the sub-family or genus, as is done in Kirby's great Catalogue of Diurnae.

It would be difficult to find two species more closely allied than Myops and Astylus, and if they are to be placed in different genera, then all we have to do is to give each of the Smerinthi a genus to itself, and in order to let posterity know to whom they are indebted for such a great work, the authors should perpetuate the genera thus created by bestowing on them their own names and those of their patrons.

The two species perhaps most dissimilar in appearance, Ocellata and Populi, are known to hybridate, which fact ought to be of some value in establishing intimate relationship.

Altogether, I do not think a much more compact group exists, and any attempt at division can only be made with violence and result in the increase of worse than useless synonyms. And if separation were to be insisted on, the shape of the wings ought to be the last point considered, for were that taken as a generic base Lepidopterology would be lost amidst a host of endless Princeps—Heroicus, Laertias, Zetides, Achilliades, Iliades, Menelaides and Fiddle-de-de-de-dees, the same as disfigured the great works of Hubner, and from which we can but appeal in the language of the Litany and pray "Good Lord deliver us."

Sept. 1873.

SMERINTHUS JUGLANDIS. Abbot & Smith.

Insects of Georgia, Vol. I, p. 57, (1797.)

(PLATE VII, FIG. 12 ♂, 13 ♀.)

Male. Expands 2½ inches.

Head and thorax flesh colored, a brown dorsal ridge on the latter; palpi brown; abdomen brown with edges of segments flesh color, the anal segment with a terminal and side tufts.

Upper surface, flesh colored; primaries with transverse brown lines and shades, and a small discal mark of like tint. Secondaries have two transverse lines accompanied by brownish shades, and are heavily clothed at the base with pale yellowish hair.

Under surface, warm reddish brown; primaries with two parallel transverse lines, between which and the external margin is an irregular band of flesh color; costa and apical part same tint. Secondaries, two median lines with the space between them flesh colored; ciliae white and brown.

Female. Expands 2½ to 3 inches.

*The genus differs from Paonias in the shape of the secondaries, and from Smerinthus in antennal structure. Grote, Bull. Buff. Soc. Nat. Sc. I. p. 2.
†Metamorphis Insectorum Surinamensium.

SMERINTHUS JUGLANDIS.

Color more dull than in male, much wanting in the pinkish tint, more inclined to ochraceous or brown; markings same as in that sex.

Habitat. Canada, and the United States generally as far westward as the Mississippi.

The larva is 2¼ inches long; has a pointed head; is of a pale green color with lateral stripes and granulated with white; caudal horn much granulated. In Abbot & Smith's work t. 29, the larva is ferruginous in color, but the text further states that "the Caterpillar is sometimes green." It feeds on the black walnut (Juglans Nigra,) and Hickory (Carya Alba,) also on the Iron Wood (Ostrya Virginica,) on which latter Mr. Lintner found the mature larvae.* Mr. Packard's assertion that it "lives on the Wild Cherry" is erroneous.†

The pupa is brown and has the three terminal segments flattened beneath.

This is by no means a common insect though rather wide spread. There is some variation in the color of the female, some examples being of a decided ochrey or brownish shade, whilst others approximate more to the flesh color of the male. This in common with the other Smerinthi varies much in size in different examples.

In Abbot's plate the colorist performed some funny work, in that the primaries in both sexes are colored pretty close to nature but the secondaries are painted yellow, and the worst feature of this is that in the text this difference is mentioned as a fact, and brought directly to notice by the author making some remarks in connection with this and a somewhat analagous European species,‡ it will be seen by my figs. 12, 13, as well as by the description, that this has no foundation, except in the fancy of the person who colored the plates, who doubtless imagined that a little variety introduced would improve the natural plain appearance of the insect.

I have commenced the Catalogue which heads this paper with this species, believing the grounds for retaining it in a separate genus to be entirely insufficient.

SMERINTHUS PALLENS. Nov. Sp.

(PLATE VII, FIG. 14 ♀.)

FEMALE. Expands 2½ inches.

Same form as Juglandis; wings a little broader in proportion than in that species.

Upper surface, uniformly ochraceous; primaries with two light brown sub-basal lines; two other lines, parallel with the exterior margin, traverse the wing from inner margin to costa, near and at which they curve inwardly towards the base; on the disc, between the 1st and 2nd median nervules, is a very faint greyish shade. Secondaries have two pale brown median lines which follow the curve of the outer margin, there are also faint indications of a third line nearer the base.

Under surface, same color and markings as above, perhaps a possible shade paler, with the exception of the sub-basal lines of primaries which are here wanting; emarginations whitish.

Described from a unique ♀ example received from Texas near the Mexican boundary. I would like to say something further regarding this rather curious insect, but as the above embodies all I really know concerning it, I will spare my readers, and not attempt to make more verbiage supply the paucity of fact.

SMERINTHUS EXCAECATA. Abbot & Smith.

Insects of Georgia, Vol. I, p. 49, t. 25, (1797.)

PLATE VII, FIG. 1 ♂, 2 ♀.)

MALE. Expands 2¼ to 2½ inches.

Head and body fawn colored; a broad dorsal thoracic patch widest near the abdomen, narrow in the middle and terminating in a line on the head; a dorsal stripe on the abdomen.

Upper surface, primaries fawn colored with, (in fresh examples,) a faint violet shade, a broad brown

*Proc. Ent. Soc., Phila., Vol. III, p. 666.
†Guide to Ent., p. 274.
‡ "Mr. Abbot very justly remarks the affinity between this and Sphinx Populi, the Antennae which in the male are also, in a degree pectinated; but there is not much affinity in the markings of the wings, nor is there so great a difference of hue between the fore and hind wings of S. Populi as in that now before us." Abbot & Sm. Ins. Ga., Vol. I, p. 51, (1797.)

somewhat triangular median space crossed with shades of darker hue, from this outwards are various transverse wavy lines and shades of brown; a small black discal spot; emarginations acutely dentate and white. Secondaries, base and middle rose color; costal and exterior margins fawn color, on the former some paler lines; a large black ocellus with a single blue pupil; a pale brown shade crosses the wing transversely from the apex to the ocellus, and beyond this to the anal angle, where it assumes a darker color, and with which it connects the latter; emarginations white.

Under surface; primaries, basal half rose colored, exterior to this is a pink transverse band, traversed and edged with brown lines, beyond which to the exterior margin the space is chocolate brown with an uneven pink transverse line widest in the middle, where it joins the inner band of same color; a dark yellow patch at posterior angle. Secondaries, chocolate brown with a pink median band, which is joined outwardly in the middle by a triangular spot of the same color; dark yellow apical and anal dashes.

FEMALE. Expands 3 to 3¼ inches.

Marked as in the male but paler in color, and the rosy hue of upper surface of secondaries extends to the exterior margin.

The larva is light green, palest on the back, with white granulations, pale yellow stripes on the sides, and green caudal horn. It feeds on the leaves of various kinds of Apple, and according to Abbot on the Wild Rose (Rosa Carolina). Mr. Lintner also found it on Maple.

It is the commonest of our species and occurs throughout the Atlantic States as well as in Ohio, Kentucky, Indiana, &c., &c.

Some Lepidopterists consider Geyer's figures, 835, 836 in the "Zutrage," as representing a distinct species, this theory is advocated by Grote & Robinson in the notes appended to their Cat. of 1865,* and in Grote's Cat. of 1873, to both of which I have referred in the introductory remarks to this paper; I cannot imagine how any one after seeing Geyer's figures could for a moment suppose them to represent anything else than Excaecata, they are certainly more recognizable than Abbot's,† which has the primaries painted rose color like the secondaries, besides, Geyer's descriptive remarks‡ are pertinent and to the point; he even mentions that the female (of which he gave no figure) is generally larger and less bright in color; he gives its locality as Pennsylvania.

The most astounding revelation that we find regarding this species is where Prof. Packard tells us in his Guide that the ocellus or eye-like spot of hind wings has "two or three blue pupils!"§ such utterly erroneous and culpably careless assertions are the more lamentable as the book in question was sought after more particularly by beginners and those who had not yet acquired the knowledge sufficient to discriminate between the chaff and the good grain.

SMERINTHUS MYOPS. ABBOT & SMITH.

Insects of Georgia, Vol. I, p. 51, t. 26, (1797.)

(PLATE VII, FIG. 9, ♂.)

MALE AND FEMALE. Expand 2¼ to 2½ inches.

Head and body chocolate and purplish brown; a golden yellow dorsal ridge on thorax. On each side of abdomen a row of irregularly shaped yellow spots; a dark brown dorsal line.

Upper surface; primaries purplish brown ornamented with lines and shades of rich chocolate; a brown discal dot; a bright yellow spot near the inner angle and another not far from the apex. Secondaries yellow, broadly bordered with chocolate at the costa and outer half of exterior margin; a yellow spot at apex; two pale transverse lines on brown margin near and at the costa; a black ocellus, with blue centre, between which and the inner margin the color is pale brown and purplish.

Under surface; primaries, basal half yellowish and plain; outer half marked and colored as above, but more brilliantly. Secondaries with a median band, composed of alternate pale and dark lines, succeeded outwardly by a somewhat broken one of rich yellow, the space beyond which to the margin is chocolate.

---

* SMERINTHUS PAVONINUS: A hitherto unidentified, and, since Geyer wrote, unnoticed species of Smerinthus, which the author mentions having received from Pennsylvania. It seems allied to S. Excaecatus, while Geyer compares it with the European S. ocellatus; compared with the former Geyer's figures offer too many points of distinctiveness to allow us to consider it as the species intended." Grote & Robinson, Proc. Ent. Soc., of Phila., Vol. V, p. 189.

† Abbot & Smith, Insects of Georgia, t. 25, (1797.)

‡ PAONIAS PAVONINA: Die t oberer Spk. Ocellata sehr verwandt art ist jedoch durch die eckigere gestalt der vorderflugel, ihre durchaus braune Grundfarbe und deutlichere Zeichnung, so wie durch den hinten Augendeck der Hinterflugel, von der angeführten xonogenü unterschieden. Das Weib is gewohnlich vielgrosser, und in der Farbung weniger bunt. Heimath, Pennsylvanien, von Herrn Grimm." Geyer, Zutrage zur Sammlung Exotischer Schmetterlinge, p. 15, (1837.)

§ S. Excaecatus. Smith has the hind wing rosy on the inner angle. The "ocellus" or eye-like spot is black, with two or three blue pupils." Packard's Guide to the study of Insects, p. 275. (1869.)

Larva feeds on the Wild Cherry, and is pale green, with transverse oblique yellow and red lines or bands on the sides, the last of these extending up the caudal horn; stigmata red.

This is found in the same localities as Excaecata, Juglandis, &c., but is very rare.

Chenu* has on the plate between pages 4 and 5 a male figure which, notwithstanding the exaggeration of the apical and outer portion of primaries, is easily recognizable as Myops, beneath we are informed it is "Smerinthe Ocelli, femelle," which latter intelligence does not quite take our breath, as the first shock is experienced on looking at plate 1 of the same work where the tailless, plain antennaed female of Saturnia Isabella is figured with the title of "Attacus Isabellæ male." †

SMERINTHUS ASTYLUS. Drury.
Illustrations of Exotic Entomology, Vol. II, p. 45, t. 26, (1773.)

(PLATE VII, FIG. 10, ♂.)

MALE. Expands 2¼ to 2¾ inches.

Head and body cinnamon colored, a yellowish red dorsal ridge on thorax. Abdomen with a brown dorsal line, not very distinct, sides somewhat yellow.

Upper surface; primaries, flesh colored basal patch; rest of wing cinnamon colored, with the inner margin blue grey, a mere line at the pale basal patch, but beyond that it becomes wider and is abruptly terminated not far from the inner angle by a yellow spot; there is also another yellow spot at the apex; a submarginal flesh colored line, and several smaller ones at the costa; a brown discal spot. Secondaries, inner part yellow, which about half way in becomes merged in cinnamon red, which color occupies the balance of the wing; near the costa are two pale lines; a round black spot with blue pupil near the anal angle.

Under surface very much the same as in S. Myops, with, however, more of a reddish cast throughout.

Female differs only from the male in being a little larger, and paler in color.

To the kindness of Prof. Meyer, of Brooklyn, who discovered the larva, and who is so far, I believe, the only person who has bred this rare species, I am indebted for colored drawings in which it is represented as being 1¾ to 2 inches in length; of a pale green color, beautifully variegated with dorsal and lateral yellow and red stripes and spots, somewhat in the manner of S. Myops. Its food plant is the tall Whortleberry or Huckleberry, (Vac. Corymbosum.)

Habitat. N. York, N. Jersey, Pennsylvania; of exceeding great rarity.

SMERINTHUS GEMINATUS. Say.
American Entomology, Vol. I, p. 25, t. 12, (1824.)

(PLATE VII, FIG. 6, ♂, 7, ♀.)

MALE. Expands 2¼ inches.

Head pale grey; palpi dark brown; thorax pale grey with a large brown dorsal patch; abdomen greyish brown.

Upper surface, primaries, pale grey, of a somewhat pinkish tint, with various transverse brown lines and shades; a dark reddish brown patch in the median space interiorly; a pale discal mark; a dark reddish brown lunate spot, edged interiorly with pale gray, at apex, and another on interior margin near the inner angle. Secondaries, deep rose color, somewhat broadly margined exteriorly and on costa with clay color; near the anal angle a black spot containing two blue marks, from whence the insect derives its name; this ocellus is prolonged into a hook-like black mark, which connects it with the anal angle.

* Chenu.—Encyclopedie d'historie Naturelle. Papillons Nocturnes.
† The ♂ of Saturnia Isabella, Grsells, (Ann. Soc. Fr. p. 241, 1859,) has the secondaries tailed like our Luna, (though belonging to a different group,) and broadly pectinated antennæ.

SMERINTHUS GEMINATUS.

Under surface; primaries, basal half rosy; a dark reddish brown median line, the space from thence to exterior margin marked nearly as on upper side, but more obscure. Secondaries, brown; a white discal mark; a red brown transverse line, half-way between this and posterior margin are two parallel transverse white lines, with faint indications of a third one between them; a dark brown patch at anal angle.

FEMALE. Expands 2½ to 3½ inches, and is marked and colored like the male.

Larva, 2 inches in length; pale green, lightest above, with yellow lateral granulated stripes; caudal horn violet; stigmata red. It feeds on the willow.

Habitat. Mass., N. York, Penna., Md., Va., Ill., Ky., &c.

Mr. Lintner has perfected the history of this species in his Entomological Contributions,* where he has followed it through all its stages with his usual conscientious and exhaustive exactness, an example which cannot be sufficiently commended. He there states, the egg is slightly flattened, and of pale green color. His observations also establish the fact of this species being double brooded, the two broods occurring in June and August. From eggs deposited June the 12th the larvae issued on the 19th, and by the 26th all had undergone their first molt, on the 30th the second took place, and on July the 4th they patriotically commenced to throw off their last old garment; eight days later, after only three brown moltings, they went into the ground; the first imago emerged on the 30th of July, and the last on the 10th of August. He obtained from thirty-six larvae thirty-one imagines, among them a female of the variety "Jamaicensis," the description of which follows this.

In juxtaposition to the accurateness of the paper above alluded to, we would refer to page 211 of Morris' Synopsis, published in 1862, where we find the following description of Geminatus, which includes every word that is there mentioned of the secondaries; "Posterior wings rosy, along exterior and terminal border yellowish gray; ocellus black, emitting a short broad line to inner angle, and with two or three blue pupils;" this is quoted from Dr. Clemens, and his name is appended thereto by the conscientious compiler, who in a measure saved himself thereby, but in a measure only, for Dr. Morris was too old an entomologist not to know the characteristics of this species, and too good a scholar not to know better than that Geminatus, as its name indicates, would have but two marks or pupils, and two only, and in his case we can only ascribe it to sheer carelessness. But this ridiculous error was not to stop here, for on turning to page 275 of Packard's Guide, (1869,) in the article already referred to in our remarks on S. Excaecata, we read, "S. Geminatus, Say, is so called from the two or three blue pupils in the black ocellus, the hind wings are rosy;" this is all the author says of Geminatus, (except that "the pupa has been found at the roots of willows,") and it is certainly enough, and to spare, of the kind; although there is neither authority or quotation marks given by Dr. Packard in the above, we still apprehend that he derived the information regarding the three pupils from the same source; might it not, perhaps, have been as well for the author, ere he commenced writing, to have given the subject at least a little superficial attention, or to have even taken a mere glance at examples of two of our commonest species, which are to be found in every schoolboy's collection, to ascertain that Excaecata had but one pupil, and that Geminatus had only two, and not "two or three;" in that event the student would probably not have been enlightened with the rather original information that the insect derived its name from having two or three pupils in the ocellus, and that the term geminate could with equal propriety be applied to things trinal as well as binate.

Var. JAMAICENSIS, DRURY, Illust. Exotic Entomology, Vol. II, p. 43, t. 25, (1773.)

(PLATE VII, FIG. 8, ♀.)

Color and ornamentation nearly as in the ordinary form, with the exception of the black ocellus of secondaries, which encloses but one blue spot instead of two.

I have seen but two examples, the first, (which is the original of my fig. 8,) was captured near Baltimore, by Mr. J. P. Wild, about fifteen years since, and is at present in my possession; from the time I first received it I regarded it as Drury's Jamaicensis, but, of course, could not determine its position as a true species, or as merely a variety of Geminatus, although I always inclined to the latter opinion, which was at length confirmed, and the mooted question, of what Drury's figure was meant to represent, was at last put to rest by the careful observations of Mr. Lintner, who had the rare fortune, before referred to, of raising an example † from eggs deposited by a pinned specimen of Geminatus; this example, which I have had the opportunity of examining and comparing with mine, is a little larger, being about 2½ inches in expanse; and "just below the first median nervule" the two bands which cross the middle of the wing are nearer to each other than in my example or in examples of Geminatus; in this respect, as in most others, it accords with Drury's figure. The blue

---

* Entomological Contributions No. II, by J. A. Lintner, in the Twenty-fourth Annual Report of the New York State Museum of Natural History, p. 139-127. (1870.)

† Variety.—Among the above imagines was a female, having but a single blue pupil on the black ocellated spot of the secondaries." Lintner, Ent. Contributions in the 24th Report N. York State Mus. Nat. Hist., p. 122. (1870.)

pupil in Mr. Lintner's example is in the upper part of the black spot, in mine it occupies the centre, in Drury's figure the right hand pupil is in the centre, and the left hand more towards the lower part.

Drury's figure in the original edition of 1773 has the thorax, abdomen and primaries flesh or fawn-colored, and the costal and outer margins of secondaries yellow, of which latter color no mention whatever is made in the description with which Mr. Lintner's example and mine agree exactly. Drury mentions particularly in his preface that he hopes any inaccuracies in the illustrations will be credited to their proper source, the artists, and it is evident that in this instance the artists laid on gamboge with the intention of improving nature, for which they doubtless considered themselves fully competent. In Westwood's edition of Drury, published in 1837, the bright yellow had grown in such favor in the sight of the artists, that in addition to the margins of secondaries, they extended operations and laid the favored pigment also over the abdomen.

Drury's figures represent a male, Mr. Lintner's example and mine are females.

Grote and Robinson contended that Drury's figure represented a distinct species; they say "*Smerinthus Jamaicensis*, Drury sp., seems to us, judging from Drury's figure and description, quite distinct from the northern species from the Atlantic District,"* and Mr. Grote expresses the same belief in his last Catalogue,† where he thinks it must be conceded that Mr. Lintner's reasons are partly speculative when he refers Drury's figure to S. Geminatus; we do not think anything of the sort must be conceded, when we are fully informed of the fact that Mr. Lintner's specimen, which agrees with Drury's description and figure, with the exception of the false coloring on secondaries of latter, was produced from ova deposited by Geminatus.

Fabricius thought Drury's figure was intended for S. Ocellata, and as Smith says, "quotes it as such without any scruple."‡

With regard to Drury's locality of Jamaica, it is scarcely necessary to state that the earlier writers, owing to the want of precise information, frequently gave erroneous localities; thus, Cramer cites Dys. Boreus as a native of India, whilst its true home is Surinam, and to come nearer home, Donovan figures our Anth. Genutia in his "Insects of India." I would further remark the well attested fact, that so far there is no authentic instance of any species of Smerinthus having yet been found in the West Indies, or South and Central America.

We think a critical comparison of our figure with Drury's illustration and description will convince the most skeptical of their identity.

That we cannot summarily dispose of the name *Jamaicensis*, so utterly inapplicable to a form indigenous to New York and Maryland, is much to be deplored, but according to the same law of priority that allows the stability of such a name as *Schmidtiiformis*, § it will have to remain so.

SMERINTHUS OPTHALMICUS. BOISDUVAL.

Ann. Soc. Ent., France. (1855.)

(PLATE VII, FIG. 4, ♂, 5, ♀.)

MALE AND FEMALE. Expand 3 to 3½ inches.

Head and palpi brown; thorax pale grey, with a large dark brown dorsal patch; abdomen brown.

Upper surface; primaries, a large pale grey basal patch, edged outwardly with dark brown, which latter extends obliquely across the wing outside of the white discal lune to the costa; beyond this are several other brown, undulate transverse lines and shades; the space from these to outer margin is brown, with an irregular grey band extending from the inner angle to within a short distance of the apex; a grey dash at the apex. Secondaries rose-color; outer margin clay-color; costal and inner margins yellowish white; ocellus black, enclosing a blue iris which encircles a black pupil; this is connected with the anal angle by a short black band.

Under surface; primaries, basal half rosy, with a narrow white discal mark; outer half brown, with some rather indistinct whitish transverse lines; costa whitish. Secondaries brown, traversed by a broad pale median band and several dark brown lines; a white discal mark; costa white.

Of the larva I am able to say nothing; as far as I am aware, it is as yet unknown.

Habitat. California, Washington Ty., Lake Superior.

There is between the male and female examples from California but little difference in colouration; but a female from Lake Superior (of which fig. 5 is a representation) has the markings of primaries of a less de-

* Proc. Ent. Soc., Phila., Vol. V, p. 148. (1865.)
† Bull. Buff. Soc. Nat. Sc., Vol. I, p. 23. (1873.)
‡ Abbot & Smith, Insects of Georgia, Vol. I, p. 43. (1797.)
§ Morta Schmidtiiformis, Freyer. (1856.)

cided character, and the color is not gray or ashen, but of a general pale reddish brown or umber tint, on both upper and lower surfaces; the costa of primaries in this example is much more rounded than in any I have seen from California.

This rare insect is the nearest American analogue of the European S. Ocellata, L., a fact alluded to by Dr. Boisduval in his very short description.* The species is so rare, that, until recently, but few opportunities have offered for the entomologist to examine it in nature. Dr. Clemens, who had evidently never seen an example, thought it might possibly be a variety of S. Geminatus,† into which supposition he was doubtless led by Dr. Boisduval's remark in the description above alluded to.

The specimen referred to by Mr. Grote,‡ as coming from the Isthmus, is a female of this species; the party from whom I obtained it had collected in Costa Rica, but before coming east he visited California, and sojourned there awhile, receiving additional material from that state, which he was by no means careful to keep apart from his more southern collections. This example has been the victim of a series of atrocious abuses, the first of which was perpetrated by the thundering fool who captured it, and who merits the unmitigated contempt of all scientists on earth, and torments unspeakable hereafter, in Hades; this talented individual came across the poor thing just after it had emerged from the pupa, and killed it before the wings had expanded to one-fourth of their proper size. When it came into my possession the abdomen had been left somewhere in California, but the conscientious collector, in order to give *quantum sufficit*, had replaced it with one of Arachnis picta.

SMERINTHUS CERISYI. Kirby.

Fauna Boreali-Americana, Vol. IV, p. 302, t. 4, fig. 4. (1837.)

(PLATE VII, FIG. 3, ♂.)

MALE. Expands 2½ to 3 inches.

Head and palpi brown; thorax pale ashen, nearly white, with a large dark brown dorsal patch; abdomen brownish grey above, pale ashen beneath.

Upper surface; primaries, pale ash-colored with numerous brown, undulate, transverse lines and shades; a white discal mark, which color is continued along the median nervure to the pale basal patch; joining this latter exteriorly, and between the median nervure and interior margin, is a brown patch or cloud. Secondaries rose-colored, but of a less lively tint than in any other species; towards the exterior margin the rose color is tinged with greyish; costal and interior margins white, or nearly so; ocellus black, containing a blue iris which almost encircles a black pupil; the blue does not quite unite, opposite the inner margin, in surrounding the pupil; the ocellus is prolonged towards, and connects with the anal angle.

Under surface; primaries, basal half dull rose-colored; outer half marked as on upper side, but paler and less distinctly. Secondaries white, with pale brown, undulate, transverse bands.

Of the female nothing is known.

This is certainly the rarest of all the heretofore described N. American Sphingidæ; but three authentic examples, all male, are known; the first was figured and described by Kirby, in 1837, § who did not know in what precise locality it was captured; this example, perhaps, may still be preserved in the British Museum, otherwise it is probably lost; the second one was taken by the late Robt. Kennicott at Rupert House, in British America, and is at present in the Museum of Comp. Zool. at Cambridge; this is the largest specimen of the three, expanding about three inches. The third and last, the original of figure 3, I received in a small collection of things from near Providence, Rhode Island.

---

* "Le S. *Ophtalmica* assez rapproché de notre *scellata*, plus voisin de *Geminatus* de Say, mais l'œil n'est pas doublé et il diffère de toutes les espèces du même groupe par sa large bande brune, anguleuse, qui traverse le milieu des ailes supérieures." Ann. Soc. Ent. Fr., t. III, 3me ser. xxxii. (1855.)
† Jrl. Acad. Nat. Sc., Phila., p. 164. (1859.)
‡ "I learn from Mr. Strecker that a specimen referable to this genus has been 'received from the Isthmus.'" Grote, in Bull. Buff. Soc. Nat. Sc., Vol. I, p. 25. (1873.)
§ "Body ash-colored; thorax with a large trapezoidal brown spot dilated next the abdomen; primaries angulated, ash-colored, with a transverse series of brown, sub-marginal crescents in a paler band, between which and the posterior margin is another obsolete paler one; above the crescents is a straight, whitish band, and a linear angular forked one, under the internal sinuses of which the wings are clouded with dark brown; underneath, the above markings of the wings are very indistinct; the secondaries are rose-color, paler at the costal and posterior margins; underneath they are dusky, clearer with a whitish band coinciding with that of the primaries, a transverse series of crescents and a dentated lesser side band, all rather indistinct; but the most conspicuous character of the secondaries is a large eyelet situated at the anal angle, consisting of a black pupil, nearly, but not quite surrounded by a blue iris, and situated in a black triangular spot or atmosphere which extends to the anal angle, and is surmounted by some blue scales; the abdomen above is dusky ash-colored.

This insect appears to be the American representation of *S. Ocellatus*, from which, however, it differs considerably. It comes very near *S. Geminatus*, (Say, Am. Ent. i. t. xli.) but in that the eyelet has two blue pupils. Taken in North America, locality not stated."—Kirby, Fauna Boreali-Americana, Vol. IV, p. 302. (1837.)

Mr. Grote mentions that Mr. Stph. Calverly of N. York, (of whom we have heard nothing for some years,) once informed him that he had raised this species from the larva; but as there is no record of the particulars of so interesting an event, we may be pardoned for suggesting that perhaps Mr. C. may possibly have been mistaken in the species, which, of course, can yet be determined if his examples are still extant.

SMERINTHUS MODESTA. Harris.

Cat. N. Am. Sphingidae, Sill. Jul. Art & Sc., Vol. 36, p. 292, (1839.)

(PLATE VII, FIG. 11.)

MALE AND FEMALE. Expand 4 to 5 inches.
Head and body pale grey.
Upper surface; primaries, basal third very pale grey, with faint transverse shades; a broad olivaceous median band, within which is a small white discal spot; adjoining this is a pale transverse shade, and a narrow undulate band; the space from these to the exterior margin is olivaceous. Secondaries dusky rose-color; costal and abdominal margins very pale grey; exterior margin olivaceous; near the anal angle is a bluish grey patch surmounted by a curved black streak.
Under surface; pale olivaceous grey, broadly margined exteriorly with a somewhat darker shade; base of primaries dusky rose-color, on which the pale discal spot is visible.
Habitat. Canada, Lake Superior Region, New England and Middle States. Very rare.
One can scarcely understand why Dr. Harris should have designated this noble species, the prince of its genus, by so humble an appellation, unless he labored under the fallacious idea that greatness and modesty are inseparable, which may have been the case with his generation, but in our day it is precisely the reverse. We may, however, have yet to fall back on Walker's more appropriate name of Princeps, for should Dentatus, Cram.,* and Modesta, Fabr.,† be eventually determined as distinct from each other, of which there is every probability, then Harris' name will long have been preoccupied, and Walker's would have to be retained in its place.

SMERINTHUS HYBRIDUS. Westwood.

Humphrey's British Moths, t. 1, (1845.)
Menetries, Wien. Ent. Monatschrift, Vol. II, p. 197, (1858.)
Staudinger, Cat. Lep. des Eur. Faunengebiets, p. 37, (1871.)

HIBRIDA EX S. OCELLATA ET S. POPULI.

(PLATE VII, FIG. 15.)

Same size as S. Populi. Head and body brown, ground color of primaries pinkish, same as S. Ocellata; markings brown, and same style as in S. Populi. Secondaries brown, with a reddish basal patch; an obscure grey spot replaces the ocellus near the anal angle.
Under surface is a complete compound of the colors and markings of both Ocellata and Populi, favoring, however, the former the most.
This monstrosity, an offence against nature and local collectors, is, nevertheless, as Menetries says, at all events very remarkable,‡ and, although not of our fauna, I have figured it as a curiosity, as well as for its affording an illustration of the close affinity of the Smerinthid species with each other, for there can scarce be any species more unlike in appearance than the European S. Ocellata and S. Populi, of which this abnormity is the product.

* Cramer, Papillons exotiques, Vol. II, t. 125, p. 42, (1779.)
† Fabricius, Entomologia Systematica, Tom. III, pars. 1, p. 356, (1793.)
‡ "Der in den Transactions der Londoner entom. Gesellschaft abgebildete Bastard von Sph. Ocellata und Populi. Ist allerdings sehr merkwurdig." Wien. Ent. Mon., Vol. II, p. 197.

Of the following species I am anxious to obtain examples, either by exchange or purchase; any Naturalists having duplicates of any of them will confer a great favor by communicating with

HERMAN STRECKER,
Box 111 Reading, P. O.,
Berks Co., Pennsylvania, U. S. of N. America.

Ornithoptera Hippolytus, Cram.
Ornithoptera Amphrysus, Cram.
Ornithoptera Helena, Linn.
Ornithoptera Croesus, Wall.
Ornithoptera Brookiana ♀, Wall.
Papilio Evan, Doubl.
Papilio Pericles, Wall.
Papilio Blumei, Boisd.
Papilio Macedon, Wall.
Papilio Philippus, Wall.
Papilio Arcturus, West.
Papilio Phorbanta, Linn.
Papilio Homerus, Fabr.
Papilio Garamas, Hub.
Papilio Caiguanabus, Poey.
Argynnis Rudra, Moore.
Argynnis Oscarus, Evers.
Argynnis Cuilia, Feld.
Argynnis Jerdoni, Laug.
Argynnis Dexamene, Boisd.
Argynnis Jainedeva, Moore.
Argynnis Ruslana, Motsch.
Argynnis Anna, Blanch.
Argynnis Childreni, Gray.
Argynnis Aruna, Moore.
Acherontia Satanas.
Papilio Ascanius, Cram.
Cœrous Chorinæus, Fabr.
Parthenos Tigrina, Voll.
Charaxes Epijasius Reiche.
Charaxes Kadenii, Feld.
Charaxes Jupiter, But.
Charaxes Etheocles, Cram.
Catagramma Excelsior, Hew.
Papilio Wallacei, Hew.
Papilio Slateri, Hew.
Papilio Eudoxus Boisd.
Dyctis Biocalatis, Guerin.
Romaleosoma Sophron, Dbldy.
Romaleosoma Pratinus, Dbldy.
Romaleosoma Arcadius, Fabr.
Nymphalis Calydonia, Hew.
Saturnia Epimethea, Dru.
Calinaga Buddha, Moore.
Papilio Icarius, West.
Papilio Elephenor, Dbldy.
Papilio Dionysus, Dbldy.
Diadema Boisduvalii, Dbldy.
Pavonia Aorsa, West.
Any species of Phyllodes.

Papilio Gundlachianus, Feld.
Dynastor Napoleon, Doubl, Hew.
Argynnis Sagana ♂, Doubl, Hew.
Zeuxidia Aurelius, Cram.
Urania Rhipheus, Cram.
Urania Sloanus, Cram.
Nyctalemon Oroates, Linn.
Nyctalemon Cydnus, Feld.
Erasmia Pulchella, Hope.
Actias Mænas
Saturnia Derecto, Msn.
Saturnia Argus, Drury.
Any Asiatic species of Parnassius.
Citheronia Phronima.
Castnia Dædalus, Cram.
Pyrameis Gonerilla, Fabr.
Pyrameis Abyssinica, Feld.
Pyrameis Dejeanii, Godt.
Pyrameis Tameameo, Esch.
Vanessa v. Hygiæa, Hdrch.
Vanessa v. Elymi, Rbr.
Dasyopthalmia Rusina, Godt.
Morpho Phanodemus, Hew.
Any species of Agrias.
Any species of Callithea.
Pandora Chalcothea, Bates.
Pandora Hypochlora, Cates.
Pandora Divalis, Bates.
Colius Viluiensis, Men.
Colias Ponteni, Wallengr.
Buiaea Deroyllei, Thom.
Buiaea Phædusa, Dru.
Rinaca Zuleica, Hope.
Papilio Disparilis, Herr-Sch.
Acræa Perenna, Dbldy.
Ossiphanes Boisduvalii, Dbldy.
Clothilde Jægeri, Herr-Sch.
Pieris Celestina, Boisd.
Euphœa Eurypon, Hew.
Paphia Panariste, Hew.
Limenitis Lymire, Hew.
Io Beckeri Herr-Sch.
Eacles Kadenii, Herr-Sch.
Smerinthus Tinæsia, Stoll.
Smerinthus Panopus, Cram.
Sphinx Substrigilis, West.
Saturnia Larissa, West.
Eusemia Victrix, Bellatrix, Amatrix, Dentatrix, West.

These are a few of the very many of the rarer species that I am eager to procure; of course there are numberless others from all parts of the world, equally desirable and coveted by me.

North Atlantic Express Co.,

NEW YORK,
OFFICE, 57 BROADWAY.

Chartered by Special Act of Incorporation.

CENTRAL EUROPEAN OFFICE:
5 RUE SCRIBE, PARIS.

PRINCIPAL OFFICE IN GREAT BRITAIN:
4 Moorgate St., London, E. C.
B. W. & H. HORNE, AGENTS.

BRANCH OFFICES: Golden Cross, Charing Cross; George & Blue Boar, High Holborn; 108 New Bond Street; 474½ New Oxford Street.
OFFICE IN LIVERPOOL: 5 Knowsley Buildings Tithebarn Street.
CONTINENTAL OFFICES: 5 Rue Scribe, Paris; 82 Rue d'Orleans, Havre; 88 Rodingsmarkt, Hamburg; 20 Bahnhofs Strasse, Bremen.

Merchandise, specie, bullion, stocks, bonds, or other valuables and packages and parcels of every description, personal effects, baggage, etc., forwarded to and from Europe and all parts of the United States, the States and Territories of the Pacific Coast, British Columbia and the Canadas included, *at fixed Tariff rates, with no extra charges whatsoever for Cartage, Storage, commission, delivery, etc., etc.,* the shipper or receiver being under no other care or expense than the stipulated freight from the point of shipment to place of delivery, and the amount of duties and government fees actually paid at the Custom-house of the United States or Europe.

For the convenience of shippers, where agencies of the Company are not established, packages or heavy goods may be forwarded to either of the offices or agencies of the Company, by either of the express or transportation companies in the United States, or by post, by railway, through the parcel delivery companies, or forwarding houses in any part of Great Britain or the Continent of Europe.

All packages, bonds, or parcels forwarded by this Company will be landed on arrival simultaneously with the mail, and for immediately thereafter and will be entered at the Custom-house, duties paid and delivered to the parties to whom addressed in any part of Europe, the United States, the Canadas, or British Columbia, with the greatest possible dispatch. Transportation charges and duties collected on delivery, or may be prepaid, at the option of shipper.

Insurance against marine risk taken by the Company, when desired by the shipper, at the lowest current rates; premiums payable in all cases in advance.

Shippers to or from any part of America, and Americans traveling in Europe, will find this the quickest, cheapest and most reliable medium of transportation, the business of this Company being conducted upon the well known prompt American express system, which has become so great a commercial necessity and convenience throughout the United States.

Purchases made, and collections and communications in every part of Europe and the United States promptly executed.

☞ Circulars sent on application to

S. D. JONES,
Manager,
71 BROADWAY, N. Y.

No. 8. PRICE 50 CENTS.

LEPIDOPTERA,

RHOPALOCERES AND HETEROCERES,

INDIGENOUS AND EXOTIC;

WITH

Descriptions and Colored Illustrations,

BY

HERMAN STRECKER.

Reading, Pa., 1874.

Reading, Pa.:
OWEN'S STEAM BOOK AND JOB PRINTING OFFICE, 515 COURT STREET,
1874.

PAPILIO COPANÆ. Reakirt.

Proc. Ent. Soc., Phil., Vol. II, p. 141, (1863).
Kirby, Cat. Diurnal Lep., p. 521, (1871).

(PLATE VIII, FIG. 1, ♀.)

FEMALE. Expands 4½ inches.
Antennæ and head black; thorax black, spotted with yellow on the sides. Abdomen dark shining green above, black on the sides and beneath, three rows of yellow streaks on each side, and two rows of white spots and one of yellow streaks below.

Upper surface; primaries blackish brown, the basal and interior part glossed with shining green which changes in depth and shade in different lights; five sub-marginal yellow, dart-shaped dashes, the first, between the discoidal nervules, is almost obsolete, the second is the largest, extending almost to the discoidal cell, the one nearest the anal angle is double; a narrow yellow dash on edge of costa.

Secondaries blackish brown, with green reflection much more noticeable than on primaries; a mesial band of seven yellow spots, the one nearest the anal angle composed of a few atoms, the next large and rhomboidal in shape, the succeeding four larger and more or less oval, the seventh extends from the first sub-costal nervule to the costa, and is concave on both outer and inner edges; a row of indistinct marginal lunules; exterior margin dentate, with yellow emarginations.

Under surface; primaries, basal half blackish brown, disk pale brown; four of the spots of the upper surface reproduced and more distinct, but not so yellow; the largest of these spots extends into the discoidal cell.

Secondaries shining brown, varying in shade in different positions; a marginal row of red spots bordered narrowly with black, and, as the original description very aptly says, "resembling chevrons in form."

The male I have never seen, nor do I know if it be at all known.

Habitat. Guatamala.

The type from which the above description and accompanying figures were taken, came from near Copan, and is in my cabinet; the only other example I know of is in the Mus. of the Am. Ent. Soc.

This species differs entirely from the others of its group,* in the number and arrangement of the lateral and ventral rows of streaks (spots we can scarcely call them, as they are parallelogramic in form,) on the abdomen, of which there are in all nine, one yellow and two white below, and three yellow on each side.

We have such poor facilities in this country for properly studying the exotic species, that it was a bold venture of Mr. Reakirt to describe this as new, especially as it belongs to a group so replete with varieties; nevertheless, I have a strong conviction that it will not share the fate of so many of that author's species. It is here truthfully figured and coloured from the type, and those abroad who have the advantages of larger material for comparison, can pronounce their verdict as to its genuineness.

In remarks at the close of my friend Reakirt's original description, he must have been carried away a little by his enthusiasm, when he said "the lustrous brilliancy of its upper surface is alone surpassed by the Morphidæ." I can only see that its lustrous brilliancy exceeds a little that of P. Polydamus, and is not equal to that of P. Latinus.

PIERIS NAPI, Linnæus.

Napi, Linnæus, (Papilio N.) Faun. Succ., p. 271, n. 1037, (1761); Syst. Nat. I, 2, p. 760, (1767). Seba, Rer. Nat. Thes., Vol. IV, t. 2, (1765). Esper, Schmett., Vol. I, I. t. 3, (1777). Hubner, Eur. Schmett., Vol. I, f. 406, 407, (1798–1803).
Pieris Napi, Godart, Enc. Meth., Vol. IX, p. 161, (1819). Boisdwval, Sp. Gen. 1, p. 518, (1836). Staudinger, Cat. Lep. Eur. I, p. 3, (Jan., 1871). Kirby, Cat. Diurnal Lep., p. 453, (March, 1871).

---
* Such as composed Hubner's genus Rhobolus, viz.: P. Hyperion, Hub., P. Polydamus, L., P. Crassus, Cram., &c.

PIERIS NAPI.

Pontia Napi, Duncan, Nat. Lib. Ent., Vol. III, p. 121, t. 9, (1835).
Tachyptera Napi, Berge, Schmetterlingsbuch, p. 94, t. 30, f. 4, (1842).
Pieris Venosa, Scudder, Proc. Bost. Nat. Hist. Soc., VIII, p. 182, (1861). Morris, Synopsis,
    p. 320, (1862). *Weidemeyer*, Proc. Ent. Soc., Phila., Vol. II, p. 151, (1863). *Kirby*,
    Cat. Diurnal Lep., p. 454, (1871). *Edwards*, Syn. N. Am. Lep., p. 4, (1872).
Pieris Nasturtii, Boisduval, Lep. Cal., p. 38, (1869).
Var. BRYONIAE, Ochsenheimer, (*Papilio B.*) Schmett., Eur. I, 2, p. 151, (1808). *Pieris B.*,
    Godart, Enc. Meth., Vol. IX, p. 162, (1819). *Staudinger*, Cat. Lep. Eur. I, p. 3,
    (1871). *Kirby*, Cat. Diurnal Lep., p. 453, (1871).
    Papilio Napi, Esper, Schmett., I, 2, t. 64, (1743). *Hubner*, Eur. Schmett., Vol. I, f. 407,
    (1798–1803).
Var. SABELLICAE, Stephens, (*Pontia S.*) Ill. Brit. Ent. Haust, I, p. 21, t. 3, (1827). *Duncan*,
    Nat. Lib. Ent., Vol. III, p. 123, t. 8, f. 3, (1835). *Pieris S.*, *Kirby*, Cat. Diurnal
    Lep., p. 453, (1871).
Var. NAPAEAE, Esper, (*Papilio N.*) Schmett., I, 2, t. 116, f. 5, (1800). *Hubner*, Eur. Schmett.,
    Vol. I, f. 664, 665, (1803–1818). *Pontia N.*, *Duncan*, Nat. Lib. Ent., Vol. III, p.
    122, (1835). *Pieris N.*, *Staudinger*, Cat. Lep. Eur. I, p. 3, (1871). *Kirby*, Cat.
    Diurnal Lep., p. 453, (1871).
Var. PALLIDA, Scudder, Proc. Bost. Nat. Hist. Soc., VIII, p. 183, (1861). *Morris*, Synopsis,
    p. 321, (1862). *Weidemeyer*, Proc. Ent. Soc., Phila., Vol. II, p. 151, (1863). *Kirby*,
    Cat. Diurnal Lep., p. 455, (1871). *Edwards*, Syn. N. Am. Lep., p. 5, (1872).
Pieris Ibridis, Boisduval, Lep. Cal., p. 39, (1869).
Pieris Castoria, Reakirt, Proc. Acad. Nat. Sc., Phila., p. 238, (1866). *Kirby*, Cat. Diurnal Lep.,
    p. 454, (1871). *Edwards*, Syn. N. Am. Lep., p. 4, (1872).
Pieris Reseda, Boisduval, Lep. Cal., p. 39, (1869).

PLATE VIII, FIG 2, PIERIS VENOSA, Scudder, ♂, ♀.
    FIG. 4, PIERIS PALLIDA, Scudder, (*P. Castoria*, *Reakirt*,) ♂, ♀.

I present the figs. 2–5 to my friends, the Lepidopterists that they may for themselves judge whether the insects represented should occupy positions as distinct species, or whether they be, as I firmly believe, only forms of P. Napi.

Figs. 4 and 5 have been drawn from Reakirt's original types of P. Castoria, which Mr. Scudder informs me is identical with his P. Pallida,* the description of which appeared five years previous to Mr. Reakirt's.

In the obsolescence of the dark scales, which in ordinary forms define the neuration on the under surface of P. Napi, it approaches closely the var. Napaeae, in fact the only difference observable is the absence of the two black spots on under side of primaries, but this is not a specific distinction as one-third of my European examples are also destitute of those spots on under surface, and in some examples on upper surface also. Moreover the absence or presence of these spots is not a peculiarity confined particularly to this species (Napi), for in the common form of P. Rapae these spots are in some instances almost obsolete, and in others entirely wanting; in the var. Ergane, found in Dalmatia and Turkey, they do not occur at all, and in the var Manni are quite indistinct and often entirely absent.

In fact, there is no more difference between P. Pallida (Castoria) and the typical P. Napi, than between the latter and some of its European and Asiatic varieties and aberrant forms, and the more I have studied the many examples at my command the more am I convinced that P. Pallida is but a form of P. Napi.

P. Venosa can scarcely be considered even as a variety; it resembles the ordinary P. Napi to such a degree that we are forced to believe in their identity.

I have a ♀ from Japan in nowise differing from the California examples.

May not, perhaps, P. Venosa and P. Pallida be the spring and summer generation, thus accounting for the depth of markings in the former, as is the case with P. Napi and Napaeae, P. Vernalis and P. Protodice, Pap. Ajax and Pap. Marcellus, Van. Levana and Van. Prorsa, &c. Our California friends can best tell us if such be the case, or whether they both emerge from the chrysalis at the same season of the year.

But neither form is by any means constant; I have them from the almost immaculate examples of P. Pallida, in regular gradations, to the heaviest marked P. Venosa, and where the one ceases and the other

---

*In speaking of P. Pallida and P. Marginalis, I always do so with the understanding that the former is the same as P. Castoria, and the latter as P. Yreka, which Mr. Scudder informed me was the case, after inspecting the types of the two latter; I have not seen Mr. Scudder's types of either P. Pallida or P. Marginalis, therefore I make this explanation, although I have implicit reliance on Mr. Scudder's acquaintance with his own species, and write accordingly.

begins it is, indeed, difficult to determine, but that Venosa is identical with Napi I am sure; Pallida may, perhaps, have the benefit of the shadow of a doubt, but eventually, I believe my opinion will be substantiated as to the identity of all three; both P. Pallida and P. Venosa are common in California and adjacent territory, and could the larvæ be discovered, their status as species or varieties could be then defined.

Figs. 2 and 3 are the ordinary form of P. Venosa, neither the darkest or the lightest marked; I have others much heavier marked, and, as I before said, between these and P. Pallida, all the intermediate grades.

Time will prove that there are much fewer true species of Lepidoptera than are at present supposed to exist; at first the old authors, owing to the science being in its infancy and consequent want of opportunities for observation, in many instances described males and females of the same species as distinct; especially was this the case with the tropical Lep., where the sexes, in numerous instances, are entirely dissimilar in appearance. But through the labors of Horsfield, Bates, Wallace and others, the majority of these errors have been corrected, and latterly, though occasionally some naturalist, through negligence or inability, makes male and female out of one sex, it has ceased to be a common offence, but in lieu thereof, every microscopic variation of tint or marking is seized upon with avidity in order to create a new species, and equally often is the same result attained through the student's negligence in obtaining the proper material for comparison, or in his haste to outstrip some other unfortunate in foisting an old species with a new title on the world, that will be honored by having the abbreviation of his name, like an antient tin-pan, dangling to its tail.

Though not so palatable to the advocates of multiplicity of species, how much better would it be to endeavor to define the true status of species already described, than to be eternally grinding out new ones, and only giving to after generations the trouble of undoing what has been done, and earning for themselves few thanks and much ridicule.

PIERIS RAPÆ, Linnæus.

Rapæ, Linnæus, (*Papilio R.*) Faun Suec. p. 270, (1761); Syst. Nat. 1, 2, p. 759, (1767). *Esper*, Schmett. I, 1, t. 3, f. 2, (1777). *Hubner*, Eur. Schmett. 1, f. 404, 405, (1798–1803).
Pieris *Rapæ, Godart*, Enc. Meth., Vol. IX, p. 161, (1819). *Boisduval*, Sp. Gen. 1, p. 520, (1836). *Staudinger*, Cat. Lep. Eur. I, p. 3, (1871). *Kirby*, Cat. Diurnal Lep., p. 454, (1871). *Edwards*, Syn. N. Am. Lep., p. 4, (1872).
Pontia *Rapæ, Duncan*, Nat. Lib. Ent., Vol. III, p. 117, t. 7, (1835).
Tachyptera *Rapæ, Berge*, Schmetterlingsbuch, p. 94, t. 30, (1842).
Pieris *Marginalis, Scudder*, Proc. Bost. Nat. Hist. Soc. VIII, p. 183, (1861). *Morris*, Synopsis, p. 321, (1862). *Weidemeyer*, Proc. Ent. Soc. Phila., Vol. II, p. 151, (1863). *Kirby*, Cat. Diurnal Lep., p. 454, (1871). *Edwards*, Syn. N. Am. Lep., p. 5, (1872).
Pieris *Yreka, Reakirt*, Proc. Acad. Nat. Sc. Phila., p. 238, (1866).
Var. Nelo, Borkhausen, (*Papilio N.*) Eur. Schmett. 1, p. 127, (1788).
Var. Metra, Stephens,)*Pontia M.*) Ill. Brit. Ent. Haust. 1, p. 19, (1827). *Duncan*, Nat. Lib. Ent., Vol. III, p. 119, t. 8, (1835). Pieris *M., Westwood, Humphrey*, Brit. Butt., p. 26, t. 5, (1841) *Kirby*, Cat. Diurnal Lep., p. 454, (1871).
Var. Ergane, Hubner, (*Papilio E.*) Eur. Schmett., 1, f. 904–907, (1827 ?). Pieris *E., Staudinger*, Cat. Lep. Eur., 1, p. 3, (1871). *Kirby*, Cat. Diurnal Lep., p. 454, (1871).
Pontia *Narcæa*, Freyer, Beit. Eur. Schmett., 1, t. 43, (1828).
Var. Mannii, Mayer, (*Pontia M.*) Stett. Ent. Zeit., p. 151, (1851), Pieris *M., Staudinger*, Cat. Lep. Eur. I, p. 3, (1871). *Kirby*, Cat. Diurnal Lep., p. 454, (1871).
Var. Leucotera, Stefanelli, Bull. Ent. Soc. Ital. 1, p. 147, (1869).
Var. Novangliæ, Scudder, (*Ganoris N.*) Can. Ent., Vol. IV, p. 79, (1872).

PLATE VIII, FIG. 6, 7, PIERIS MARGINALIS, Scudder, (*P. Yreka, Reakirt,*) ♂.
FIG. 8, PIERIS NOVANGLIÆ, Scudder, ♂.

How Mr. Scudder first, and afterwards Mr. Reakirt, could have imagined the examples, which they respectively dubbed P. Marginalis and P. Yreka, were new species, and distinct from the old P. Rapæ, is beyond my ken. I even yet think that Mr. Scudder must have been mistaken when he pronounced the types of P. Yreka identical with P. Marginalis; then again, it can scarce be possible that he would not know his own species! With regard to Mr. Reakirt's determinations there can be no dispute; the figs. 6, 7, were drawn from his two original types of P. Yreka which he described in 1866, and which are now in my cabinet.

The very first words he says, after "Nov. Sp.," are "Size and form of *Pieris Rapæ*, L."!! and well could he say so, for the one which he described as the ♀ is the common ♂ form of that species; the other is also a ♂, with the dark apical patch of primaries represented by a few scales only; examples exactly like this one I also have from Germany.

But whenever I think of this comedy of errors, an uncontrollable desire overcomes me to lie back and indulge in a glorious guffaw over the fallibility of us poor humans, for in our beloved studies it is as Butler says, in Hudibras, of religion, " still be doing, never done; as if religion were intended for nothing else but to be mended."

The two types of P. Marginalis are in Mus. Comp. Zool. at Cambridge; the ♂ came from Crescent City, Cal., and the ♀ from Gulf of Georgia.

The two types of P. Yreka are from California.

There is a curious thing in connection with these western examples of Rapæ, (Marginalis and Yreka,) that they should have been found in California and Washington Territory four or five years before the species was introduced into Canada and the United States from Europe; this fact furnishes material for some reflection.

I trust my readers will forgive me right cheerfully for not going into elaborate griseous, luteous, cyaneous descriptions of these common species; it may even be considered waste of time and material to have figured them, but I have done so with the purpose that all might see for themselves that P. Marginalis and P. Yreka are nothing but P. Rapæ, and P. Pallida, P. Castoria, and P. Venosa are but forms of P. Napi, and not have to depend on my determinations alone.

Fig. 8 illustrates the curious ♂ variety of P. Rapæ, first described by Mr. Scudder, under the name of "Novanglie," in 1872, and which has no analogue in the old world; it is not of unfrequent occurrence, and some of my friends inform me that these yellow males are from larvæ, which feed on Mignionette, but if that be the case, why are the females produced from larvæ feeding on that plant not likewise lemon yellow, or do only the male larvæ affect that food, perhaps there may be females of like yellow colour, but I have never yet seen or heard of any such, all that have come under my observation were males, and I am of the opinion that they bear to the normal form, the same relation that Colias Helichta† does to C. Eriate,‡ whatever that relation may be. But it is really wonderful that within the few years that have elapsed since the time of P. Rapæ's introduction from Europe, there should have arisen a variety which is so entirely unlike anything found in the old country.

ANTHOCHARIS OLYMPIA. Edwards.

Trans. Am. Ent. Soc., Vol. III, p. 266, (1871).

(PLATE VIII, FIG. 9, ♂.)

MALE. Expands 1½ inches.

Body black above, beneath white.

Upper surface white, with black at base of all wings. Primaries have a black apical patch broken with white ; a black discal spot.

Secondaries with a small black spot on costa near the apex, and a minute black discal point.

Under surface white. Primaries, a small yellowish grey spot on costa not far from the apex, also a few specks of greyish extend in a broken line from this spot to the exterior margin ; discal spot enclosing a white line.

Secondaries with three irregular bands of greenish grey, the second and third ones connected on the median nervure by a cross-band ; on these greenish bands are a number of round, white dots.

The female I have not yet had the opportunity of examining, but Mr. Edwards, in his description, says it is " similar to male."

Habitat. Virginia, Texas, In Mus. Comp. Zool., Am. Ent. Soc., W. H. Edwards.

This fine insect is distinct both in form and ornamentation from all other known American species of Anthocharis ; though smaller, it forcibly reminds us of the beautiful Zegris Eupheme§ of Russia and Syria, the markings of under side of secondaries bear a wonderful resemblance to those of that species, as does also

*Scudder, Canadian Entomologist, IV, p. 79, (1872).
†Lederer, Verh. Zool. Bot. Ges., II, p. 33, (1853).
‡Esper, Schmett., I, 2, t. 112, f. 3, (1806).
§ Esper, Schmett., I, 2 t. 113, (1800).

the whole outline of wings; but in A. Olympia there are no indications of the orange coloured apical spot that is one of the adornments of Eupheme.

I am led to believe that in reality this species is not of much greater rarity than A. Genutia,* but it is doubtless owing to its colour that it has heretofore enjoyed immunity from capture, as our white butterflies generally are such common species that the collector passes them by in his search for others more desirable; but my urgent advice would be, that in early spring these white butterflies be not despised, for, by diligence and patient watching, the careful observer may be right nobly rewarded by the capture of examples of this hitherto rare species. As the types were captured in company with A. Genutia in Virginia, and also were taken in Texas, and coupling this with the fact that A. Genutia is found in the United States east of the Mississippi generally, it is but reasonable to suppose that A. Olympia will be found in the same localities, and should any one be fortunate enough to get a surplus of examples thereof, I will be most happy to exchange for one or more such indigenous or exotic species as may be acceptable.

MELITAEA PICTA. Edwards.

Proc. Ent. Soc., Phila., Vol. IV, p. 201, (1865);
Syn. N. Am. Butt., p. 17, (1872).
Kirby, Cat. Diurnal Lep., p. 171, (1871).

(PLATE VIII, FIG. 10, ♂.)

MALE. Expands 1½ inches.
Body, black above; pale yellow beneath.
Upper surface, black with red and yellow spots and marks arranged much as in the allied species, Phaon,† &c.; fringes white with black at the terminations of veins.
Under surface; primaries whitish-yellow and red, a black patch at inner angle, another on middle of interior margin, a black elbowed bar extends from this to the costa, half way between this bar and the exterior margin is another black dash extending from costa, where it is widest, to nearly the middle of wing, where it terminates in a line.
Secondaries, entirely whitish-yellow, two black specks on costa, and two small clusters of grey scales at exterior margin; fringes as above.
Habitat. Colorado, Nebraska, Mexico. It is, as yet, represented in but few collections.
The smallest of the N. American species yet discovered, and easily distinguished from all others by the almost immaculate under side of secondaries.

MELITAEA MATA. Reakirt.

(Eresia M.) Proc. Ent. Soc., Phila., Vol. VI, p. 142, (1866).
Phyciodes Mata, Kirby, Cat. Diurnal Lep., p. 177, (1871).
Melitaea Mata, Edwards, Syn. N. Am. Butt., p. 17, (1872).

(PLATE VIII, FIG. 11, ♀.)

FEMALE. Expands 1½ inches.
Body blackish above; beneath white.
Upper surface, blackish brown with yellowish white markings. Primaries, a marginal row of lunules, the one between first and second median nervules much larger than the others; these are succeeded by two broad bands extending from costa to interior margin; within the dark space betwixt these bands and base of wing are two other narrower white bands.

*Fabricius, Ent. Syst. III, 1, p. 193, (1793).
† Proc. Ent. Soc., Phila., Vol. II, p. 505, (1864).

Secondaries with a marginal row of lunules; two broad bands, separated by a dark line, and the exterior one enclosing a row of small brown crescents, occupy the outer half of wing; within the discoidal cell is a large white spot divided by a dark line.

Under surface white; primaries have three slight brown dashes, one at posterior angle, one at middle of interior margin and the other extends from middle of costa to first median nervule.

Secondaries with markings of upper surface faintly repeated in very pale brown and yellow.

Habitat. Rocky Mts. of Colorado.

Mr. Reakirt described this from a unique ♀ example, I know of no other in any collection.

The peculiarity of colouration is remarkable, though not without precedent as in the case of Eresia Leucodesma,* E. Myia,† E. Ofella,‡ and some others, where the ornamentation is white on a dark ground. Mr. Reakirt's impression was that the example was faded, in which conclusion he was incorrect, as the portion of the secondaries which is overlapped by the primaries, proved on examination to be exactly the same colour as the exposed parts, and the under side is equally pale with the upper; besides the example was never exposed, having passed from the collector's hands, who had his specimens in papers, direct to Mr. Reakirt and finally to me, in no instance was it ever exposed to the continued action of light. I at first thought it might be an albino variety of some species or other, but on a rigid comparison with the analogous species I cannot in the least identify it with any of them, and Mr. Hewitson, the greatest living authority on Diurnal Lepidoptera, to whom I sent a careful drawing of it assures me it is "quite a stranger to him."

SATYRUS HOFFMANI.

Page 31, t. 4, fig. 8, ♀ June (1873.)

(PLATE VIII, FIG. 12, ♂.)

This species or variety,§ as the case may be, was described on page 31 of this work, and t. 4, fig. 8, represents the ♀; I did not at that time figure the ♂, considering the ♀ the most remarkable on account of its conspicuous white under surface, but Mr. W. H. Edwards, on the receipt of that No. of this work, wrote a few lines, informing me I had re-described his species, S. Wheeleri, the description of which was printed in advance sheets of Trans. Am. Ent. Soc., and were distributed end of June, 1873.

This description of S. Wheeleri I copy below, and accompany it by that of S. Hoffmani, ♂ and ♀, and I trust, that after a comparison of the descriptions and figures of the latter with the description of the former, but little further need be added to prove that they are not the same.

"Satyrus Wheeleri, n. sp.

MALE. Expands 2.3 to 2.5 inches. Upper side light yellow-brown, clouded with dark brown, especially on the disks of each wing, the dark portion forming a broad band on primaries, a narrow one on secondaries well defined outwardly but within fading insensibly into the ground colour; hind margins edged by a pale

Satyrus Hoffmani, n. sp.

MALE. Expands 2 inches. Upper surface uniform brown of no deep a shade throughout as in the darkest examples of S. Alope, S. Nephele or S. Sylvestris. On the primaries are two ocelli, black with small white pupils, the one nearest the costa is geminate, being joined with a smaller one at its lower edge. On

---
* Felder, Wien. Ent. Monat., Vol. V, p. 103, (1861).
† Hewitson, Exot. Butt., Vol. III, Eresia t. 3, (1864).
‡ Ib.
§ I held that S. Alope, Nephele, Pegala, Boopis and Hoffmani are but forms of one and the same, the stem of which was S. Alope. Between the darkest examples of Nephele and Boopis there is really no difference in appearance whatever; they are both the same colour, both have the ocelli on upper surface primaries of female surrounded with a cloud of paler colour, both are marked alike beneath, neither are restricted to the six ocelli of under surface of secondaries, both sometimes are devoid of all these ocelli, or have only one or more up to the full completeness, as the case may be. Between S. Pegala and Nephele are all grades in the width of the yellow band of primaries, which is found from the merest shade surrounding the ocelli up to the broad band of S. Alope, and from thence to the broader and still more conspicuous one of S. Pegala; my remarks regarding under surface of S. Nephele and Boopis apply equally well to that of S. Alope. In S. Pegala and S. Hoffmani, where the forms (one in the west, the other in the east,) appear to have reached the highest standard, the six ocelli of under surface, secondaries, as far as my observation of many examples goes, are always present; regarding the spots of upper side of primaries, they seem to be subject to no very particular rule, (except in the case of S. Hoffmani, where there are always two, the upper one of which is geminate;) half of the examples of S. Pegala, ♂ ♀, before me have one spot, the upper one, only on primaries, the remaining half, ♂ ♀, differing in no other respect, have two equally large and precisely like the northern S. Alope; as to the latter, I have it marked on upper surface, primaries, with two spots, big spots, little spots, and with no spots at all. On examples of one ? variety from California, allied to S. Boopis, on the upper surface the ocellus of primaries nearest the posterior angle is double, though not joined together, but distinctly separated by a dark line, the lowermost of the two is always the smaller, and is unrepresented on the under surface.

The ocellus on upper side of secondaries, near anal angle, is in all the forms mentioned, (except Hoffmani,) regardless of sex, either entirely wanting, or a mere speck, or from that on to the yellow ringed ocellus equal in size to those of the primaries; nor can I find in any examples of the many I have examined, any indications of a second or third smaller spot accompanying it, except in Hoffmani, where there is always a second, and sometimes a third one.

line, preceded by a dark one, and at some distance by a dark common stripe, sometimes macular on secondaries; primaries have a broad extra-discal band as in Alope, but pale brown, in which are two large black ocelli, the upper one (in all cases under inspection sub-pyriform, as if two spots of unequal size had been compressed into one, and encloses two white points; the lower ocellus is larger, rounded, with white points; both are enclosed in narrow yellow rings; secondaries usually have three ocelli, but sometimes the one next anal angle is wanting; those spots are placed on the sub-median and two next preceding interspaces; the middle one is round, about one-tenth inch in diameter, with white point and yellow ring; the others are usually more black dots.

Under side whitish, covered with abbreviated brown streaks, most dense from base to middle of disk on primaries, but equally distributed over whole secondaries, both wings being crossed by an irregular extra-discal brown stripe, besides which secondaries have a second similar stripe nearer the base; the ocelli of primaries as on upper side, but surrounded by broader and paler rings which coalesce; secondaries have three ocelli near costa, the two outer ones small, round, the others oval all with white dots and yellow circles; the three ocelli next anal angle distinct, also with dots and circlets.

Body yellow brown; legs same; palpi darker; antennæ brown with fine white annulations; club ferruginous.
FEMALE. Not known.

From nine males taken by the naturalists of Lieutenant Wheeler's Expedition, 1871, between Cascade and Rocky Mountains, but the precise locality not indicated. This fine species is one of the largest, equalling the largest specimens of Alope, and may at once be distinguished from any other North American Satyrus yet known by its pale color and clouded surface, and by the whitish color of under surface, and conspicuous brown transverse stripes. The Ocelli also are different from those of our other larger species, having small white pupils, and the upper one on primaries being peculiar in shape and doubly pupillated."

secondaries towards the outer margin, between first and second median nervules is another ocellus about one-fourth the size of those on primaries; between the second and third median nervules is a small black dot or point. On all wings are indistinct marginal and sub-marginal lines which except the ocelli and usual fenny sexual transverse dash of primaries, are the only marks on the upper surface, and were it not for the double uppermost ocelli it could not possibly be distinguished in appearance from a dark example of S. Nephele.

Under surface brown with some grey scales mixed, which gives it almost the exact tint of the under side of S. Pegala, although it does not look as smooth as in that species, more squamose in appearance; primaries have ocelli as on upper side but surrounded by broad yellow rings, the one encircling the lower ocellus has a small yellow spot emanating from its upper edge; narrow marginal and sub-marginal lines; an irregular narrow central brown band; the whole surface, but especially the inner half, covered with short transverse lines. Secondaries have six ocelli arranged as in S. Pegala, in two rows of three each; the middle one of the row nearest outer angle, is oblong and produced into a point inwardly; the others are round, all are encircled narrowly with yellow and pupilled with white; whole surface marked with short brown streaks which tend towards segregation near the outer margin; an irregular much broken band or line crosses the disc and a shorter one occurs half way in from this towards the base.

Body dark brown, same colour as upper surface.

FEMALE. Expands 2.3 to 2.5 inches. Upper surface pale yellowish brown. Primaries, outer half, corresponding to the yellow band of S. Alope, very pale and contains two large ocelli, the uppermost one geminate as in male, and all surrounded with yellow rings. The darker basal half where it joins the outer paler part is well defined, but on the inner side it becomes lost in the general ground colour, without any line of demarkation. Secondaries with sub-marginal lines as on primaries; an irregular narrow brown mesial band; a black ocellus accompanied by a black spot as in male, sometimes a third small black spot is between the ocellus and inner margin near anal angle.

Under surface white, ocelli and other markings precisely as in male but sharper and better defined throughout.

From one male and ten females taken by a naturalist of Lieutenant Wheeler's Expedition in 1871, at Owen's Lake Nevada.

After presenting the above two descriptions I thought I would have nothing to say, further, regarding them; but as I light my pipe and cast the approving glance, peculiar to self-satisfied quill drivers, at what I have written, like a flash it suggests itself, all at once, in the twinkling of an eye, as I look crosswise from Mr. Edwards' description to mine, and from mine to Mr. Edwards', that this might be a most curious case of crossed gynandromorphism, or mimicry, (of which the exotic species furnish so many instances,) where the male of Mr. Edwards' species mimics the female of mine, and I suppose when the female of Mr. Edwards' species (at present "unknown") turns up, it will be found to look like the male of mine! "Wonderful, indeed, are the works of nature," as the philosopher, who daily sweeps the street crossings, observed whilst he gazed wonderingly and admiringly at the anatomy of a cat which he held aloft by the nape of the neck; "wonderful, indeed," he repeated, "for if the holes in the cat's skin had been but a little higher up, or a little lower down, the cat's eyes would have been covered, and, consequently, the cat would have been unable to see."

GRAPTA GRACILIS. Grote & Robinson.

Ann. Lyc. Nat. Hist. N. Y., VIII, p. 432, (1867).
Edwards, Syn. N. Am. Butt. p. 20, (1872).
Vanessa Gracilis, Kirby, Cat. Diurnal Lep., p. 182, (1871).
Grapta C. Argenteum, var. Scudder, Proc. Ess. Inst., III, p. 169, (1862).

(PLATE VIII, FIG. 14, ♀.)

MALE AND FEMALE. Expand 2 inches.

Has the exact size, shape and markings of the western G. Zephyrus,* but on the upper surface the colour is much deeper, being about the same tint as G. Progne, the marginal band of secondaries, as well as primaries, is dark brown, whereas in its western representative the dark marginal band is confined to the primaries alone.

Under surface marked also exactly line for line as in G. Zephyrus, but the colour of inner half of all wings is darker than in the latter, and of a maroon or reddish brown shade inclining to claret on its outer edge, the darkness of this inner part of wings as well as of the exterior margin sets the intervening white mesial band out in bold relief making it, certainly, the most beautiful of all our species or varieties, for which it is time and observation will have to show far it is not yet given to earthly beings to know where these species begin or end. My belief is, though, that this is the same as G. Zephyrus, for there is no difference save in depth of colour, our eastern examples being much the darker.

G. Gracilis is at home on Mt. Washington, New Hampshire, G. Zephyrus is from Rocky Mountains of Colorado, Nevada, &c.; there is nothing remarkable in these two being identical as I believe them to be, when we consider G. Gracilis is found in the same locality as Chionobas Semidea and Plusia Hochenwarthii,† both of which are also found in the Rocky Mountains of Colorado.

Plate VIII, Fig. 13, is the immature larva of Papilio Anticostiensis, a representation of which I promised on page 49; it is from an alcoholic example brought by Mr. Couper from Anticosti Island last summer.

Dec., 1873.

ENTOMOLOGICAL NOTES.

EUDRYAS.—In Notes on Zygaenidae,‡ Dr. Packard refers Eudryas to that family and associates it with Alypia, and later, in his "Guide," he still maintains the same position, and to further confirm his theory, he states that he received "a piece of wood burrowed by E. Grata," and also, "as E. Unio is now known to burrow in the stems of plants, our opinion that Eudryas is allied to the Castniidae would seem to be confirmed," &c.; on the same ground, then, Cryptophasa and Cossus should be placed with Castnia, as they are borers to a fearful extent, but much, I doubt, if any amount of boring would ever bring them to the Castniidae, bore they ever so wisely.

In 1863 a third species from Texas was described by Mr. Grote, and for which he erected the genus Ciris, calling it Ciris Wilsonii,§ it differed from the typical Eudryas in having the antennae pectinated. In 1868 he described another, also from Texas, calling it Euscirrhopterus Gloveri,¶ this one has different antennae like Unio and Grata, these new species as well as the two old ones he has placed in sub-familia Castniinae, à la Packard, near Alypia in his and Robinson's Catalogue, before referred to in these pages. At first he appears to have had some misgivings, and (on p. 321, Vol. IV, Proc. Ent. Soc. Phila.,) makes some very sensible observations regarding the habits of Eudryas so entirely different from those of Alypia, but even there stating that the former is nocturnal, &c., he cannot, however, get out of the Zygaenidae with it; it still must sick there in such discordant company, but, it may go a little lower down near Ctenucha and allies, and finally he gave in, and adopted Dr. Packard's views, on page vi in the catalogue before mentioned.

Dr. Harris placed Eudryas in the Bombycidae near Scudonota, and Mr. Walker assigned it to the Noctuae. To tell the truth the beautiful things do not appear to fit in anywhere very well, but to place them with the Zygaenidae is too absurd! it is almost equal to putting Smerinthus with Colias, and could only be the result of ideas picked up during a sojourn in Schlaraffen Land.

That their place is with the Noctuae there ought not to be the least doubt, and their probable position is near Miselia or Hadena.

*Edwards, Trans. Am. Ent. Soc., Vol. III, p. 16, (1870).
†Plusia Hochenwarthii, Hochenwarth. Beitrage zur Insectengeschichte 1785, 335, T. 7, f. 2, P. Divergens, Fabr. Mant. 162, (1787). P. Ignea, Grote, Proc. Ent. Soc. Phila., II, p. 274, (1863).
‡Proc. Essex Inst., IV, (1864).
§Heaven forbid the bans.
‖Proc. Ent. Soc., Vol. II, p. 65, (1863).
¶Trans. Am. Ent. Soc., Vol. II, p. 185, (1867).

ENTOMOLOGICAL NOTES.

ANTICOSTI LEPIDOPTERA.—In Paper XVI of the Bull. Buf. Soc. Mr. Grote describes certain Diurnal Lep. from the Island of Anticosti, he mentions ten species, eight of which he examined and two more (Graptas) he heard about, "making ten species in all known from the Island"; he is however in error, there were in all fourteen, viz.: Pap. Turnus, Pap. var. Anticostiensis, Pieris Oleracea var., Colias Pelidne var., Colias Philodice, Argynnis Atlantis, Phyciodes Tharos, Grapta Comma, Grapta Progne, Vanessa Atalanta, Lycaena Lucia, Ly. ? Pembina, Ly. Scudderii, and Carterocephalus Paniscus.

The examples of P. Turnus, all ♂ and ♀, are small, expanding only three inches; the bodies are almost wholly black; all the black bands of wings heavy; but little of the blue or grey which is so conspicuous on the marginal band on under side of secondaries, in ordinary forms.

The Colias Philodice ♂ and ♀ are in nowise different from those of other districts.

Colias Pelidne, these are peculiar, the ♂ and ♀ both being entirely lemon yellow on upper and under surface, without the heavy grey powdering at base of wings on upper side, and of whole under surface of secondaries; in all other respects they resemble the typical Pelidne of North Labrador, of which I have some examples, among the latter there also occurs rarely a yellow ♀, and there is one instance of a ♂ in which the discs of the wings are orange, after the manner of some of the paler examples of C. Eurytheme (Keewaydin). I thought at first these Anticosti Pelidne might be Scudder's C. Interior, but that gentleman after inspecting these during a late visit here, says they are different, and were unknown to him; but for my part, I believe them to be nothing more then a form of C. Pelidne, precisely analogous to that of Palaeno,† found in the Alps, in which both ♂ and ♀ are yellow, whilst in examples of that species from all other localities the ♀ is always white.

Grapta Comma.—One ♂, small, expanding but 1¾ inches, differs in nothing else from the common form.

Lycaena Scudderii likewise resembles those from other localities.

I suppose by Cyclopides Manduc‡ Mr. Grote meant Carterocephalus Paniscus§ which is found in the higher latitudes of Europe, Asia and N. America.

Of the Lycaenas, which I believe to be Pembina, but which he described as new and named "Glaucopsyche Couperii Grote," he says "this species differs from Lygdamus and Pembina in having a much broader dark margin to the wings," &c., on examining ten examples, five ♂ and five ♀, taken by Mr. Couper in the two previous summers, I find that all the males are, on upper surface, in all respects exactly like Lygdamus, I cannot find any difference while examining them side by side; of the females some are nearly all blue but of much less brilliancy, and the border fades insensibly into the blue without any distinct line of separation between the two colours, others have the blue and dark grey or brown equally divided, and one has the blue restricted to a few basal scales only. The under surface of both sexes is paler than in Lygdamus, but the arrangement of the spots is precisely the same, i. e. on primaries a sub-marginal row of six spots, the one nearest the inner angle sometimes double, a discal bar or spot. On secondaries an irregular sub-marginal row of eight spots, the seventh sometimes geminate in such cases making nine in all; a discal bar, a spot near costa and another opposite to it, within the discoidal cell, between which and the interior margin is another minute spot, there last two are often obsolete, example A ♂ in my cabinet has all the spots, except these too, almost as large and pupilled as in Lygdamus. E ♂ has all the spots like this, but smaller, and the two nearest anal angle of secondaries are without pupils. C ♀, all spots of upper wings pupilled, but small; secondaries have the discal bar, but traces of all the other spots are scarcely discernable, except as closest examination, when a few minute white points may be observed, the whole wing looks plain grey with a white (discal) mark in middle; the other two males are nearly like B ♂. D ♀, all spots as in A ♂, well defined and pupilled; this example has the upper surface nearly all bluish. E ♀, ground colour very dark, all spots as in the one last described, upper surface dark, blue confined to basal parts. F ♀, spots of primaries large and pupilled, of secondaries small, the third, south, fifth, sixth and seventh, from costa, have minute black points in centre, this example is small, 1 inch in expanse, and the upper surface is like the ordinary ♀ examples of Lygdamus, blue and dark about equally divided. G ♀ marked below F ♀, but is nearly all blue on upper side, and expands 1 3-16 inches. H ♀, spots small, those nearest costa and inner angle, on primaries, almost obsolete, pupils quite small, secondaries have sub-marginal ocelli represented by six minute white spots, the one nearest the anal angle being wanting, no evidence of black centres, a white discal bar; this example comes in appearance close to L. Pierce,* but the spots of secondaries are smaller and the ground colour of wings darker.

I have eight examples, male and female of a species from the west and north-west, these are what I always supposed to be L. Pembina, Edwards, they are about the same size, colour and ornamentation as those I have just described from Anticosti and Labrador. The upper surface of males are like the preceding, perhaps a shade more towards violet, the ♀ like the darker forms, I will however go into a few details concerning the under surface; A ♂ is marked like A ♂ of preceding species and Lygdamus. B ♂ and C ♂ the same, but the spots all smaller. D ♂ is a small example, expands 1 inch. E ♂ has spots all large and well defined like E ♀ of the Anticosti examples. F ♀, all spots pupilled, but all smaller than on the last. G ♀, H ♀ like F ♂ of Couper's, except that the spot nearest anal angle of secondaries is absent. These examples just described, come from California and some from British America, present no particular points of difference from those taken in Anticosti, more than there is a shade difference in the blue of upper surface of males, and the females have less blue on upper surface than in some of those from that locality; should these be L. Pembina, which I think is the case, then Couperii is undoubtedly the same; I had the first examples, taken in 1872, a year prior to Mr. Grote's obtaining them, and although the examples, twelve in number, male and female, were faultless or nearly so, I could not bring myself to describe them as new, believing then, as I still do, that they were Edwards' L. Pembina. Mr. Scudder, in Article XX,** says the nine examples that Mr. Grote's description were based upon, included only one male, and "all excepting one female were more or less rubbed, and their determination was a matter of no small difficulty"; and further, "the wings of the male are rubbed so that it is impossible to assert positively that their border was any broader or less well defined than in those specimens from which it was believed to be specifically distinct"; this can, however, be no excuse for Mr. Grote, for if the specimens were in the desolate condition described, it would have been better to have left them alone, even if the one great aim of having "Grote" behind the name did not in this instance "obtain".

Mr. Scudder says in same paper, "the name Couperi will nevertheless stand for this species, for Mr. Edwards has recently called my attention to the fact that in describing Pembina he stated it to be allied to the Californian Piercea, Boisd., while in the same connection he described a Californian butterfly (Behrii) as belonging to a distinct series of which Lygdamus, Doubl., was the type"; with equal propriety might Mr. Grote give to Argynnis Nokomis another name because Mr. Edwards indulged in some little phantasy about the relation or similarity of that species to A. Diana, which belongs to a distinct group or series. Mr. Edwards in his Synopsis N. Am. Butt. (1872) has placed the Lycaenas referred to thus, No. " 50, Behrii,†† 51, Anticosti,‡‡ 52, Lygdamus,§§ 53. Pembina, so if he did think

*Scudder, Proc. Bost. Nat. Hist. Soc., p. 108, (1862.)
† Herr-Sch. Schmett. Ent. I, (4), 42, (1843).
‡C. Manduc, Edwards, Proc. Ent. Soc. Phila., Vol. II, p. 20, (1863), is the same as C. Paniscus, Fabr.
§Fabr. Syst. Ent., p. 531, (1775).
∥Boisduval, Ann. Soc. Ent. Fr., p. 297, (1852).
¶ Proc. Acad. Nat. Sc. Phila., p. 22, (1852).
** Bull. Buff. Soc., Vol. I, " Notes on the species of Glaucopsyche from Eastern N. Am." (1873).
†† Edwards, Proc. Acad. Nat. Sc., Phil., (1862).
‡‡ Boisduval, Ann. Ent. Soc. Fr., p. 290, (1852).
§§ Doubleday, Ent. I, p. 209, (1842).

ENTOMOLOGICAL NOTES.

when he described Pembina that it was allied to Pheres, (in which he was not so very far out of the way,) he has since changed his idea, for between the four species (Behrii, &c.,) and Pheres he has placed thirteen others.

Mr. Edwards, who of course knows his own species, has received also of these Antiocoli examples, and the shortest plan to determine if they are Lycaena Pembina, Edward. or Glaucopsyche Casperi GROTE, is for him to say which, at his early convenience, and much oblige a great many of us uninitiated, for as a matter of course we cannot expect Mr. Grote to do so, his time being so much taken up in attacking the life-long labours of Mr. Walker, of the British museum, that he really is unable to give the requisite attention to other matters, thus he is, how unfortunate, compelled to suffer, or as his own beautiful and touching language expresses it, " I elect to suffer through an injustice rather than countenance an apparent wrong." !

LIMENITIS PROSERPINA, Edwards, is doubtless a form of L. Arthemis bearing the same relationship to that species as does the aberrant L. Tremulae* to L. Populi† in Europe.

PIERIS BECKERII,‡ Edwards, from Nevada, is certainly identical with P. Chloridice,§ found in Turkey, Sarepan, Siberia, &c.; several Lepidopterists, as well as myself, have compared the two, and excepting that the Nevada examples are a little larger, cannot detect any differences that might warrant the retaining of the latter as a distinct species.

In what Mr. Grote's new Catocala Meskei ‖ differs from Unijuga, Walker, I am at a loss to perceive; Mr. Meske had the goodness to send me Grote's type for examination, but I cannot with my God will pronounce it anything else than Unijuga; I have since sent to Mr. Meske three examples ♂ and ♀ which he pronounces the same species as the one which Grote described as Meskei, and so they are, but nevertheless they are at the same time Unijuga too, and as we are supposed to go in nomenclature by the law of priority, it will still have to stand as Unijuga, although Mr. Grote doubtless imagines, after the manner of "the Ingenious Gentleman Don Quixote," that he has achieved a glorious victory over Mr. Walker and has thereby acquired the right of annihilating that author's species as spoils of war.

N. B.—If Catocala Arizonia, described in same paper **with C.** Meskei, does not turn out to be **one or the other of Dr.** Von Behr's species, others than myself will be much surprised.

I trust that, though the species figured in this No. are not conspicuous or showy, it still may not be devoid of interest. The next No. (IX) will be devoted to N. Am. Catocalæ.

* Esper, Schmett., I, 2, t. 114, (1800).
† Linn. Faun. Suec., p. 277, (1761).
‡ Edwards, Butt. N. Am. Pieris, t. 1, (1871).
§ Hubner, Eur. Schmett., 1, fig. 712, 713, (1803-1818).
‖ Canadian Entomologist, Vol. V, p. 161, (1873).

Of the following species I am anxious to obtain examples, either by exchange or purchase; any Naturalists having duplicates of any of them will confer a great favor by communicating with

HERMAN STRECKER,
Box 111 Reading, P. O.,
Berks Co., Pennsylvania, U. S. of N. America.

Ornithoptera Hippolytus, Cram.
Ornithoptera Amphrysus, ♀ Cram.
Ornithoptera Helena, Linn.
Ornithoptera Croesus, Wall.
Ornithoptera Brookiana ♀, Wall.
Papilio Evan, Doubl.
Papilio Pericles, Wall.
Papilio Blumei, Boisd.
Papilio Maccdon, Wall.
Papilio Philippus, Wall.
Papilio Arcturus, West.
Papilio Phorbanta, Linn.
Papilio Homerus, Fabr.
Papilio Garamas, Hub.
Papilio Caiguanabus, Poey.
Argynnis Rudra, Moore.
Argynnis Oscarus, Evers.
Argynnis Cnidia, Feld.
Argynnis Jerdoni, Lang.
Argynnis Dexamene, Boisd.
Argynnis Jaincleva, Moore.
Argynnis Ruslana, Motsch.
Argynnis Anna, Blanch.
Argynnis Childreni, Gray.
Argynnis Aruna, Moore.
Acherontia Satanus.
Papilio Ascanius, Cram.
Cœrous Chorineus, Fabr.
Parthenos Tigrina, Voll.
Charaxes Epijasius Reiche.
Charaxes Kadenii, Feld.
Charaxes Jupiter, But.
Charaxes Etheocles, Cram.
Catagramma Excelsior, Hew.
Papilio Wallacei, Hew.
Papilio Slateri, Hew.
Papilio Endochus Boisd.
Dyctis Biocnlatis, Guerin.
Romalæosoma Sophron, Dbldy.
Romalæosoma Pratinus, Dbldy.
Romalæosoma Arcadius, Fabr.
Nymphalis Calydonia, Hew.
Saturnia Epimethea, Dru.
Calinaga Buddha, Moore.
Papilio Icarius, West.
Papilio Elephenor, Dbldy.
Papilio Dionysos, Dbldy.
Diadema Boisduvalii, Dbldy.
Pavonia Aorsa, West.
Any species of Phyllodes.
Smerinthus Dentatus, Cram.

Papilio Gundlachianus, Feld.
Dynastor Napoleon, Doubl, Hew.
Argynnis Sagana ♂, Doubl, Hew.
Zeuxidia Aurelias, Cram.
Urania Rhipheus, Cram.
Urania Sloanus, Cram.
Eudæmonia Semiramis, Cram.
Nyctalemon Cydnus, Feld.
Erasmia Pulchella, Hope.
Actias Mænas
Saturnia Dercceto, Mssn.
Saturnia Argus, Drury.
Any Asiatic species of Parnassius.
Citheronia Phronima.
Castnia Dædalus, Cram.
Pyrameis Gonerilla, Fabr.
Pyrameis Abyssinica, Feld.
Pyrameis Dejeanii, Godt.
Pyrameis Tameamea, Esch.
Vanessa v. Hygiæa, Hdrch.
Vanessa v. Elymi, Rbr.
Dasyopthalmia Rusina, Godt.
Morpho Phanodemus, Hew.
Any species of Agrias.
Any species of Callithea.
Pandora Chalcothea, Bates.
Pandora Hypochlora, Cates.
Pandora Divalis, Bates.
Colias Viluiensis, Men.
Colias Ponteni, Wallengr.
Bunæa Deroyllei, Thom.
Bunæa Phædusa, Dru.
Rinæa Zuleica, Hope.
Papilio Disparilis, Herr-Sch.
Acræa Perenna, Dbldy.
Opsiphanes Boisduvalii, Dbldy.
Clothible Jægeri, Herr-Sch.
Pieris Celestina, Boisd.
Euplœa Eurypon, Hew.
Paphia Panariste, Hew.
Limenitis Lymire, Hew.
Io Beckeri Herr-Sch.
Eacles Kadenii, Herr-Sch.
Smerinthus Timesia, Stoll.
Smerinthus Panopus, Cram.
Sphinx Substrigilis, West.
Saturnia Larissa, West.
Eusemia Victrix, Bellatrix, Amatrix, Dentatrix, West.
Smerinthus Modesta, Fab. (nec. Harris.)
Smerinthus Tartarinovii, Brem.

These are a few of the very many of the rarer species that I am eager to procure; of course there are numberless others from all parts of the world, equally desirable and coveted by me.

North Atlantic Express Co.,

NEW YORK,
OFFICE, 57 BROADWAY.

Chartered by Special Act of Incorporation.

CENTRAL EUROPEAN OFFICE:
5 RUE SCRIBE, PARIS.

PRINCIPAL OFFICE IN GREAT BRITAIN:
4 Moorgate St., London, E. C.
B. W. & H. HORNE, AGENTS.

Branch Offices: Golden Cross, Charing Cross; George & Blue Boar, High Holborn; 108 New Bond Street; 474½ New Oxford Street.
Office in Liverpool: Old Castle Buildings Tithebarn Street.
Continental Offices: 5 Rue Scribe, Paris; 82 Rue d'Orleans, Havre; 88 Rodingsmarks, Hamburg; 29 Bahnhofs Strasse, Bremen.

Merchandise, specie, bullion, stocks, bonds, or other valuables and packages and parcels of every description, personal effects, baggage, etc., forwarded to and from Europe and all parts of the United States, the States and Territories of the Pacific Coast, British Columbia and the Canadas included, *at fixed Tariff rates, with no extra charges whatever for Custom-House brokerage, commission, delivery, etc., etc., the shipper or receiver being under no other care or expense than the stipulated freight from the point of shipment to place of delivery, and the amount of duties and government fees actually paid at the Custom-houses of the United States or Europe.*

For the convenience of shippers, **where agencies of the Company are not established, packages or heavy goods may** be forwarded to either of the offices **or agencies of the Company,** by either of the express or transportation companies in the United States, or by post, by railway, through the parcel delivery companies, or forwarding houses in any part of Great Britain or the Continent of Europe.

All packages, trunks, or parcels forwarded by this Company will be landed on arrival simultaneously with the mails, or immediately thereafter and will be entered at the Custom-house, duties paid and delivered to the parties to whom addressed in any part of Europe, the United States, the Canadas, or British Columbia, with the greatest possible dispatch. Transportation charges and duties collected on delivery, *or may be prepaid, at the option of shipper.*

Insurance against marine risk taken by the Company, when desired by the shipper, at the lowest current rates; premium payable in all cases in advance.

Shippers to or from any part of America, and Americans traveling in Europe, will find this the quickest, cheapest and most reliable medium of transportation, the business of this Company being conducted upon the well known prompt American express system, which has become so great a commercial necessity and convenience throughout the United States.

Purchases made, and collections and communications in every part of Europe and the United States promptly executed.

☞ Circulars sent on application to

S. D. JONES,
Manager,
57 BROADWAY, N. Y.

No. 9.   Price 50 Cents.

LEPIDOPTERA,

RHOPALOCERES AND HETEROCERES,

INDIGENOUS AND EXOTIC;

WITH

Descriptions and Colored Illustrations,

BY

HERMAN STRECKER.

Reading, Pa., 1874.

Reading, Pa.:
Owen's Steam Book and Job Printing Office, 515 Court Street,
1874.

CATOCALA ROBINSONII. Grote.

Trans. Am. Ent. Soc., Vol. IV., p. 20, (1872).

(PLATE IX, FIG. 1, ♂.)

Expands 2½ inches.
Head and thorax, above, whitish grey; abdomen blackish. Beneath, thorax white, abdomen white, powdered with grey atoms.
Upper surface; primaries whitish grey; transverse shades very faint; transverse lines narrow and black; reniform ordinary size; sub-reniform open.
Secondaries black, greyish at base; fringes white.
Under surface; primaries black; a small white basal patch; a narrow white, sub-terminal band which becomes almost obsolete towards the interior margin; a white median band extending from costa to half way between the latter and interior margin.
Secondaries, basal third white, rest black with a narrow white mesial band. Fringes on all wings white.
Habitat. New England, Middle and Western States to the Mississippi.
Somewhat rare, and distinguishable from all other black-winged species by the peculiar pale frosted appearance of the upper surface of primaries.

CATOCALA RETECTA. Grote.

Trans. Am. Ent. Soc., Vol. IV, p. 4, (1872).

(PLATE IX, FIG. 2, ♂.)

Expands 2½ inches.
Head and thorax, above, grey; abdomen blackish grey. Beneath white.
Upper surface; primaries light grey, transverse lines black and distinct; reniform surrounded by a double line; black sub-apical dash; two other black dashes, one crossing the transverse anterior line, and the other the transverse posterior line, towards the interior margin.
Secondaries black, with broad white fringes.
Under surface nearly the same as in C. Desperata, to which this species, though smaller, is closely allied.
Habitat. New England, Middle, and doubtless other of the United States.
Compared with C. Desperata and C. Flebilis, it differs in the ground colour of primaries, being brighter and of a less bluish cast; the black shades sharper, not so spread or suffused; and in the greater depth of the white fringe of secondaries.
It is a rare species, and has generally, heretofore, been confounded with C. Desperata.

CATOCALA FLEBILIS. Grote.

Trans. Am. Ent. Soc., Vol. IV, p. 4, (1872).

(PLATE IX, FIG. 3, ♂, 4, variety ♂.)

Expands 2½ inches.
Head and thorax, above, grey; abdomen blackish. Beneath white.
Upper surface; primaries bluish grey; transverse lines black and well defined; reniform brown and not very distinct; sub-reniform open; the sub-apical dash is continued directly across the whole length of the wing to its base.

CATOCALA FLEBILIS.

Secondaries black, with white fringe.

Under surface; primaries black; a small white basal patch, a white spot or space in cell, and a very narrow, half obsolete, white sub-marginal band.

Secondaries white; a broad black marginal and narrower mesial band, the white space between these two is very narrow.

Found in same localities as C. Desperata, but by no means as common.

FIG. 4 is a variety, occasionally occurring, in which the broad central longitudinal dash is broken in the middle at the reniform and sub-reniform.

For the original of this figure (4) I am indebted to friend Angus, of West Farms, N. Y., who captured near that village, at various times, examples of this variety, and to whose goodness I have been again and again indebted for valuable additions to my cabinet, as well as many other acts of kindness.

CATOCALA AHOLIBAH. Nov. Sp.

(PLATE IX, FIG. 5, ♀.)

Expands 3 inches.

Head and thorax, above, dark brown, with scattered white or grey scales; abdomen brown. Beneath light brownish grey.

Upper surface; primaries dark brown frosted and intermixed with white and grey; a white space adjoining the reniform inwardly; reniform indistinct; sub-reniform very small, white, surrounded with black, and entirely disconnected with the transverse posterior line.

Secondaries crimson, with brownish hair at the base; median band rather narrow and regular, and continued to within a short distance of the abdominal margin, where it turns upwards and is lost in the brownish hair that clothes that part.

Under surface; primaries crossed by three black bands, none of which join or merge with each other; the spaces between the base and sub-basal band, and between the latter and the median band, are orange coloured inclining a little to crimson at the interior margin; the space between the median and marginal bands is white; fringe white, with black at terminations of the veins.

Secondaries; inner two-thirds crimson, a little paler than on upper side, rest white; marginal band tinged with grey at and near the costa; median band terminates about one line from abdominal margin; slight indications of a discal crescent connecting with the median band; fringe white.

Habitat. California.

The above description and accompanying figure were taken from the single ♀ example contained in the collection of Mr. James Behrens, of San Francisco, to whose practical and extended labors in Entomology we are indebted for our knowledge of many of the Pacific species, and who, in order to enable me to present the species, had the almost unprecedented generosity to rob his own fine cabinet of the only example it contained of this insect. He says, in reference to it, that "it is a frequenter of the deepest, darkest gulches and glens of the higher mountains of California," and further, that it flies in July and August, and was the wildest animal he ever saw.

This species closely resembles C. Sponsa* and its ally, C. Dilecta†; the primaries, on upper surface, have a striking similarity, especially to Sponsa, and the ground colour of secondaries is the same, but there the resemblance ceases; the black bands of secondaries are different, and in the under surface of primaries of the two European species the black bands are broader, and the sub-basal and median at the inner half of the wing are connected, and the median and marginal are almost confluent at and towards the interior margin, and the narrow spaces between all these bands are entirely white. In the secondaries the crimson extends much nearer to the costa, and there is a large black discal lune or spot. I have been thus particular in my descriptive remarks of the above analogous European species, inasmuch as, no matter how careful a drawing be made, the student does not of course feel that certainty whilst comparing his example with it, and is often apt to think, if the differences are not very strongly marked ones, that they may be the result of the artists not being exhaustively accurate, and is, consequently, sometimes thereby led to erroneous conclusions. But the shape of the black bands on upper surface of secondaries, and the spaces between the black bands on under surface of primaries

* Linné. Syst. Nat., 841, (1767).
† Hubner, Sam. Eur. Schmett., 328, (1793-1827).

which are open and clear from costa to interior margin, and which are also orange coloured between the median and sub-basal, and the latter and base, are points that are so distinctive as to preclude all idea of the identity of our species with either of its European allies alluded to. The fact of the red on under side of primaries being of an entirely different tint from that of secondaries is very remarkable; I do not believe it exists in any other known Catocala.

March 1st, 1874.

CATOCALA MARMORATA. Edwards.
Proc. Ent. Soc., Phil., Vol. II, p. 508, (1864).

(PLATE IX, FIG. 6, ♀.)

Expands 4 inches.
Head and thorax light grey; abdomen is wanting in the single example so far known.
Upper surface; primaries pale grey and white, more or less powdered with dark grey or blackish atoms, (and bear a superficial resemblance to those of the European C. Fraxini*); transverse lines black; beyond the transverse posterior line, a brown band, succeeded outwardly by another which is much narrower and pure white; reniform dark, and shape not well defined; sub-reniform joined by a line to, not formed by, a sinus of the transverse posterior line; fringe white.
Secondaries scarlet of a lovely shade; mesial band narrowed in the middle, and extends almost to the abdominal margin; fringe white.
Habitat. Yreka, California.
A regal insect, exceeding in size all known American species; the unique type from which the annexed figure was drawn is in the Museum of the Am. Ent. Soc.; its sex can not be determined, as, unfortunately, the abdomen, as I before stated, is non est, but from general appearances I should suppose the example in question to be a ♀.
One can but regret that so little concerning this fine species is known; the original description contains no further remarks than "from Yreka, California," and we can only hope that time, which "at last sets all things even," will enable us to receive specimens, and learn more concerning this superb insect.

CATOCALA ULTRONIA. Hubner.
(Eventis U.) Sam. Exot. Schmett., II, 26, f. 347, (1793-1827).
Catocala U., Guenée, Noct. III, 89, (1852).
Catocala U., Packard, Guide, p. 317, t. 8, fig. 4, (1869).

(PLATE IX, FIG. 7, ♀.)

Expands 2 to 2½ inches.
Head and body brown above, greyish white beneath.
Upper surface; primaries pale ash-coloured, a broad, longitudinal, rich deep brown space covers the lower one-third of the wing to the interior margin; a broad, suffused, sub-apical dash of the same colour; reniform small, generally almost obsolete; sub-reniform open.
Secondaries deep red; mesial and marginal bands regular, and extending to abdominal margin; fringe white.
Under surface; primaries, base black, between this and the median band the space is red, between the median and marginal bands it is yellowish white.

* Linne. Syst. Nat., 512, (1758).

Secondaries red, greyish near the costa; mesial band irregular in width and extends to inner margin; a black discal lune joins the mesial; fringes white and black.

Habitat. Canada, and the United States generally east of the Mississippi.

A common and very pretty species which, by the peculiar appearance of the primaries, can be easily known from all others.

CATOCALA PIATRIX. Grote.

Proc. Ent. Soc., Phila., Vol. III, p. 28, t. III, (1864).
Proc. Ent. Soc., Phila., Vol. III, p. 532, (1864).
Trans. Am. Ent. Soc., Vol. IV, p. 16, (1872).

(PLATE IX, FIG. 8, ♂.)

Expands 2¾ to 3 inches.

Head and thorax brown, with darker lines; abdomen yellowish brown. Beneath pale ochraceous.

Upper surface; primaries brown, varied with darker basal, median and sub-apical shades; transverse lines black; reniform large and enclosed in a dark shade; sub-reniform open and pale, from this to the costa, interior to the reniform, is a paler space.

Secondaries yellow, base clothed with brownish hair; bands rather broad, but narrowing towards their termination at the abdominal margin.

Under surface * of all wings pale ochraceous, darker at interior margins; primaries have three transverse bands, the sub-basal and mesial black and distinct, the marginal pale, much suffused with yellow, especially towards the exterior margin; on secondaries the mesial band is irregular in width, narrow towards the costa, broader on disc, and is terminated some distance from inner margin; marginal band darkest near the anal angle and becomes almost obsolete as it nears the apex and costa.

Habitat. New England, Middle and Southern States.

A rather common species, belonging to the same group as Subnata and Neogama, in company with which it occurs in many localities.

CATOCALA MULIERCULA. Guenée.

Noct., Vol. III, 97, (1852).

(PLATE IX, FIG. 9, ♂.)

Expands 2½ to 2¾ inches.

Head and body dark brown above; beneath yellowish grey.

Upper surface; primaries dark, rich reddish brown, with none of the markings very distinct; reniform small; sub-reniform pale, space immediately interior to the reniform also a little paler; transverse lines black.

Secondaries deep yellow clothed with brown hair at base and abdominal margin; marginal and mesial bands extend to interior margin; fringe blackish, except near apex, where it is white.

* The original description of the under surface, and still more, the second one by the same author, which followed it a few months later, would lead one to expect, instead of a very ordinary looking Catocala, some gorgeous insect rivaling the richness of Erasmia or Eterusia. The first description says: " Under surface pale luteous, pale ochraceous brownish along external margins, orange coloured at base of posterior wings, median bands black, slightly iridescent." Proc. Ent. Soc., Phil., III, 82. The second description, in regard to which the author says, " I allow the present description to supersede the one given by me on page 82 of the present volume," is as follows: " Under surface of both pair pale grayish ochraceous, iridescent, irrorate basally and sub-discally tinged with an orange shade; anterior wings crossed by three, posterior pair by two black transverse bands." Proc. Ent. Soc., Phil., III, 533. In the third description of this species, in Trans. Am. Ent. Soc., IV, 16, all the splendour of description is transferred to the upper surface, the author doubtless considering that he had conscientiously performed his duty towards the under side in the preceding two descriptions, allows it "presently" to rest in peace.

Under surface yellow, darkest at and near inner margin of secondaries; three black bands on primaries, the sub-basal and median connected near the inner margin; the two black bands of secondaries extend from costa to abdominal margin.

Habitat. Middle, Western and Southern United States.

Is an exceedingly rare species with us, but occurs more frequently in Georgia and Florida, from which latter states I have occasionally received it.

CATOCALA CONSORS. Abbot & Smith.

Phalæna Consors, Lepid. Georgia, Vol. II, p. 177, t. 89, (1797).
Catocala Consors, Guenee. Noct. III, 99, (1852).

(PLATE IX, FIG. 10, ♂.)

Expands 2¼ to 2½ inches.

Head and thorax, above, smoky grey; abdomen yellowish brown, beneath yellowish grey.

Upper surface; primaries dark smoky **grey**; **transverse lines black**, dull, and not deeply dentate; reniform brownish; sub-reniform small.

Secondaries; deep yellow, with brown hair at base and abdominal margin; marginal **band with three deep** indentations interiorly, mesial band very irregular; the shape of these bands cause the **wing to have a chequered** appearance, **one** point of the marginal band almost touching another of the mesial, **on the disc**; fringes black and yellow.

Under surface dusky yellow; three broad black bands on primaries, mesial and marginal bands of secondaries much as above.

The larva, according to Abbot, is found on the Bastard Indigo (*Amorpha Fruticosa* L.).

Habitat. From Maryland to the **Gulf**.

An exceedingly rare species, or **at least** difficult to obtain, as it is represented in but few American collections; the example from which fig. 10 was drawn is in the collection of Mr. Chas. Blake, of Philadelphia, well known by his extensive labours on the N. Am. Mutillidae,* and to whose uniform goodness I am indebted for innumerable favours, far more than, with my best will, I ever shall be able to repay.

CATOCALA NEBULOSA. Edwards.

Proc. Ent. Soc. Phila., Vol. II, p. 510, (1864.)
Catocala Ponderosa, Grote & Robinson, Proc. Ent. Soc. Phila., Vol. VI, p. 23, t. 4, (1866). *Grote*, Trans. Am. Ent. Soc., Vol. IV, p. 11, (1872).

(PLATE IX, FIG. 11, ♀.)

Expands **3** to 3½ inches.

Head **and body** brown above, and yellow beneath.

Upper surface; primaries, ground color greyish yellow, heavily clouded with maroon **or** dark reddish brown, which has on fresh examples a perceptible bluish sheen, especially on that **portion from** the transverse anterior line to the base, which is so dark as to appear almost black, the **transverse** anterior line widens unequally from its middle upwards to the costa where it is very broad; transverse posterior line deeply sinuated; reniform moderately large and doubly annulated; sub-reniform connected with the transverse posterior line.

Secondaries rich yellow; marginal band broad, space between this and the median band narrow; the portion of the **wing** from **the** median band to the base almost entirely covered with heavy brownish hair, giving the wing much **the same appearance** as in Cerogama, Guen.; a yellow apical spot; fringe yellow.

Under surface **yellow**; **primaries with** three purplish black bands, the sub-marginal, which **is narrowest, and** the median, **extend from costa to interior margin**, the sub-basal reaches only **to** the sub-median **nervure**;

* In the Trans. Am. Ent. Soc., Vol. III, (1871).

none of these bands are connected with each other. On secondaries the mesial and marginal bands extend to abdominal margin. Fringes on all wings yellow.

Habitat. Middle, and Western States to the Mississippi; rather rare.

In the original description of this species we have another instance of how utterly valueless, aye, worse than valueless, are such things unaccompanied with figures; Mr. W. H. Edwards' description (in Proc. Ent. Soc., Phila., 1864,) is better than nine-tenths of such things generally are, and, moreover, is written in language that can be understood, nevertheless, after a lapse of two years, so little had this description been recognized, that Messrs. Grote & Robinson re-described this, one of the largest of our Catocalæ, and one so prominently unlike all others, as a new species, and even made remarks comparative concerning the difference between Edwards' Nebulosa and their Ponderosa, but I had better quote their own words literally and in full, which follow their technical description: "Several specimens examined. Resembles the description of *C. Nebulosa*, Edwds., but differs in several important particulars, the color of the ordinary spots, conformation of the median band on the under surface of the secondaries and the general aspect of these on the upper surface seem to be different, while some of the minor details, such as the color of the scales clothing the nervules, etc., will not apply properly to *C. Ponderosa*, nobis."* I believe Mr. W. H. Edwards published no protest, perhaps he cared nothing about it, or it may be that their description was as unintelligible to him as his was to them, for theirs was a third longer and infinitely more abstruse and grandiose, and, in consequence, he may not have been aware of the identity of his Nebulosa and their "Ponderosa, nobis." Six years later Mr. Grote again described it under the name of Ponderosa, giving Nebulosa as a synonym; after his technical description comes the following (quoted in full): "Mr. Edwards compares the secondaries quite wrongly with those of C. Cerogama,† which C. Ponderosa in nowise resembles. The specific name chosen by Mr. Edwards had already been used five times in the family;"‡ by this we understand that he has at last became acquainted with the fact that Nebulosa, Edwds., and Ponderosa, Grote & R., are the same; but in this instance it appears that the law of priority must succumb, in order that the G. & R. may still obtain, at all events the G., for were that stricken off in all instances where it is attached to synonyms, the taint of synonymy would be removed from the great bulk of N. American Heteroceres. No! Ponderosa must stand because Nebulosa "had already been used five times in the family." Now, how has it been used five times? it has been applied to an Agrotis,§ a Mamestra,‖ a Hadena,¶ a Dryobota** and a Taeniocampa,†† the latter, however, is Nebulosus, not Nebulosa, and in these five the name only holds for one, *Mamestra Nebulosa*, Hufnagel; as regards the others, they are only synonyms, and no longer used to designate the species. The connection between Mamestra Nebulosa, Hufn., and Catocala Nebulosa, Edwds., is about as intimate as between Papilio Philenor, L.,‡‡ and Parnassius Cladius, Men.,§§ and they resemble each other about as much as do those two diurnals. Mr. Grote's own words will, however, support me in retaining Mr. Edwards' prior name of Nebulosa, for he says, (in speaking of another species, C. Marmorata,) "with regard to the specific name, this is already used in the Noctuidæ for a species of Hadena. It has been hitherto the custom to reject such names, but this should not be done where, as in the present case, there is no danger of confusion."‖‖ The Hadena alluded to is a small affair found in N. E. Labrador, Greenland, Iceland and northern Scotland; it is a little smaller than Mamestra Nebulosa, has brownish primaries and smoky secondaries; its true name, however, is not H. Marmorata, but H. Exulis,¶¶ the former title was bestowed on it by Dr. Herrich-Schæffer much later; but, in either case, to object to a species of the large, brilliant, semi-Geometrid genus Catocala, occupying a position at the termination of the great family of the Noctuidæ, being designated by the same name as one of the small obscure moths comprised in the widely different genera of Mamestra or Hadena which stand near the head of that family, seems to be as useless as it is inconsistent.

* Proc. Ent. Soc., Phila., Vol. VI, p. 25, (1866).
† This incorrect statement of Mr. Grote's, regarding Mr. Edwards comparing C. Nebulosa to C. Cerogama, he corrected about nineteen months later, in the Canadian Entomologist, Vol. V, p. 162, (Sept., 1873).
‡ Trans. Am. Ent. Soc, Vol. IV, p. 12, (Jan., 1872).
§ Agrotis Decora, Hub., 45; Nebulosa, Hub., 402, Sam. Eur. Schmett.
‖ Mamestra Nebulosa, Hufnagel, Berlinisches Mag., Vol. III, 418, (1767); *Bimaculosa*, Esper.
¶ Hadena Basilinea, Fab. Mantissa. Ins. 183, (1787); *Nebulosa* Viewg, Tab. Verz., T. I, (1789); *Sordens* Werneburg, Beitrage zur Schmett., I, 251, (1864).
** Dryobota Protea, Borkh., Nat. Ges. Eur. Schmett., IV, 336, (1792); *Nebulosa* Walch, Naturforscher, XIII, p. 29, (1779).
†† Taeniocampa Incerta, Hufn., Berlin. Mag., Vol. III, 298, 424, (1767). *Nebulosa* Haworth, Lep. Britannica, p. 120, (1803-1820).
‡‡ Linnæus, Mantissa, I, p. 535, (1771).
§§ Menetries, Cat. Mus., St. Petersberg, Lep. I, p. 73, (1855).
‖‖ Grote, Trans. Am. Ent. Soc., Vol. IV, p. 7, (1872).
¶¶ Hadena Exulis, Lefèbvre, Ann. Soc. Ent. Fr., 392, (1836). *Marmorata*, Herrich-Schæffer.

CATOCALA AMASIA. Abbot & Smith.

Phalæna Amasia, Lep. Georgia, Vol. II, p. 179, t. 90, upper figure, (1797).
Catocala Amasia, Duncan, Nat. Lib. Ent., Vol. VII, p. 205, t. 26, (1841).
Catocala Amasia, Guenee, Noct., Vol. III, 103, (1852).

(PLATE IX, FIG. 12, ♂.)

Expands 1½ to 1¾ inches.
Head and thorax pale grey and white, with black marks; abdomen yellowish; beneath yellowish white.
Upper surface; primaries white, transverse lines black and distinct, reniform and sub-reniform distinctly defined by black lines, space from the transverse posterior line to the exterior margin brownish, traversed from costa to inner margin by a narrow white zig-zag band.
Secondaries yellow, marginal band broken about two-thirds in from the costa, but replaced with a spot at the anal angle; median band narrow and nearly straight, and discontinued some distance from the abdominal margin.
Under surface yellow, darkest at bases and at inner half of secondaries; a marginal and median band of ordinary width extending from costa to inner margin; of the sub-basal band, an almost imperceptible shade is all that is noticeable, at least in the examples I have or have access to; perhaps in large suites there may occur examples in which this band may be more distinct. Bands of secondaries same as on upper side.
Habitat. Virginia, Georgia, Florida, and other of the Southern States. Rare.
According to Abbot, the caterpillar is grey, with darker lines laterally, and its food various kinds of oaks, but that it also was found on the Pride of China, (Melia Azedarach, L.), that it spun the beginning of May and came out the end of the same month.
On the lower part of Abbot's plate 90, where this insect was first represented, there is another species which purports to be its female, and which is found not only in the south, but as far north, to my knowledge, as Rhode Island; it is a species of the same size as Amasia, and was described as C. Formula*; in a succeeding plate it will also be delineated.
The nearest European representative of Amasia is C. Nymphagoga,† but the similarity exists principally in size and markings, as the upper side of primaries in the latter are dark, whilst in our species they are white, but the style of ornamentation, arrangement of bands, etc., are very similar.
But few examples of C. Amasia find their way into collections, owing to the non-residence of collectors or Lepidopterists in the Southern States, and, however speculative and enterprising a people the Americans may be, they have not yet found a way to make the natural sciences pecuniarily remunerative; and in this respect, as well as in some others, we need not be ashamed to learn something from the old country.

A FEW WORDS ON THE CATOCALA NOMENCLATURE.

I have a sort of old-fashioned respect for the way the fathers of science used to name these things; for instance, the Catocalæ all had amatory names, relating to love or marriage, Amatrix, Cara, Relicta, etc., etc. Of course these terms would soon be exhausted, and, in fact, have been; then, names that would in a great measure keep up the connection would naturally be next selected, and the most appropriate ones for the purpose would be those of women famous in ancient history for their lust or talents, or both combined, as in the case of C. Messalina,‡ C. Helena § and C. Briseis,‖ of later authors, and it might be well to continue in the same plan. Of upwards of forty species found in Europe and Siberia, none had the names of any scientist, ancient or modern, bestowed upon them, though such names as Lederer, Felder, Hewitson and Moschler will, nevertheless, stand whilst printing or science endure. But to us progressive Americans it is owing that the harmony of the Catocala Nomenclature has been broken; Edwards first, with his C. Walshii, and then Grote

* Grote & Robinson, Proc. Ent. Soc., Phil., Vol. VI, p. 27, (1866).
† Esper, Schmett., 105, 5, (1787).
‡ Catocala Messalina, Guenee, Noctuelites, III, p. 105, (1852).
§ Catocala Helena, Everamann, Bull. Mos., II, (1856).
‖ Catocala Briseis, Edwards, Proc. Ent. Soc., Phil., II, (1864).

with C. Clintonii, C. Robinsonii, etc.; it is, however, done, and irrevocably so, and we can only in sadness submit. I can not, however, refrain from thinking that there is a great deal in the appropriateness of a name, for I never yet knew one of your George Washington Smiths, or John Quincy Adams Warrens, or Michael Angelo Joneses, leaving any very perceptible foot-prints on the sands of time, and vividly I remember, whilst walking, years ago, through a plantation in S. Carolina, that every third field hand was Julius Cæsar Agamemnon, or Mark Antony Aurelius, and one burly fellow carried, in addition to about 300 pounds adipose tissue, the fearful additional load of Clarence Theophrastus Columbus Porcher Barton. In the case of these overloaded unfortunates, the grandeur of the name was, like the helmet in the "Castle of Otranto," crushing instead of adorning. In the case of the beauteous and wonderful works of nature it is just the contrary, their loveliness and marvelous structure are such that the grandest names of science, art and history seem almost too feeble to apply to them, whilst names of lesser note cannot be exalted by the association, but serve only as a blot to deface the beautiful. I believe that all that is great and sublime in nature and art is more or less intimately connected, but now, in Heaven's name, what grandeur, or historical or poetical idea can we associate with such names? It is true, they may answer the purpose of identification, but so would Catocala No. 1, Catocala No. 2, etc., for that matter equally as well, but how different when we gaze on the gorgeous Priamus Butterfly* what a flood of thought it suggests! the court of the old Trojan King arises and is "followed fast and followed faster" by each varied scene of the Iliad; the Golden Crosus† reminds in an instant of the magnificence of the Lydian monarch and the death of the hapless Atys; and the splendid Sardanapalus,‡ of the sumptuousness of that prince; and Humboldtii,§ though any to whom science is dear scarce need a reminder, of one far exceeding in rank all of earth's potentates, one of whom a monarch of Europe once said, "Der groesste mann seit Noah."‖

ENTOMOLOGICAL NOTES.

POLAR LEPIDOPTERA.—During a recent visit to Washington I had the opportunity of examining, at the Smithsonian Institution, the few entomological examples collected by Dr. Emil Bessels, of the unfortunate "Polaris Expedition," at Polaris Bay, N. Lat. 81°, 83°. There are three species of Lepidoptera, Het., which I identified as follows:

Dasychira Rossii, (Læria R.) Curtis, Ross's 2nd Voy. App. Nat. Hist., p. 76, l. A. (1835), one pair of ♀♀, also the web with eggs surrounded by the hair of the larva. This species has been found also in N. E. Labrador.

Anarta Richardsonii, (Hadena R.) Curtis, Ross's App., p. 72, l. A. (1835). A. Algida, Lefebvre, Ann. Soc. Fr. 395, Pl. 10, 6, (1836). Two examples. Occurs also in Labrador and Northern Norway, and I have seen one example taken on Mt. Washington, New Hampshire.

Cidaria Salini, (Psychophora S.) Curtis, Ross's App., p. 73, l. A, (1845), five or six examples. The later described C. Polysitaria, Gn., found in Lapland, is doubtless identical with this species.

There are also several examples of a Hymenopterous insect, Bombus Kirbiellus, Curtis; and a Diptera, Tipula Arctica, Curtis, both figured and described in the same work as the Lepidoptera above.

After my examination of these entomological treasures, still having some time to spare, I strolled through various other departments of the Museum of the Institution; on reaching the upper apartment, devoted mainly to casts and remains of pre-Adamite animals, and whilst gazing on these stupendous relics of a period swept in obscurity almost equal to that of futurity itself, I was roused from my musings by the sound of a succession of raps on some evidently hard substance, when on turning my head I saw two animals of the present era, ♂♀, with artificial coverings of the texture and appearance of broadcloth and silk, communing together, and at short intervals striking, the one with a cane, the other with the end of a parasol, the cast of the Glyptodon; every rap caused a white mark to appear, the result of the striking bone of the pigment from the plaster which it covered; I much fear I had little regard for etiquette or the rules of well-bred society, for without a moment's reflection I expressed to these dignified Yahoos my unqualified opinion of their Vandalic conduct, which, of course, like all opinions unsolicited, was by no means graciously received; nor was my opportunity further resented, after the departure of these poor mindless things, by perceiving on the frontal plate or bone of the same Glyptodon, that some wretches had scrawled their pitiful, miserable, unknown, degraded names! But bidding farewell to the thoughts of these debased creatures, not one tithe as noble as the monster whose semblance or remains they contaminated, I left the apartment and wended my way towards other objects of interest. Ere I close I cannot fail to express my appreciation of the uniform kindness and attention I received from the various scientific gentlemen connected with the Institution, as well as those of the neighboring Museum of the Agricultural Department, the latter almost solely the creation of the untiring, indefatigable Prof. Glover.

Finally, I can scarce avoid mentioning, among the vast number of examples of nature and art accumulated in the Museum of the Smithsonian Ins., the splendid specimen of the great Rocky Mt. Goat, an animal so rare as almost to have led one to the belief that it was apocryphal; the cast of the shell of an immense ? Chelonian which measures nearly three paces in length and two in width, and is about four feet in height; a huge Octopus (the Devil-Fish of Victor Hugo's "Toilers of the Sea,") in alcohol, which we should judge to measure, with arms extended from tip to tip some ten feet or more, and a single arm of another much larger; the numerous and most curious wood carvings, etc., etc., of the Alaska Indians, their Masque of Death, the Bird that brought their fathers from the Lord only knows where. In the Geological and Mineralogical Department, under the supervision of my fellow-townsman, Dr. Endlich, is a huge mass of native copper, weighing I don't know how much, and surmounted by a famous acolite of fabulous proportions. Good friends, I must close, or I do not know when I might stop; you will perhaps say this is not Lepidopterology, why should it be here introduced? true,

---

* Ornithoptera Priamus, Linnaeus, Mus. Lud. Ul. Reg., p. 182, (1764).
† Ornithoptera Croesus, Wallace, Proc. Ent. Soc., Ser. II, Vol. V, p. 70, (1859).
‡ Agrias Sardanapalus, Bates, Proc. Ent. Soc., Ser. II, Vol. V, p. 111, (1860).
§ Tithorea Humboldtii, Latr., Perisama Humboldtii, Guer.
‖ The greatest man since the flood.

but each page of God's great book is connected with the other, bound in its mighty cover, the Universe, and we cannot admire one without admiring the other; we do not love our mistress's hand alone, but also her brow, hair and eyes, her whole beautiful form, the entire faultless work.

NORTHERN LEPIDOPTERA.—I here give a list of the Heterocerous Lepidoptera received by me sometime since from Mr. Couper, who took them in S. Labrador and Anticosti Island in the summers of 1872 and 1873; there are still several Agrotis and Crambis that I am not quite certain of, and which I must defer attending to until I receive the few Polar species that are yet lacking, to my cabinet, for comparison.

Alypia Octomaculata, Fabricius, (*Zygaena A.*) (1793), appears to have been common, as I received twelve examples; they present no particular differences from those found elsewhere. The opinion has been expressed, though I doubt its accuracy, that these Anticosti specimens are the ♂ of A. Langtonii, which latter has but one yellow spot on the secondaries, whilst Octomaculata, as we all know, has two white ones.

Alypia Langtonii, Couper, five examples, presenting no variation from some which I obtained in Luzerne County, Pennsylvania. This species appears to be very closely allied to the Californian A. Sacramenti, Boisd., and to judge from Mr. Stretch's figures on Plates 1 and 6 in his admirable work, I should consider them to be identical; Sacramenti I have not yet seen in nature.

Deilephila Gallii, Rott., (1775). *D. Chamaenerii*, Harris, one example; this is a species common both to Europe and America.

Setia Ruficaudis, Kirby, Faun. Bor. Am., IV, p. 303, (1837). *Hemorrhagia Uniformis*, Grote & Robinson. Five examples.

Thyatira Pudens, Guenée, Noct. I, 13, (1852). The single example differs from those found in Pennsylvania, in that the white spot on the middle of costa, on primaries, is indistinct on the inner edge, where it is much broken and merged into the grey ground colour, this latter is more or less freckled with white throughout.

Agrotis Chardinyi, Boisduval, Eur. Lep. Ind. Meth., p. 94, (1829). *A. Hetera*, Zetterman. Bull. Mos., p. 35, (1837), two examples, a trifle larger, but agreeing exactly in all other respects with the typical examples from Central Russia and Siberia. I believe this is the first instance of the capture of this species in the Western Continent. It belongs to the same group as the European A. Fimbria, A. Orbona, A. Pronuba, etc., commonly known in England as yellow underwings, the secondaries being yellow with a plain black margin.

Agrotis Porphyrea, Hubner. Two examples, present no difference whatever from those found in Piedmont and other parts of Europe.

Agrotis Clandestina, Harris, one example; this species is found as far north as Greenland.

Agrotis Confusa, Treitschke, Schmett. Eur., VI, 1, (1827). Three; do not present any obvious points of difference from examples from Iceland in my possession.

Agrotis Fennica, Tauscher, Mem. Mosc., (1806). One example; in the British Museum are examples credited to Trenton Falls and Nova Scotia, but the one alluded to above is the first and only one I ever saw that was taken in N. America; it is a handsome species, expanding about 1¼ inches, primaries are dark purplish grey margined with pale flesh colour along interior margin, reniform and orbicular also flesh colour; secondaries white, outwardly greyish or smoky.

Mamestra Condita, Guen., Noct. II, 78, (1852). One example.

Hadena Rurea, Fabr. Syst. Ent., 615. (1775). One example, differing in no particular from those received from various parts of Europe.

Lemonia.—apparently L. Albilinea, Hubner, but the single example is in too wretched a condition to speak of with any certainty.

Drasteria Erechtea, Guen., two small sized examples.

Hyperetis Alienaria, H-S. Three.

Metrocampa Perlata, Guen. Sixteen examples, all smaller than those found in the United States.

Acidalia Frigidaria, Moschler, Wiener. Ent. Monatschrift, Vol. IV, p. 373, t. 10, (1860). Two examples.

Cidaria Hastata, L., var. Gothicata, Guen. Nine examples, one with secondaries entirely black, like many of those found commonly in Pennsylvania, N. York, etc., the others are nearer to the European stem-form Hastata, having a much white in them as in many of the latter; some agree exactly with Mr. Moschler's fig. 4 on t. 10, Wien. Ent. Monats, Vol. IV.

Cidaria Tristata, Lin. Syst. Nat., X, 526, (1758). One example, identical in every particular of size, colour and ornamentation, with those from Europe.

Cidaria Obductata, Mosch., Wien. Ent. Monats., Vol. IV, p. 375, t. 10, (1860), three examples, all agreeing with the excellent figure cited. Mr. Moschler, in his original description, ventures the suggestion * that perhaps this mae be a polar form of C. Luctuata, Hb., a species common in most parts of Central Europe; so sure am I that this surmise will prove correct, that when I first received the examples, before I was acquainted even with the figure and description of Mr. Moschler, I placed them in my collection below C. Luctuata as a variety of that species; the principal difference in Obductata, on the upper surface, is the absence of the white mesial band of secondaries, beneath it is greater, the prevalent colour being black.

Baptria Albovittata, Guen. Seven specimens.

Sericoris **Glaciana**, Mosch., Wien. Ent. Monats., Vol. IV, p. 380, t. 10, (1860). One example.

NEMEOPHILA PLANTAGINIS.—Of this species and its varieties, Hospita, etc., I have seen in various collections, and have myself received many examples from Colorado, Nevada and California, as well as some melanotic forms which are unrepresented in the old world, one of which is the New. Petrova of Walker. The synonymy of this species is:

NEMEOPHILA PLANTAGINIS, Linnaeus (*Phalaena P.*) Systema Naturae, 501, (1758); Fauna Suec., 301, (1761); *Bkh.*, English Moths and Butt., t. 50, (1773); *Espr.* Schmett., 36, (1777–1794); *Donovan*, Nat. Hist. Brit. Ins., IV, t. 134, (1792–1836); *Hubner*, (*Bombyx P.*) Samm. Eur. Schmett., 127, 128, (1793–1827); (*Pyrausta P.*) Verz. bek. Schmett., 181, (1816); *Ochsenheimer*, Schmett. Eur., III, 312, (1810); *Godart*, Hist. Nat. Lep. Fr., III, 33, (1821–1824); *Stephens*, (*Nemeophila P.*) Brit. Ent., (1827–1835); *Duncan*, Nat. Lib. Ent., IV, 216, (1836); *Freyer*, N. Beit., 612, (1831–1858). *Borgs*, (*Bombyx P.*) Schmetterlingsbuch, 68, t. 15, (1842). *Saalmuller*, (*Nemeophila P.*) Cat. Eur. Lep., 56, (1871).

Nemeophila Cosyttis, Grote & Robinson, Trans. Am. Ent. Soc., 1, p. 337, t. VI, (1868). ib. IV, 428, (1873).

Nemeophila Uchorti, Grote & Robinson, Trans. Am. Ent. Soc., 1, p. 338, t. VI, (1868). ib. IV, 428, (1873).

---

*"Cidaria Obductata, Moschl., Taf. 10, Fig. 3 (an *luctuata* var.?) Zwei von Labrador erhaltene uebereinstimmende Exemplare wage ich nicht mit Bestimmtheit von *luctuata* zu trennen, denn obwohl dieselben auffallende unterschiede zeigen, wäre es doch möglich, dass sie als nordische Varietäten zu jener art gehörten." Moschler, Wien. Ent. Monats., Vol. IV, p. 375, (1860).

ENTOMOLOGICAL NOTES.

Var. MATRONALIS, Freyer, N. Beitr., 405, (1831–1858). *Hubner, (Plantaginis)* Sam. Eur. Schmett., 238, (1793–1827).
Var. HOSPITA, Schifermuller, Syst. Vera. 310, (1776). *Ochsenheimer*, Schmett. Eur., III, 314, (1810). *Esper, (Plantaginis)* 36, (1777–1794). *Hubner*, Sam. Eur. Schmett., 126, (1793–1827).
Var. PETROSA, Walker, Cat. Brit. Mus., [I], 625, (1855).
Euspychana Geometrica, Grote, Proc. Ent. Soc. Phila., **IV**, p. 318, t. II, (1865).
Euspychana Geometroides, Grote & Robinson, List N. Am. **Lep.**, p. vii, (1865).

I would here say a few words, more or less, regarding our American examples; the single types of Cæspitis and Cichorii, which were taken in California by M. Lorquin, present no differences from some of the endless variations found in Europe; Cæspitis is like one of the common varieties that has the basal half of the secondaries black, and Cichorii, really, has no points in particular to distinguish it from the ordinary European examples; not even its size will save it, as I have trans-Atlantic examples equally as small, and one daunting still smaller; it may not be out of place here to quote in full the author's remarks which follow his technical description of Cichorii. "This species is smaller than N. Cæspitis, and however variable in ornamentation it may prove to be, will be readily distinguished by the black fringes and clear yellow bands of the upper surface of primaries. The larvæ of these species are stated to be quite distinct, and to be found on different food plants." * The black fringes may distinguish it from the single example which served as the foundation for Cæspitis, but they won't separate it from any number of European examples, one of which, now before me, has the fringes on all wings black, another has the fringes black and yellow, according as these colours on the surface extend to the margin, the like colours also prevail on the fringes; the same applies to the var. Hospita, both European and American examples. As regards difference of larvæ and food plant; if the student chooses to confine himself to closet study, entirely neglecting to see nature under more favourable circumstances, he must not be disappointed if error is the result; I thought that the omnivorous appetite of the Arctian larvæ was too well known for any one to base specific distinctions on what they eat; I have had them to feed on anything from grass to an old green pasteboard box, and I doubt if a green thing exists that they would not attempt to digest if you give them a chance. Too much stress is also often laid on difference in appearance of caterpillars, and that too in the face of the fact that some species in various genera are produced from larvæ presenting most remarkable differences of colouration; it is needless to enumerate such; it will be sufficient to refer to Euc. Imperialis, Thyreus Abbotii, various Giaptas, etc., etc., nor can I well **see** why there should not be variation **in** the larvæ of the same species, as well as in the imago.

We now come to var. Petrosa, examples of which I have as yet seen none from Europe, though I have little doubt but that they may occur there; this is the form rediscribed later by Mr. Grote, as Geometrica, who allied it to Cucuscha and created the genus Eupsychona for its reception, placing it in the Zygænidæ.† That Mr. Walker should have considered it a distinct species is not so much a matter of surprise, he probably not having seen the many intermediate varieties, but to create, as Mr. Grote did, a new genus for a Nemeophila, and place it with the Zygænidæ, is about out-Heroding Herod. Why the specific name was changed to Geometroides, in G. & R's List N. Am. Lep., I do not know; In the original description and plate it is Geometrica, but whatever name was meant to be retained "is comparatively of little moment," since this Zygænid Arctian ally of Eudryas must loose its pretensions and fall back **to** Stephen's genus Nemeophila and Walker's name of Petrosa, and stand thus: Nemeophila Plantaginis L. var. Petrosa, Wlk.

The wonderful and countless variations occurring among the Arctians are too well known to need more **than a passing notice,** but I cannot refrain from citing a few; on t. 5, Illustrations Zyg. &c. Bombyx, by R. H. Stretch, are 16 figures representing nine varieties of Leptarctia Lena, Boisd., and they are most astonishingly dissimilar, some having primaries grey and secondaries yellow with plain black margin, some have secondaries spotted in various ways, some have them red, others have secondaries all black, and white spots **or** bars on primaries, and in my possession are eighteen examples received from the author of the above work, all of which are different, more or less, from his figures; one has all the wings entirely black on upper surface. On t. 3 of same work are three figures of Epicalia Virginalis, Boisd., one with rather secondaries having broken black bands, one with black secondaries with ochraceous spots, and **the** third, with the exception of a few small spots, has the secondaries entirely black, and in the eight examples in my cabinet **are** all sorts of intermediate forms between these. Of Arctia Caja L., the varieties are almost endless; they have red hind wings, orange ones and yellow ones, with three spots, five spots, six spots, spots and bands, spots connected and spots isolated, one example from British Am. has the upper wings almost entirely brown, the white being reduced to fine lines; and there are examples in which the upper wings are entirely brown and the lower one entirely black. But to return to Plantaginis; I have received, at various times, of European and American examples, twenty-seven of the ordinary form in many variations, besides of var. Hospita, six from Europe and five from Colorado and Nevada, of var. Petrosa nine from Colorado, Nevada, etc.; some of those latter have the secondaries entirely black, and with three white, disconnected marks on primaries; others have a white anal spot on secondaries, and four pale marks on primaries, connected (all except the spot within the cell, which is always free), in some instances and in others not, one example has the two of the white marks connected in one of the primaries, whilst on the opposite wing the same marks are not united; **in** some there is so much pale patching that it becomes hard to say to which variety they belong, whether to Hospita or Petrosa.

Of Hospita, I believe the first examples found on this continent were taken by Mr. Mead, who captured quite a number of both that var. and Petrosa in Colorado; of the latter I also received specimens taken by Mr. Drexler many years since, and by the Wheeler Exped. of 1871, as well as from others at **various** times. I noticed also an example, **among** a number of unspread Rocky Mt. Lep., in the coll. of Mr. Schonborn, in Washington; this also was from the Rocky Mts., and is **very** close to the type of Petrosa.

March 17, 1874.

PARNASSIUS SMINTHEUS, Dbldy.—I was formerly a strong advocate of the distinctness of this form from the Alpine P. Delius, Esp., but this will only serve as another illustration of the folly of arriving at such conclusions without the fullest material for comparison, for having lately received examples of P. Intermedius, Men., from the Altai Mts., S. W. Siberia, which is by all European authorities considered to be but a variety of P. Delius, I can only add that our Rocky Mt. P. Smintheus is also but a form of Delius, as between the examples of Smintheus from Colorado and Montana, and the lately received Intermedius from Altai, there is simply no difference whatever, they are identical; and so sure were the trans-Atlantic Lepidopterists of this fact, that in the great Catalogues of both Staudinger and Kirby, Smintheus is cited as a variety of Delius; and Mr. Hewitson has repeatedly expressed to me the same opinion.

* Trans. Am. Ent. Soc., I, p. 336, (1868), Grote & Robinson.
† "A. Zygænoid genus allied to *Cucuscha* and presenting some analogies in the neuration to Eudryas." Grote, Proc. Ent. Soc., Phil., IV, p. 317, (1865).

Of the following species I am anxious to obtain examples, either by exchange or purchase; any Naturalists having duplicates of any of them will confer a great favor by communicating with

HERMAN STRECKER,
Box 111 *Reading, P. O.,*
Berks Co., Pennsylvania, U. S. of N. America.

Ornithoptera Hippolytus, Cram.
Ornithoptera Lydius, Feld.
Ornithoptera Helena, Linn. ♀
Ornithoptera Croesus, Wall.
Ornithoptera Brookiana ♀, Wall.
Papilio Evan, Doubl.
Papilio Pericles, Wall.
Papilio Blumei, Boisd.
Papilio Macedon, Wall.
Papilio Philippus, Wall.
Papilio Arcturus, West.
Papilio Phorbanta, Linn.
Papilio Homerus, Fabr.
Papilio Garamas, Hub.
Papilio Caiguanabus, Poey.
Argynnis Rudra, Moore.
Argynnis Oscarus, Evers.
Argynnis Cnidia, Feld.
Argynnis Jerdoni, Lang.
Argynnis Dexamene, Boisd.
Argynnis Jainadeva, Moore.
Argynnis Ruslana, Motsch.
Argynnis Anna, Blanch.
Argynnis Childreni, Gray.
Argynnis Aruna, Moore.
Acherontia Saturnus.
Papilio Ascanius, Cram.
Cœrœus Chorinœus, Fabr.
Parthenos Tigrina, Voll.
Charaxes Epijasius Reiche.
Charaxes Kadenii, Feld.
Charaxes Jupiter, But.
Charaxes Etheocles, Cram.
Catagramma Excelsior, Hew.
Papilio Wallacei, Hew.
Papilio Slateri, Hew.
Papilio Eudoxius Boisd.
Dyctis Bioculatis, Guerin.
Romalæosoma Sophron, Dbldy.
Romalæosoma Pratinus, Dbldy.
Romalæosoma Arcadius, Fabr.
Nymphalis Calydonia, Hew.
Saturnia Epimethea, Dru.
Calinaga Buddha, Moore.
Papilio Icarius, West.
Papilio Elephenor, Dbldy.
Papilio Dionysus, Dbldy.
Papilio Gundlachianus, Feld.
Daidema Boisduvalii, Dbldy.
Pavonia Aorsa, West.
Any species of Phyllodes.
Smerinthus Dentatus, Cram.

Dynastor Napoleon, Doubl, Hew.
Argynnis Sagana ♂, Doubl, Hew.
Zeuxidia Aurelias, Cram.
Brahmæa Whitei.
Brahmæa Certhia.
Urania Sloanus, Cram.
Eudæmonia Semiramis, Cram.
Nyctalemon Cydnus, Feld.
Erasmia Pulchella, Hope.
Aetias Mænas
Saturnia Derceto, Msn.
Saturnia Argus, Drury.
Any Asiatic species of Parnassius.
Citheronia Phorunea.
Castnia Daedalus, Cram.
Pyrameis Gonerilla, Fabr.
Pyrameis Abyssinica, Feld.
Pyrameis Dejeanii, Godt.
Pyrameis Tameamea, Esch.
Vanessa v. Hygiæa, Hdrch.
Vanessa v. Elymi, Rbr.
Dasyophthalmia Rusina, Godt.
Morpho Phanodemus, Hew.
Any species of Agrias.
Any species of Callithea.
Pandora Chalcothea, Bates.
Pandora Hypochlora, Cates.
Pandora Divalis, Bates.
Colias Vilniensis, Men.
Colias Ponteni, Wallengr.
Bunea Deroylici, Thom.
Bunea Phaedusa, Dru.
Rinaca Zuleica, Hope.
Papilio Disparilis, Herr-Sch.
Acræa Perenna, Dbldy.
Opsiphanes Boisduvalii, Dbldy.
Clothilde Jägeri, Herr-Sch.
Pieris Celestina, Boisd.
Euplœa Eurypon, Hew.
Paphia Panariste, Hew.
Limenitis Lymire, Hew.
Io Bœkeri Herr-Sch.
Eacles Kadenii, Herr-Sch.
Smerinthus Timesia, Stoll.
Smerinthus Panopus, Cram.
Sphinx Substrigilis, West.
Saturnia Larissa, West.
Eusemia Victrix, Bellatrix, Amatrix, Dentatrix, West.
Smerinthus Modesta, Fab. (nec Harris.)
Smerinthus Tartarinovii, Brem.

These are a few of the very many of the rarer species that I am eager to procure; of course there are numberless others from all parts of the world, equally desirable and coveted by me.

North Atlantic Express Co.,

NEW YORK,
OFFICE, 57 BROADWAY.

Chartered by Special Act of Incorporation.

CENTRAL EUROPEAN OFFICE:
5 RUE SCRIBE, PARIS.

PRINCIPAL OFFICE IN GREAT BRITAIN:
4 Moorgate St., London, E. C.
B. W. & H. HORNE, AGENTS.

BRANCH OFFICES: Golden Cross, Charing Cross; George & Blue Boar, High Holborn; 108 New Bond Street; 474½ New Oxford Street.

OFFICE IN LIVERPOOL: Old Castle Buildings Tithebarn Street.

CONTINENTAL OFFICES: 5 Rue Scribe, Paris; 82 Rue d'Orleans, Havre; 88 Rodingsmarks, Hamburg; 29 Bahnhofs Strasse, Bremen.

Merchandise, specie, bullion, stocks, bonds, **or** other valuables and packages and parcels of every description, personal effects, baggage, etc., forwarded to and from Europe and all parts of the United States, the States and Territories of the Pacific Coast, British Columbia and the Canadas included, at *fixed Tariff rates*, with *no extra charges whatever for Custom-House brokerage, emmisions, delivery, etc., etc.*, the shipper or receiver being under **no** other cost or expense than the stipulated freight from the point of shipment to place of delivery, and the amount of duties and government fees actually paid at the Custom-house of the United States or Europe.

For the convenience of shippers, where agencies of the Company are not established, **packages or heavy goods** may be forwarded to either of the offices or agencies of the Company, by either of the **express or transportation** companies in the United States, or by post, by railway, through the parcel delivery companies, or forwarding houses in any part of Great Britain or the Continent of Europe.

All packages, trunks, or parcels forwarded by this Company will be landed on arrival simultaneously with the mails, or immediately thereafter and will be entered at the Custom-house, duties paid and delivered to the parties to whom addressed in any part of Europe, the United States, the Canadas, or British Columbia, with the greatest possible dispatch. Transportation charges and duties collected **on** delivery, or may be prepaid, at the option of shipper.

Insurance against marine risk taken by the Company, when desired by the shipper, at the lowest current rates; premiums payable in all cases in advance.

Shippers to or from any part of America, and Americans traveling in Europe, will find this the quickest, cheapest and most reliable medium of transportation, the business of this Company being conducted upon the well known prompt American express system, which has become so great a commercial necessity and convenience throughout the United States.

Purchases made, and collections and communications in every part of Europe and the United States promptly executed.

☞ Circulars sent on application to

S. D. JONES,
Manager,
57 BROADWAY, N. Y.

No. 10.   Price 50 Cents.

LEPIDOPTERA,

RHOPALOCERES AND HETEROCERES,

INDIGENOUS AND EXOTIC;

WITH

Descriptions and Colored Illustrations,

BY

HERMAN STRECKER.

Reading, Pa., 1874.

Reading, Pa.:
Owen's Steam Book and Job Printing Office, 515 Court Street,
1874.

The N. American Species of the genus Lycæna.

I can find no sufficient grounds for retaining the genus Chrysophanus or Polyommatus for the copper-coloured species, as there really seems to be, in the Lycænidæ of this country and Europe, no particular characteristics that are sufficiently constant to separate the red and the blue species into different genera.

The colour and ornamentation amounts to but little; in some species the males are blue and the females red or brown, as in *Saepiolus* and *Heteronea*; in others both sexes are brown, as *Agestis*,* *Eurypilus*,† etc., or, again, both are blue, as *Lucia*, *Argiolus*,‡ and many others; and *Heteronea*, though the male is blue, is certainly much nearer to the copper-coloured *Sirius* and *Gorgon* than it is to such other blue species as *Lucia*, *Lygdamus* or *Comyntas*.

Neither is the presence or absence of a tail to the secondaries of the least moment, as these appendages are found in some of the fiery-coloured species as well as in many of the blue ones; as instances of the former, I would mention *Arota*, *Virginiensis* and *Lampon*,§ and of the latter, *Comyntas*, *Tejua*, *Balkanica* ‖ and *Theophrastus*.¶ In some species the spring brood is tailless, whilst the summer generation of the same insect is provided with those ornaments.

In good truth I cannot see why all the N. American and European species, except the few contained in *Eumaeus*, Hub., should not be embraced within one genera, even including the Theclas, for on examination of these latter we find the same diversity of form and colour as in the others, some tailed, others destitute of those appurtenances, some brown, others blue, etc.; between *Arota* or *Virginiensis* and *Niphon* ** there is certainly no more difference than between *Niphon* and *Melinus* †† or between *Melinus* and *Grunus*.‡‡

Lederer retained the two groups, *Polyommatus* and *Lycæna*, but arranged under the former the fiery or copper-coloured species, and such blue ones as *Optilete*,§§ *Aegon*,‖‖ *Battus*,¶¶ and in the latter the Theclas and such other blue or brown ones as *Corydon*,*** *Damon*,††† *Telicanus*,‡‡‡ etc.

Hubner divided them into many groups or genera, not always placing the most closely allied together; for whilst he has his genus *Eumaeus*, (containing *E. Minyas*,§§§) placed in the same sub-family and immediately preceding *Nomiades*, which contains *Damon*, *Alsus*,‖‖‖ and allies, he has, not very felicitously, placed between this and *Chrysophanus* (copper species) not only nine genera, but has even put the latter in another sub-family.

At a time when comparatively few species were known, there might have appeared plausible grounds for separating the red from the blue species, but since the many later perplexing and curious intermediate forms have been discovered in Asia Minor, Persia and California, the frail foundation on which the distinction was founded has not been equal to the task of sustaining it; and the Lycæna, like the great genus Papilio, will not bear disruption without violence.

I subjoin a list of all the described species of N. America.

Those that are unknown to me in nature are prefixed with a ?.

Those that are wanting to my collection are designated by a *.

Such as I possess the author's original types of, are denoted by a ‡.

The numbers over some species are the numbers attached to the figures of same species on plate X, thus " 29 ♂, 30 ♀, Sirius, Edwards."

To such as I have figured I have added no descriptions, as whether there be figures or not, the descriptions of such things are little better than waste of time, although to such as I have no other knowledge of I have quoted the author's diagnosis in full.

---

* Lycæna Agestis, Hubner, (Papilio A.) Eur. Schmett. I, f. 303–305, (1798–1803).
† Lycæna Eurypilus, Freyer, Neuere Beitrage, VI, t. 573, f. 4, (1852).
‡ Lycæna Argiolus, Lin., (Papilio A.) Fauna Svecica, p. 284, (1761).
§ Lycæna Lampon, Lederer, (Polyommatus L.) Hor. Soc. Ent. Ross. VIII, p. 8, t. 1, (1870).
‖ Lycæna Balkanica, Freyer, Neuere Beitrage, V, t. 421, (1844).
¶ Lycæna Theophrastus, Fabricius, (Hesperia T.) Ent. Sys. III, 1, p. 281, (1793).
** Thecla Niphon, Hubner, (Lica N.) Zutr. Ex. Schmett. f. 203, 204, (1823).
†† Thecla Melinus, Hubner, (Strymones M.) Zutr. Ex. Schmett. f. 121, 122, (1818).
‡‡ Thecla Grunus, Boisduval, Ann. Soc. Ent. Fr., p. 289, (1852).
§§ Lycæna Optilete, Knoch, (Papilio O.) Beit. Ins. Ges. L, p. 76, t. 3, (1781).
‖‖ Lycæna Aegon, Schiffermüller & Denis, (Papilio A.) Wien. Verz. p. 185, (1776).
¶¶ Lycæna Battus, (Papilio B.) Schiff. Wien. Verz. p. 185, (1776).
*** Lycæna Corydon, Poda, (Papilio C.) Musei Graecensis, p. 77, (1761).
††† Lycæna Damon, Schiff., (Papilio D.) Wien. Verz., p. 182, (1776).
‡‡‡ Lycæna Telicanus, Lang, (Papilio T.) Verzeichniss Schmett. p. 47, (1789).
§§§ Eumaeus Minyas, Hubner, (Eustixis Adolescens M.) Samml. Exot. Schmett., (1806–1816).
‖‖‖ Lycæna Alsus, Schiff., (Papilio A.) Wien. Verz. p. 184, (1776).

LYCÆNA. Fabr.

(PL. X, F. 3, ♂.)

‡ TEJUA, Reakirt, Proc. Acad. Nat. Sc., Phil., p. 245, (1866). *Edwards*, Syn. N. Am. Butt., p. 35, (1872).
  Cupido Tejua, Kirby, Cat. Diurnal Lep., p. 356, (1871).
  Described from a single ♂ received from Southern California.

(PL. X, F. 18, ♂.)

‡ MONICA, Reakirt, Proc. Acad. Nat. Sc., Phil., p. 244, (1866).
  Cupido Monica, Kirby, Cat. Diurnal Lep., p. 356, (1871).
  Lycæna Monica, Edwards, Syn. N. Am. Butt., p. 34, (1872).
  From same locality as the preceding. The description was taken from two males, one of which the author curiously mistook for a female.

COMYNTAS, Godart, (*Polyommatus C.*) Enc. Meth., IX, p. 660, (1823). *Morris*, Cat. Lep. N. A., p. 12, (1860). Syn. Lep. N. Am., p. 83, (1862). *Harris*, Ins. Injurious to Vegetation, Flint's Ed., p. 275, (1862).
  Argus Comyntas, Boisduval & Leconte, Lep. Am. Sept., p. 120, t. 36, (1833).
  Cupido Comyntas, Kirby, Cat. Diurnal Lep., p. 356, (1871).
  Lycæna Comyntas, Edwards, Syn. N. Am. Butt., p. 34, (1872). *Packard*, Guide, p. 265, (1869).
  The commonest of our species, found in Canada and from thence southward to the Gulf of Mexico, and westward from the Atlantic to the Rocky Mts. It is closely allied to the European *Polysperchon*, Berg.

AMYNTULA, Boisduval, Ann. Soc. Ent. Fr., p. 294, (1852). *Edwards*, Syn. N. Am. Butt., p. 34, (1872).
  Polyommatus Amyntula, Morris, Cat. Lep. N. Am., p. 12, (1860); Syn. Lep. N. Am., p. 87, (1862).
  Mr. Kirby, in his Cat., (p. 356) cites this as a variety of the preceding, which it indeed represents on the Pacific slope but with which I do not think it is identical; it is generally of much larger size; the tails are not nearly so long or slender in comparison, and there are many other minor points of difference. Common in California and adjoining territories.

PSEUDARGIOLUS, Boisduval & Leconte, (*Argus P.*) Lep. Am. Sept., p. 118, t. 36, (1833). *Morris*, Cat. Lep. N Am., p. 12, (1860); Syn. Lep. N. Am., p. 82, (1862).
  Lycæna Pseudargiolus, Edwards, Proc. Ent. Soc., Phila., Vol. VI, p. 204, (1867); Butt. N. Am., t. 2, Lyc., (1869); Syn. N. Am. Butt., p. 38, (1872).
  Polyommatus Pseudargiolus, Harris, Ins. Inj. to Veg., Flint's Ed., p. 274, (1862).
  Papilio Argiolus, Abbot & Smith, Insects of Georgia, Vol. I, t. 15, (1797).
  Lycæna Neglecta, Edwards, Proc. Acad. Nat. Sc., Phila., p. 57, (1862); Butt. N. Am., t. 2, Lyc., (1869); Syn. N. Am., Butt., p. 38, (1872). *Packard*, Guide, p. 265, (1869).
  Cupido Pseudargiolus et *C. Neglecta*, Kirby, Cat. Diurnal Lep., p. 371, (1871).
  A delicate, handsome species, expanding 1 to 1¼ inches; male is on upper surface pale azure blue, secondaries, except at outer margin, paler than primaries. Female white, blue at bases and sometimes on disc of primaries; costal and exterior parts of primaries broadly margined with black. Under side satiny white or light grey, markings sometimes tolerably well defined, and in other instances faint or nearly obsolete. Found in the Atlantic States from Canada, southwards.
  Mr. Edwards has exercised a great deal of ingenuity in his efforts to persuade the world and himself that two species were confounded under the name of *Pseudargiolus*, but his labors have not been crowned with proportionate success, in proof of which I would refer to his six figures on plate II, Lyc., in Butt. N. Am., three of which the text informs us are *Pseudargiolus*, and three *Neglecta*, for truthfulness these figures cannot be excelled, but the funniest part is that with the exception of the one being a little larger than the other, the most critical eye will fail to detect the slightest difference between them.

PIASUS, Boisduval, Ann. Soc. Ent. Fr., p. 299, (1852). *Edwards*, Syn. N. Am. Butt., p. 37, (1872).
  Polyommatus Piasus, Morris, Cat. Lep. N. Am., p. 12, (1860); Syn. Lep. N. Am., p. 89, (1862).
  Cupido Piasus, Kirby, Cat. Diurnal Lep., p. 363, (1871).
  Lycæna Echo, Edwards, Proc. Ent. Soc., Phila., Vol. II., p. 506, (1864).
  Same size as and very near in most other respects to *Pseudargiolus*, the main difference being in the blue of upper surface, which is deeper and more inclined to violet; the markings of under surface are identical with that species. Common in California and adjacent country.

LUCIA, Kirby, Fauna Boreali Americana, Vol. IV., p. 299, t. 3, (1837). *Edwards*, Syn. N. Am. Butt., p. 37, (1872).

LYCÆNA.

Polyommatus Lucia, Morris, Cat. Lep. N. Am., p. 12, (1860); Syn. Lep. N. Am., p. 90, (1862).
Harris, Ins. Inj. Veg., Flint's Ed., p. 275, (1862).
Cupido Lucia, et *Violacea*, Kirby, Cat. Diurnal Lep., p. 368, (1871).
Lycæna Violacea, Edwards, Proc. Ent. Soc., Phil., Vol. VI., p. 201, (1866); Butt. N. Am. t. 1, Lyc. (1868); Syn. N. Am. Butt., p. 37, (1872).

About 1 inch in expanse. The male above is bright shining blue, with white fringes, sometimes brown at terminations of veins. Female is blue bordered with black at exterior margin, broadest at the apex and extending inwards on the costa. Southward, in Virginia, the prevalent colour of the female on the whole upper surface is uniform dark brown; examples also occur in same locality, that are intermediate in colour between these brown ones and the common northern blue form. The under surface is greyish white and varies in depth of markings. There is a row of brown sub-marginal spots succeeded or surmounted inwardly by a row of crescents, in many examples the space between these latter and the outer margin is entirely filled with dark brown, especially on the secondaries, thus forming a scalloped border; in some specimens in addition to this latter there is on the disk of the secondaries a large brown patch; this is represented in Kirby's figure in Faun. Am. Bor.; it seems the further northward the more prominent the markings on the under side become. Found in Labrador, Canada, and Eastern United States to Virginia; I have not heard of its having occurred further southward than the last named state.

This unfortunate insect has also been a victim to the insatiable mania for manufacturing new species, which seems to be a national affliction with the majority of American Lepidopterists.

HANNO, Stoll, (*Papilio H.*) Supplement to Cramer, t. 39, (1790).
Rusticus Adolescens Hanno, Hubner, Samm. Ex. Schmett., (1806–1816).
Hemiargus Hanno, Hubner, Verz. Bek. Schmett., p. 69, (1816).
Cupido Hanno, Kirby, Cat. Diurnal Lep., p. 350, (1871).
Polyommatus Ubaldus, Godart, Enc. Meth. IX, p. 682, (1823).
Polyommatus Filenus, Poey, Cent. Lep., (1832).
Argus Filenus, Boisduval & Leconte, Lep. Am., Sept., p. 114, (1833). Morris, Cat. Lep. N. Am., p. 12, (1860); Syn. Lep. N. Am., p. 82, (1862).
Lycæna Filenus, Edwards, Syn. N. Am. Butt., p. 35, (1872).
Argus Pseudoptiletes, Boisduval & Leconte, Lep. Am., Sept., p. 114, t. 35, (1833).

The size of *Comyntas* and much the same colour on upper surface, a black spot on secondaries towards the anal angle. Under surface silky brown, with a number of spots of same colour surrounded by paler rings and arranged in broken rows; two black spots at costa of secondaries, one within the cell and another between this latter and the abdominal margin; between the second and third median nervules, near exterior margin, a large, round, black spot with a few silvery green atoms at its outer edge; between this spot and the anal angle is a small, double, silver green spot. Southern States and West Indies—very common in Florida and Georgia.

EXILIS, Boisduval, Ann. Soc. Ent. Fr., p. 295, (1852). Edwards, Syn. N. Am. Butt., p. 35, (1872).
Polyommatus Exilis, Morris, Cat. N. Am. Lep., p. 12, (1860); Syn. N. Am. Lep., p. 87, (1862).
Cupido Exilis, Kirby, Cat. Diurnal Lep., p. 357, (1871).
Lycæna Fea, Edwards, Trans. Am. Ent. Soc., Vol. III, p. 211, (1871).

The smallest of all the known N. Am. Lycæna—expands from ½ to ⅔ inch. Upper side reddish brown, darker at the margins, fringe white except towards the inner angle of primaries where it is grey or smoky. Under side whitish at base of wings on primaries, from thence to outer margin reddish striated with fine, irregular, white lines. Secondaries, on disc coloured and marked in same way, and with a marginal row of spots, the one at anal angle silver, the next four black, and the last two, at apex, silver; these spots are succeeded inwardly by a white space. California, Nevada, etc.

*†SHASTA, Edwards, Proc. Acad. Nat. Sc., Phila., p. 224, (1862); Syn. N. Am. Butt., p. 35, (1872).
Thecla Shasta, Kirby, Cat. Diurnal Lep., p. 401, (1871).

This species is unknown to me, nor am I able to identify it by the original description which I here transcribe as another illustration of the valuelessness of such things.

"Expands one inch. Male. Upper side violet blue with a pink tinge; hind margin broadly fuscous; a large black discal spot on each wing; two or three obsolete spots near anal angle, the second from the angle with a faint yellow lunule; fringe brownish white. Under side greyish white, blueish next base; primaries have a fuscous spot near base, a discal bar and transverse sinuous row of elongated fuscous spots, each edged with whitish; along the margin obsolete spots surmounted by faint lunules. Secondaries have three fuscous points near base, a discal bar and a transverse sinuous row of fuscous spots; whole hind margin bordered by small metallic blue spots, each surmounted by a blackish lunule.—Female: upper side clear brown; the obsolete spots next anal angle, surmounted by a narrow crenated yellow band, under side as in male, but the five yellow spots next anal angle are surmounted by ochrey yellow lunules, edged above with black, fringe long and fuscous at terminations of nervures. California, Dr. Behr."

*†Isola, Reakirt, Proc. Acad. Nat. Sc., Phila., p. 332, (1866). *Edwards*, Syn. N. Am. Butt., p. 35, (1872).
    Cupido Isola, Kirby, Cat. Diurnal Lep., p. 376, (1871).

This is another species with which I am entirely unacquainted, the types from which it was described were from Mexico. Mr. W. H. Edwards in his synopsis says also, Waco, Texas. I append Mr. Reakirt's description.

"*Upper surface brownish black, glossed with violet blue; a black terminal line, broadest at the apex of the fore wings, thence diminishing to the anal angle; a small rounded, submarginal black spot near the latter; fringe white. Underneath dark ash grey; primaries with two submarginal, slightly waved whitish lines; interior to these a row of six large rounded black spots, all ringed with white; two white streaks at the end of the cell. Secondaries with a submarginal row of indistinct brown spots, of which the three nearest the anal angle are black, the first and third irrorated with metallic golden-green atoms, and the third surmounted by yellowish hunule; all the others are preceded by whitish crescents; above these three is a suffused white belt, and still farther, two double rows of waved and crenulated whitish lines; a small sub-costal black ocellus near the base. A narrow terminal black line edges the outer margin of the four wings; fringe ashy white. Expanse .88 inch. Antennae black ringed with white. Hab. Mexico (near Vera Cruz).*"

*†Gyas, Edwards, Trans. Am. Ent. Soc., Vol. III, p. 210, (1871); Syn. N. Am. Butt., p. 35, (1872).

Another species of doubtful validity, the description says:

"*Male.*—Expands .95 inch. Upper side pale violet blue, immaculate except a fuscous point near anal angle. Under side pale brown with a wash of whitish; primaries have a faint, discal bar, and a straight row of spots across the wing, the second and fourth back of the line; all edged with white; on margin traces of lunules. Secondaries have a similar discal bar and a median row of spots; a small round fuscous spot in cell, two others on costa, one near middle the other near base; a faint row of spots on hind margin, of which the two next anal angle are distinct, blackish. From Arizona, taken by Dr. Palmer, and in the collection of the Agricultural Department."

Lygdamus, Doubleday, (*Polyommatus L.*) Entomologist, p. 209, (1842).
    Lycaena Lygdamus, Edwards, Butt. N. Am., t. 1, Lyc., (1868); Syn. N. Am. Butt., p. 37, (1872).
    Cupido Lygdamus, Kirby, Cat. Diurnal Lep., p. 368, (1871).

Male expands 1¼ to 1½ inches, upper side is beautiful silvery blue with narrow black margins exteriorly, and greyish fringes. Female smaller, not so bright, and the outer half of wings much suffused with grey. Both sexes, beneath, grey with black discal bars and sub-marginal rows of large, black spots, two spots near base of secondaries, one near base of primaries, all spots encircled with white. Southern United States—rare.

(PL. X, F. 10 ♂, 11 ♀.)

Pembina, Edwards, Proc. Acad. Nat. Sc., Phil., p. 224, (1862); Syn. N. Am. Butt., p. 37, (1872).
    Thecla Pembina, Kirby, Cat. Diurnal Lep., p. 401, (1871).
    Glaucopsyche Couperi, Grote, Bull. Buf. Soc. Nat. Sc., Vol. I., p. 185, (1874).

Allied to *Lygdamus* which it resembles very closely, especially on the upper surface; I have made full comparisons between the species on p. 69 of this work. Labrador, British Columbia, Oregon.

Since Mr. W. H. Edwards described this species, it very nearly had the misfortune of losing its birthright: the author having through accident lost his types; and what was equally unfortunate, his memory even when aided by the lengthy original description would not allow him to identify with any certainty, examples that were subsequently submitted to him. This was rather placing the species in a forlorn position, but at this juncture the great Species-mill gave a revolution or two or three and the Lycaena was transmogrified into Glaucopsyche, a new specific name was of course added, and the whole fabrication attached to the trade mark of the mill, which latter was of course understood to make the insect immortal, but alas! "All glory but dazzles and dies," and so was it with "Glaucopsyche Couperi Grote," for "Like the swift shadows of noon, like the dreams of the blind it vanished away as the dust in the wind," and in its place stands the Prodigal Pembina, tired of the husks and returned to its first honourable estate and title.

Antiacis, Boisduval, Ann. Soc. Ent. Fr., p. 300, (1852). *Edwards*, Syn. N. Am. Butt., p. 37, (1872).
    Polyommatus Antiacis, Morris, Cat. Lep. N. Am., p. 12, (1860); Syn. Lep. N. Am., p. 90, (1862).
    Cupido Antiacis, Kirby, Cat. Diurnal Lep., p. 371, (1871).

Not quite as large as the two preceding. Male violet blue on upper side with white fringe. Female brownish grey, a little bluish towards base—under surface in both sexes coloured and marked much as in Lygdamus. California.

Behrii, Edwards, Proc. Acad. Nat. Sc., Phil., p. 224, (1862); Syn. N. Am. Butt., p. 37, (1872).
    Thecla Behrii, Kirby, Cat. Diurnal Lep., p. 400, (1871).*)
    Lycaena Polyphemus, Boisduval, Lep. Cal., 49, (1869).
    Cupido Polyphemus, Kirby, Cat. Diurnal Lep., p. 373, (1871).

Another species closely allied to Lygdamus and Pembina, but is generally a little larger than either of them; the blue in the male is more violaceous and less lustrous, and the female on upper side is, with the

---

*) In Kirby's Catalogue there are two species confounded under one name on p. 400, thus "337, T. Behrii, Edw., Proc. Acad. Nat. Sc., Phil., 1862;" and "Trans. Am. Ent. Soc., 1870, p. 18," the first citation refers to the above Lycaena, and the second is Thecla Behrii, an entirely different thing, though it was rather ill advised in Mr. Edwards to designate species by the same name that are in groups so closely connected as the Lycaena and Thecla.

exception of a few blue scales at base, entirely brown. Beneath, both sexes are coloured and marked as in Lygdamus. Fringes white, both above and below. Common in California.

*†AMICA, Edwards, Proc. Ent. Soc., Phil., Vol. II, p. 80, (1863); Syn. N. Am. Butt., p. 36, (1872).
    Cupido Amica, Kirby, Cat. Diurnal Lep., p. 376, (1871).

As I have no acquaintance with this Arctic species, which Mr. W. H. Edwards described from the male only, I here append his original description:

"Male. Expands 1 1-10 inch. Upper side silvery-blue, brownish along the margins, with a narrow, straight discal mark on primaries; fringe white. Under side glossy greyish white; primaries have a narrow discal mark and a curved row of six minute black spots across the disk; secondaries have a nearly straight row of five minute black spots, besides two on the costa, one of which is in the middle, the other near the base, all edged with white; there is also a sub-marginal row of points and small brown lunules, sometimes obsolete. From Mackenzie's River, by Mrs. Ross."

*†MARICOPA, Reakirt, Proc. Acad. Nat. Sc., Phila., p. 245, (1866). *Edwards*, Syn. N. Am. Butt., p. 36, (1872).
    Cupido Maricopa, Kirby, Cat. Diurnal Lep., p. 377, (1871).

I do not know this species; when Mr. Reakirt's types came into my possession, this was not among them, nor have I, to my knowledge, ever seen it. His description reads thus:

"Male. Upper side brown, glossed with violet blue; a narrow terminal dark line along the outer margins; a black discal bar on the primaries, sometimes wanting, and some obsolete rounded spots on the hind margin of the secondaries. Fringe ash-coloured. Underneath ash-brown, darkest towards the base. Primaries; a large black discal bar, a sub-central, transverse, sinuated row of seven large rounded black spots all narrowly ringed with white; following these, and parallel with the margin, another series of seven indistinct spots. Secondaries; a discal bar and two spots, one within the cell, the other above it; three transverse maculose bands; the first composed of eight large rounded black spots, and bent twice at right angles, the second of smaller, and sagittiform, and in common with the third, which is almost marginal, and very indistinct, runs parallel with the border; all these markings are encircled with white, and the seventh spot of the first and second rows are sometimes confluent. Expanse 1.25-1.35 inches. Body black above, with some blueish hairs; beneath greyish; antennae black with white annulations, lower part of club whitish. Hab.—California."

*†MERTILA, Edwards, Proc. Ent. Soc., Phila., Vol. VI, p. 206, (1866); Syn. N. Am. Butt., p. 36, (1872).
    Cupido Mertila, Kirby, Cat. Diurnal Lep., p. 377, (1871).

It is almost with despair that I turn from one description to the other, always the same monotonous thing, the same stereotyped greyish under side, the same tedious "sinuous rows of spots," and the same everlasting this shape or that shaped discal bar, spot or mark. Oh! that we could but throw out every description that is unaccompanied by a figure, how our labours would be lightened, how we would be spared the maledictions of after generations for all time to come. With what boundless veneration do we look on the tomes of Cramer, Seba, Drury, Hubner, Hewitson, and Herrich-Schaeffer, no winding into countless useless descriptions in all sorts of scattered periodicals, but a great massive work—grand, compact, solid, every description accompanied by coloured figures. I never open these mighty volumes but I feel my soul expand in Hallelujahs to the Almighty that through his great goodness such intellects were allowed to sojourn here and to bequeath to us the result of their vast labours.

These thoughts were suggested by reading the description of the above cited species "Mertila" founded on a female example, and will apply to a host of other Lycaenidæ equally as well as to this probably mythical one—and although I have not a particle of faith in half the species of this author, I now copy his description of Mertila and hope my friends and enemies will forgive me for inflicting on them, and my God for wasting the time in so doing.

"Female. Expands 1 1-10 inch. Primaries long and narrow; both wings brown, with slate-coloured hairs at base and along inner margin of primaries. Under side clear cinereous; bluish at base; primaries have a single transverse sinuous row of round black spots, each circled with white, as also is the lunule in the arc; from the arc a whitish ray runs towards base. Secondaries have a row of eight small black spots in points, each circled with white; of these, two are on central margin, four nearly parallel with the hind margin; the seventh below the others and geminate; the eighth inmost, nearly concealed by the marginal hairs; between the 2nd and 3rd and the 6th and 7th the spaces are wide; on the arc a streak, and midway between this and the base a black point; on the costa above this one slightly larger, all circled with white. From California. The male of this *distinct*) species I have not seen."

*†ORCUS, Edwards, Trans. Am. Ent. Soc., p. 376, (1869); Syn. N. Am. Butt., p. 37, (1872).
    Cupido Orcus, Kirby, Cat. Diurnal Lep., p. 377, (1871).

Described from a single male specimen from California.

"Male. Expands 1.1 inch. Upper side pruinose blue, paler on costa of primaries; hind margins broadly fuscous; fringes long, cinereous. Under side grey cinereous, bluish at base; on arc of primaries a narrow black bar bent outwards, and faintly edged without by white; on secondaries a faint discal streak; both wings have a sub-marginal line of points, scarcely discernable. Beneath thorax covered with blue, grey-hairs, abdomen grey; palpi white above at base, black at tip, and cinereous below; antennae black annulated with white; club black, tipped with cinereous."

PHERES, Boisduval, Ann. Soc. Ent. Fr., p. 297, (1852). *Edwards*, Syn. N. Am. Butt., p. 36, (1872).
    Polyommatus Pheres, Morris, Cat. Lep. N. Am., p. 12, (1860); Syn. Lep. N. Am., p. 89, (1862).
    Cupido Pheres, Kirby, Cat. Diurnal Lep., p. 362, (1871).

*) These italics are mine.

Expands 1¼ inches. Male violet blue, with narrow black border on outer margins; fringe white. Female greyish brown, bluish towards base. Under side in both sexes very pale grey; primaries, a black discal spot and mesial row of six black spots, the one nearest inner angle geminate; secondaries with a mesial and sub-marginal row of white spots, also a white discal spot and another near the base; none of the spots are pupilled. California, rather common.

XERCES, Boisduval, Ann. Soc. Ent. Fr., p. 296, (1852); *Edwards*, Syn. N. Am. Butt., p. 35, (1872).
 Polyommatus Xerces, *Morris*, Cat. Lep. N. Am., p. 12, (1860); Syn. Lep. N. Am., p. 88, (1862).
 Cupido Xerces, *Kirby*, Cat. Diurnal Lep., p. 373, (1871).

1½ inch in expanse. Upper surface, male blue, female greyish brown; fringes white. Under surface, both wings with white discal spot and sinuous row of large sub-marginal white spots, all spots blind. California.

*†ARDEA, Edwards, Trans. Am. Ent. Soc., Vol. III, p. 209, (1871); Syn. N. Am. Butt., p. 37, (1872).
Unknown to me—here is a copy of the author's description:

"Male.—Expands 0.95 inch. Upper side violet blue, grey blue when seen obliquely; hind margin of primaries very narrowly edged by fuscous; of secondaries by a black line; fringes long, white. Under side fawn color, secondaries tinted with blue at base; primaries have a large black reniform discal spot, edged with white; an imperfect transverse median row of four black dots surrounded by white, those at either extremity obsolete; faint traces of a sub-marginal series of brown lunules. Secondaries have traces of a similar series, still less distinct; no median spots; a large white patch on arc. Body above coulored; beneath white; legs white; palpi white tipped with grey; antennæ annulated white and black; club black, tip ferruginous. From Nevada, vicinity of Virginia City."

*†ERYMUS, Boisduval, Lep. Cal., p. 48, (1869). *Edwards*, Syn. N. Am. Butt., p. 36, (1872).
 Cupido Erymus, *Kirby*, Cat. Diurnal Lep., p. 366, (1871).

"Oregon." Another entire stranger to me.

(Pl. X, F. 1 ♂, 2 ♀.)

‡CATALINA, Reakirt, Proc. Acad. Nat. Sc., Phila., p. 244, (1866). *Edwards*, Syn. N. Am. Butt., p. 35, (1872).
 Cupido Catalina, *Kirby*, Cat. diurnal Lep., p. 376, (1871).
 Lycæna Daunia, *Edwards*, Trans. Am. Ent. Soc., Vol. III, p. 272, (1871); Syn. N. Am. Butt., p. 50, (1872).

A rare species; the types came from or near Los Angeles, California; those that were re-described later under the name of *Daunia* were taken in Colorado.

(Pl. X, F. 16, ♂.)

ORBITULUS, DePrunner, (*Papilio O.*) Lepidoptera Piedmontana, p. 75, (1798). *Esper*, Schmett. 1, t. 112, f. 4, (1800). *Ochsenheimer*, Schmett. I, 2, 43, (1808). *Hubner*, Eur. Schmett. I, f. 841, (1818-1827).
 Agriades Orbitulus, *Hubner*, Verz. Bek. Schmett., p. 68, (1816).
 Polyommatus Orbitulus, *Godart*, Enc. Meth., IX, p. 688, (1823).
 Lycæna Orbitulus, *Staudinger*, Cat. Lep. Eur., p. 11, (1871).
 Cupido Orbitulus, *Kirby*, Cat. Diurnal Lep., p. 363, (1871).
 Papilio Meleager, *Hubner*, Eur. Schmett. I, f. 522-525, (1798-1803); f. 761, 762, (1803-1818).
 Lycæna Rustica, *Edwards*, Proc. Ent. Soc., Phila., Vol. IV, p. 203, (1865); Syn. N. Am. Butt., p. 36, (1872).
 Cupido Rustica, *Kirby*, Cat. Diurnal Lep., p. 377, (1871).
 Lycæna Tehama, Reakirt, Proc. Acad. Nat. Sc., Phila., p. 245, (1866). *Edwards*, Syn. N. Am. Butt., p. 36, (1872).
 Cupido Tehama, *Kirby*, Cat. Diurnal Lep., p. 377, (1871).
 Lycæna Cilla, *Behr*, Proc. Cal. Acad. Nat. Sc., Vol. III, p. 281, (1867). *Edwards*, Syn. N. Am. Butt., p. 33, 50, (1872).
 Cupido Cilla, *Kirby*, Cat. Diurnal Lep., p. 363, (1871).

Upper surface of female entirely brown, with a darker discal mark on each wing. Under surface precisely as in the male. The male figure on t. 10 was drawn from one of Reakirt's original types of *Tehama* now in my possession; some other examples which I have are much darker, showing none of the yellowish grey of this one which is "var. a." *) further distinguished by the distinctness of the sub-marginal spots and lunules of upper side of secondaries. Found in the Swiss Alps and Pyrenees, as well as on the higher peaks of Colorado and the Sierras of California.

AQUILO, Boisduval, (*Argus A.*) Icones, t. 12, f. 7, 8, (1832).
 Lycæna, Aquilo, *Herrich-Schaeffer*, Schmett. Eur., Vol. I., f. 24, 25, (1843); f. 343, 344, (1847). *Duponchel* I, 47, 6, 7. *Wallengren*, Skand. Dagf., p. 211, (1847). *Moschler*, Wien. Ent. Mon., Vol. IV., p. 343, (1860). *Staudinger*, Cat. Eur. Lep., p. 11, (1871). *Kirby*, Cat. Diurnal Lep., p. 363, (1871). *Edwards*, Syn. N. Am. Butt., p. 35, (1872).
 Lycæna Franklinii, *Curtis*, Ross, 2d Voy. App. Nat. Hist., p. 69, t. A, (1835).

*) In Reakirt's description.

Considered to be a Polar variety of *Orbitulus*, which it closely resembles, but is smaller, and on under surface the marks are much more sharply defined and the ground colour darker, especially on secondaries. Found in Labrador at 57° N. L. and from thence northward.

*†KODIAK, Edwards, Trans. Am. Ent. Soc., Vol. III., p. 20, (1870). Syn. N. Am. Butt., p. 37, (1872).
   Cupido Kodiak, Kirby, Cat. Diurnal Lep., p. 376, (1871).

With regard to this species I can do no more than copy the description, as I have heretofore done in the cases of such as I do not know and consider doubtful.

" Male. Expands 1.25 inch. Upper side dull violet blue; margins narrowly edged with fuscous; fringes sordid white. Under side fawn colour, bluish at base; slightly clouded with grey on secondaries; both wings have fuscous discal bars, edged with whitish; a common median row of rounded fuscous spots, all edged with whitish, that of primaries curved beyond the cell, of secondaries parallel to the margin; a common sub-marginal row of faint spots, the second from anal angle surmounted by a round spot, perhaps belonging to median row, but much posterior to the line of same; near base, a black point on cell and a second on costa. Body above blue, thorax beneath blue grey; palpi white at base, furnished with long hairs, black at tip; antennae black, annulated with white; club black above, fuscous below and at tip. Female. Expands 1.5 inch. Upper side light brown, deep blue at base, covering half the wing on primaries, fading gradually towards the hind margin, and on secondaries covering the cell and upper abdominal margin; primaries have a curved black line at extremity of cell; faint traces of fulvous spots next anal angle. Under side as in male. From Kodiak. 1 ♂ 1 ♀. Collection Dr. Behr."

(PL. X, 14 ♂, 15 ♀.)

‡RAPAHOE, Reakirt, Proc. Ent. Soc., Phila., Vol. VI, p. 146, (1866). *Edwards*, Syn. N. Am. Butt., p. 36, (1872).
   Cupido Rapahoe, Kirby, Cat. Diurnal Lep., p. 377, (1871).

The figures on t. X are drawn from the original types which were taken in the Rocky Mts. of Colorado.

*†NESTOS, Boisduval, Lep. Cal., p. 50, (1869). *Edwards*, Syn. N. Am. Lep., p. 33, (1872).
   Cupido Nestos, Kirby, Cat. Diurnal Lep., p. 365, (1871).

From Oregon. Entirely unknown to me, nor have I present access to the work in which it is described.

ICARIOIDES, Boisduval, Ann. Soc. Ent. Fr., p. 297, (1852). *Edwards*, Syn. N. Am. Butt., p. 36, (1872).
   Polyommatus Icarioides, Morris, Cat. Lep. N. Am., p. 12, (1860); Syn. Lep. N. Am., p. 88, (1862).
   Cupido Icarioides, Kirby, Cat. Diurnal Lep., p. 366, (1871).

Expands 1 1-5 inch. Male; violet blue with exterior margins of all wings narrowly bordered with brown; fringes white. Female; greyish brown tinged with violet towards the base. Underneath both sexes are almost white; on primaries a large black discal spot, a mesial row of six black spots and a sub-marginal row of smaller, fainter ones. Secondaries; three black points near base, a discal bar, a mesial and sub-marginal row of small black spots or points; all the spots circled with white. Sierras of California, evidently rare.

*†PHILEROS, Boisduval, Lep. Cal., p. 50, (1869). *Edwards*, Syn. N. Am. Butt., p. 36, (1872).
   Cupido Phileros, Kirby, Cat. Diurnal Lep., p. 366, (1871).

Probably the same as the preceding.

*REGIA, Boisduval, Lep. Cal., p. 46, (1869). *Edwards*, Syn. N. Am. Butt., p. 34, (1872).
   Cupido Regia, Kirby, Cat. Diurnal Lep., p. 366, (1871).

About an inch in expanse; the upper surface of the male is a most beautiful silvery blue, but what obviates the necessity of all further description, and distinguishes this from all other North American species, is the large orange or gold-coloured patch near inner angle on primaries. The female is unknown to me. Mountains of California, very rare.

SCUDDERII, Edwards, Proc. Acad. Nat. Sc., Phil., p. 164, (1861); Syn. N. Am. Butt., p. 34, (1872). *Morris*, Syn. Lep. N. Am., p. 329, (1862).

Expands 1¼ inch. Males dark violet blue edged exteriorly with a black line. Female brown, suffused more or less with violet near base; on secondaries a sub-marginal connected row of orange or yellow lunules, more or less distinct in different examples. Fringe in both sexes white. Under surface grey; on all wings a discal bar and mesial row of black spots, also sub-marginal row of spots each surmounted by a crescent; the space between these spots and crescents is yellow; sub-marginal spots of secondaries edged inwardly with silvery scales. The whole under surface is very like the European *Argus L.* Found in N. Labrador, British Columbia, Canada, New England States, New York and Michigan.

BATTOIDES, Behr, Proc. Cal. Acad. Nat. Sc., p. 282, (1867). *Edwards*, Syn. N. Am. Butt., p. 34, (1872).
   Cupido Battoides, Kirby, Cat. Diurnal Lep., p. 366, (1871).

Expands nearly one inch. Male, on upper side, is dark violet-blue, with blackish exterior margins; fringe smoky. Under side yellowish-grey, marked much as in Scudderii but has an additional black spot in cell of primaries; all spots much heavier than in that species, and no indications of silver or golden scales on the sub-marginal spots of secondaries. The female I have not yet seen, but she is doubtless brown on upper side. California, Colorado, etc.,—scarce.

LYCÆNA.

*†ALCE, Edwards, Trans. Am. Ent. Soc., Vol. III., p. 272, (1871); Syn. N. Am. Butt., p. 50, (1872).
Another of those doubtful affairs for which I can do no more than give the author's description:

Male expands 1 inch. Upper side brown with pinkish blue reflection, deeper blue next base; secondaries have two fuscous points in the interspace next anal angle and a round spot in the next preceding; fringes grey-white. Under side fawn color, on the outer half of both wings reticulated with whitish; primaries have a mesial series of large black rounded spots, and a concolored spot on arc, all edged with white. Secondaries have three spots on hind margin corresponding to those of upper side, velvet black with metallic green edges; two black spots on costa and two at base. Body covered with blue hairs, below grey; palpi white, last joint black; antennæ annulated black and white; club black above, fulvous below and at tip. From Colorado, taken by Mr. Mead."

GLAUCON, Edwards, Trans. Am. Ent. Soc., Vol. III., p. 210, (1871); Syn. N. Am. Butt., p. 34, (1872).
⅜ inch in expanse. Male resembles closely in colour and markings Battoides, Behr, already described. Female is brown on upper side, beneath same as male. Nevada.

CALCHAS, Behr, Proc. Cal. Acad. Nat. Sc., Vol. III, p. 281, (1867). Edwards, Syn. N. Am. Butt., p. 34, (1871).
  Cupido Calchas, Kirby, Cat. Diurnal Lep., p. 358, (1871).
  Lycæna Nivium, Boisduval, Lep. Cal., p. 47, (1869).
Very close to the preceding, to which it bears a most alarming similarity in both sexes. California.

*†RHEA, Boisduval, Lep. Cal., p. 51, (1869). Edwards, Syn. N. Am. Butt., p. 34, (1872).
  Cupido Rhea, Kirby, Cat. Diurnal Lep., p. 367, (1871).
"California." I am unacquainted with this species, nor at present have I access to the work in which it is described.

(Pl. X, F. 4 ♂, 5 ♀.)

ANNA, Edwards, Proc. Acad. Nat. Sc., Phila., p. 163, (1861). Morris, Syn. Lep. N. Am., p. 329, (1862).
  Edwards, Syn. N. Am. Butt., p. 34, (1872).
  Cupido Anna, Kirby, Cat. Diurnal Lep., p. 358, (1871).
  ‡Lycæna Cajona, Reakirt, Proc. Ent. Soc., Phila., Vol. VI, p. 147, foot note, (1866).
  Lycæna Argyrotoxus, Behr, Proc. Cal. Acad. Nat. Sc., Vol. III, p. 281, (1867).
  Lycæna Philemon, Boisduval, Lep. Cal., p. 47, (1869).
A beautiful species, presenting on the under surface a rather different appearance from its allies. California, rare.

(Pl. X, F. 8 ♂, 9 ♀.)

MELISSA, Edwards, Trans. Am. Ent. Soc., Vol. IV, p. 346, (1873).
Resembles very much, on upper side, the preceding, with which it has sometimes been confounded. California, Nevada, Colorado, Arizona.

ACMON, Doubleday, Hewitson, Genera Diurnal Lep., t. 76, (1852); Edwards, Syn. N. Am. Butt., p. 34, (1872).
  Lycæna Antegon, Boisduval, Ann. Soc. Ent. Fr., p. 295, (1852).
  Polyommatus Acmon et Antegon, Morris, Cat. Lep. N. Am., p. 12, (1860).
  Polyommatus Antegon, Morris, Syn. Lep. N. Am., p. 87, (1862).
  Cupido Acmon et Antegon, Kirby, Cat. Diurnal Lep., p. 358, (1871).
Expands ⅞ to 1 inch. Male, upper surface violet blue, wings edged with a black line; on secondaries a row of black sub-marginal spots succeeded inwardly by a narrow orange band; fringe white. Female dark brown with orange sub-marginal band on secondaries. Beneath both sexes nearly like Anna. California, common.

*†LUPINI, Boisduval, Lep. Cal. p. 46, (1869). Edwards, Syn. N. Am. Butt., p. 34, (1872).
  Cupido Lupini, Kirby, Cat. Diurnal Lep., p. 358, (1871).
"California." I have no knowledge of this species.

*†LYCEA, Edwards, Proc. Ent. Soc., Phila., Vol. II, p. 507, (1864); Trans. Am. Ent. Soc., Vol. III, p. 273, (1871); Syn. N. Am. Butt., p. 50, (1872).
  Cupido Lycea, Kirby, Cat. Diurnal Lep., p. 377, (1871).
Of this species, which is unknown to me, the original description says:

"Male. Expands 1 2-10 inch. Upper side purplish blue, colour of Antiaris, fluid, with broad fuscous hind margins; fringe white. Under side grey white; both wings have a row of brown points representing the lunules of obsolete marginal spots; a second row of eight black spots, each circled with white; the first on costa minute, the second round, the third oval, the fourth, fifth and sixth cordate, the others round; all, except first, conspicuous; discal spot reniform. Secondaries have a second row of small spots nearly parallel with the margin; the second and third separated by a wide space; near the base three points in a line, one upon the costa, the second in the cell, the third upon the abdominal margin; all the spots circled with white; discal streak faint."
"Female. Expands 1.4 inch. Same size as male. Upper side fuscous, slightly blue at base of both wings, the discal spot of primaries appearing through the wing; under side fawn color, marked as in male. Taken in Colorado by Mr. Mead."

ENOPTES, Boisduval, Ann. Soc. Ent. Fr., p. 298, (1852); *Edwards*, Syn. N. Am. Butt., p. 35, (1872).
   Polyommatus Enoptes, Morris, Cat. Lep. N. Am., p. 12, (1860); Syn. Lep. N. Am., p. 89, (1862).
   Cupido Enoptes, Kirby, Cat. Diurnal Lep., p. 363, (1871).

   " *Upper side* violet blue, with a rather wide black border; the fringes intersected with white and black on the primaries only, entirely whiteish on the secondaries. *Under side* ashy-white, with a great number of black ocellate points; the two strix of posterior points are separated on the secondaries by a series of five yellow lunules. California."

   The above is Boisduval's description as translated in Morris' Synopsis. I have little doubt but that this species is identical with some of those since redescribed by later American authors. What a mighty reduction of species there would be if Boisduval's, Behr's, Edwards' and Reakirt's types were to meet together and undergo impartial comparison and examination.

EVIUS, Boisduval, Lep. Cal., p. 49, (1869). *Edwards*, Syn. N. Am. Butt., p. 36, (1872).
   Cupido Evius, Kirby, Cat. Diurnal Lep., p. 363, (1871).
   "California." Likewise, I regret to say, unknown to me.

HELIOS, Edwards, Trans. Am. Ent. Soc., Vol. III, p. 208, (1871); Syn. N. Am. Butt., p. 37, (1871).
   Here is a copy of the author's original description; if any one has the species, I only hope he will be able to identify it thereby:

   " *Male.*—Expands 1.1 inch. Upper side dull pruinose blue; hind margin of primaries fuscous, of secondaries edged by a black line; fringes long, soiled white, at apex of primaries partially replaced by fuscous. *Under side* grey brown, bluish at base of secondaries; both wings have a sub-marginal series of small black lunules, and a median row of rounded black spots; those of primaries large, the first on costa nearly or quite obsolete; those of secondaries minute on a white ground, but usually about half the size of the smaller on primaries and uniform; on arc of primaries a large oval black spot, of secondaries a faint streak; a point in cell and another on costal margin; all the lunules and spots faintly edged with white. Body above blue, abdomen beneath soiled white; thorax grey white; palpi same with long black hairs on front; antennæ annulated black and white; club black, tip fulvous. *Female.*—Same size. Dull blue, obscured by pale fuscous; on arc of primaries a faint streak; hind margin of secondaries bordered by indistinct brown oval spots. Under side clear drab; all the lunules and spots distinct and not edged with white. Three ♂, one ♀ from California. H. Edwards."

FULIGINOSA, Edwards, Proc. Acad. Nat. Sc. Phila., p. 164, (1861); Syn. N. Am. Butt., p. 33, (1872).
   Cupido Fuliginosa, Kirby, Cat. Diurnal Lep., p. 364, (1871).
   Lycæna Sænsa, Boisduval, Lep. Cal., p. 51, (1869).

   Expands 1¼ inch. Male, above, blue with brownish exterior margins; female greyish brown. Under surface both sexes grey, brown or black discal marks, and mesial and sub-marginal rows of black spots, on primaries the sub-marginal nearly obsolete. Mr. Edwards says in his description that the male is blackish brown on the upper side. I have seen none of these black males, all mine are blue. California and adjacent Territory.

ACHAJA, Behr, Proc. Cal. Acad. Nat. Sc., p. 280, (1867); *Edwards*, Syn. N. Am. Butt., p. 33, (1872).
   Cupido Achaja, Kirby, Cat. Diurnal Lep., p. 366, (1871).
   Lycæna Rufescens, Boisduval, Lep. Cal., p. 48, (1869).
   Expands 1¼ inch, much resembles the preceding. California.

MINTHA, Edwards, Trans. Am. Ent. Soc., Vol. III, p. 194, (1870); Syn. N. Am. Butt., p. 35, (1872).

   " *Male.*—Expands 1.15 inch. Upper side dull pruinose blue, slightly fuscous on hind margins. Under side grey brown, with a tinge of blue at base; primaries have a large reniform discal spot, a row of six large black spots, the sixth duplex, all circled with white; the row from the third spot to inner margin straight, differing from most species in this respect; a submarginal row of fuscous points. Secondaries have a large black spot on costa near base, a point on abdominal margin; a faint discal streak; a row of eight spots parallel to margin, the first six large, round, the seventh and eighth points only and back of the line, all circled with white; a submarginal row as on primaries. Body dull blue, below grey blue; legs white; palpi grey. *Female.*—Same size. Upper side fuscous; beneath a shade darker than male. From 1 ♂, 1 ♀, Nevada. Collection H. Edwards, Esq."

   Probably another of those which differ in name only from some previously described.

FULLA, Edwards, Trans. Am. Ent. Soc., Vol. III, p. 194, (1870); Syn. N. Am. Butt., p. 35, (1872).
   Appears, from the description, to be either the same or very near to *Rapahoe*, Reak. The present name had been previously used in this family by Mr. Hewitson to designate a species of Amblypodia.*)

PARDALIS, Behr, Proc. Cal. Acad. Nat. Sc., Vol. III, p. 279, (1867). *Edwards*, Syn. N. Am. Butt., p. 35, (1872).
   Cupido Pardalis, Kirby, Cat. Diurnal Lep., p. 374, (1871).
   Resembles closely *Fuliginosa*, Edw., and *Achaja*, Behr. California.

VIACA, Edwards, Trans. Am. Ent. Soc., Vol. III, p. 209, (1871). Syn. N. Am. Butt., p. 35, (1872).
   Here is a copy of the original description, which as usual, will conveniently serve to identify two or three or more dozen besides the one intended.

*) Amblypodia Fulla, Hewitson, Cat. Lyc. B. M., t. 6, (1862).

LYCÆNA.

"Male.—Expands 1.4 inch. Upper side greenish blue with a metallic lustre, somewhat obscured by fuscous on secondaries; hind margin of primaries largely bordered by fuscous, of secondaries narrowly; fringes of primaries white, black at end of nervules, of secondaries white. Under side grey-brown mottled with calcareous white; primaries have a sub-marginal series of brown lunules, not distinct apically; a median row of large round black spots, the first four from costa forming an arch, the fifth much anterior to fourth and widely separated from it; the sixth duplex; all edged with white; on the arc a sub-reniform black spot and one nearly similar in cell. Secondaries have a submarginal series of brown lunules; a median sinuous row of round black spots, less conspicuous than those of primaries, except the first, fourth and last; the second, fifth, sixth and seventh half the size of the first; on arc an indistinct bent streak; a small black spot in cell, a large one on costa and a third below cell; fringes beneath on both wings cut by brown. Body above blue, beneath thorax blue grey; legs black and white; palpi white, black at tip and on upper side; antennæ annulated black and white; club black, tip fulvous. From collection H. Edwards. Taken in the Sierra Nevada, Cal."

SAPIOLUS, Boisduval, Ann. Soc. Ent. Fr., p. 297, (1852). *Edwards*, Syn. N. Am. Butt., p. 36, (1872).
Polyommatus Sapiolus, Morris, Cat. Lep. N. Am., p. 12, (1860); Syn. Lep. N. Am., p. 88, (1862).
Cupido Sapiolus, Kirby, Cat. Diurnal Lep., p. 373, (1871).

Expands 1¼ inch. Male, above, greenish blue not very lustrous; a discal mark on primaries; blackish borders at exterior margins, broadest on primaries; fringe white. Female dark brown. Under surface in both sexes grey with discal, mesial and sub-marginal spots as in allied species. California, not scarce.

*†LORQUINI, Behr, Proc. Cal. Acad. Nat. Sc., Vol. III, p. 280, (1867). *Edwards*, Syn. N. Am. Butt., p. 36, (1872).
Cupido Lorquini, Kirby, Cat. Diurnal Lep., p. 377, (1871).

"California." I do not know this species; the name, at any rate, must give way, as Dr. Herrich-Schaeffer has already employed it in 1850 for a Mediterranean species.*)

*†DÆDALUS, Behr, Proc. Cal. Acad. Nat. Sc., Vol. III, p. 280, (1867). *Edwards*, Syn. N. Am. Butt., p. 36, (1872).
Cupido Dædalus, Kirby, Cat. Diurnal Lep., p. 366, (1871).

"California." Unknown to me.

(Pl. X, F. 17, ♂.)

GORGON, Boisduval, (*Polyommatus G.*) Ann. Soc. Ent. Fr., p. 292, (1852). *Morris*, Cat. Lep. N. Am., p. 12, (1860); Syn. Lep. N. Am., p. 86, (1862).
Lycæna Gorgon, Kirby, Cat. Diurnal Lep., p. 343, (1871).
Chrysophanus Gorgon, Edwards, Syn. N. Am. Butt., p. 33, (1872).

One of the largest N. American species; I have not yet seen the female, but Boisduval says: "Upper side of female dull brown, spotted with fulvous, as in the allied species but of a paler tint." California, rare.

EPIXANTHE, Boisduval & Leconte, (*Polyommatus E.*) Lep. Am. Sept., p. 127, t. 38, (1833). *Morris*, Cat. Lep. N. Am., p. 12, (1860); Syn. Lep. N. Am., p. 85, (1862). *Moschler*, Stett. Ent. Zeit., p. 114, (1870). *Staudinger*, Cat. Eur. Lep., p. 8, (1871).
Lycæna Epixanthe, Harris, Ins. Inj. Veg., Flint's Ed., p. 274, (1862). *Kirby*, Cat. Diurnal Lep., p. 343, (1871).
Chrysophanus Epixanthe, Edwards, Syn. N. Am. Lep., p. 32, (1872).
§ *Lycæna Dorcas, Kirby*, Faun. Bor. Am., Vol. IV, p. 299, t. 4, (1837). W. F. Kirby, Cat. Diurnal Lep., p. 343, (1871).
Polyommatus Dorcas, Morris, Cat. Lep. N. Am., p. 12, (1860); Syn. Lep. N. Am., p. 90, (1862).
Chrysophanus Dorcas, Edwards, Syn. N. Am. Butt., p. 32, (1872).

Size of Phlæas. Male, upper surface primaries dark brown, glossed with purple on discs, edge of costa orange; a black discal spot, another within the cell, and sometimes a third one between this latter and the interior margin. Secondaries, a black discal mark, midway between this and the exterior margin are two small spots; a small orange spot at anal angle continued in one or two more or less dimly defined lunules; fringe smoky. Female more of a reddish cast on the discs, no purple reflections; in addition to the spots of the male there is on the primaries an irregular mesial row of nearly confluent black spots; secondaries also with mesial row of like spots; fringes white. Under surface both sexes yellowish, spots on primaries arranged as above with the addition of a row of sub-marginal spots, the three nearest inner angle distinct, the others scarcely discernable. Secondaries have the spots of upper surface represented by mere black points or dots, a connected row of orange sub marginal lunules, the four nearest the anal angle brightest. Labrador, Canada, New England States and New York. Kirby's figure of *Dorcas* agrees exactly with Epixanthe ♀, and the wonder is that their identity was not long ago discovered. I give below the short description of *Dorcas* that commences Kirby's article; this is followed in the Fauna Am. Bor. by another, much longer, but in part more obscure diagnosis, which want of space will not allow of insertion here:

"*Dorcas* Lycæna, wings above brown-ferruginous dotted and spotted with black; beneath tawney; primaries with black spots and crescents; secondaries obsoletely dotted with black; marked at apex with obsolete orange crescents. Expansion of wings 1 inch. Taken in Lat. 54°."

*) *Lycæna Lorquinii, Herr.-Sch.*, Schmett. Eur. I, f. 442-444, (1850), VI, p. 25, (1852).

PHLÆAS VAR. AMERICANA, D'Urban, Can. Nat., V, p. 246, (1857). *Harris*, Ins. Inj. Veg., Flint's Ed., p. 273, (1862). *Kirby*, Cat. Diurnal Lep., p. 344, (1871).
   Polyommatus Americana, Morris, Syn. N. Am. Lep., p. 91, (1862).
   Chrysophanus Americana, Edwards, Syn. N. Am. Butt., p. 32, (1872).
   Polyommatus Hypophlæas, Boisduval, Ann. Soc. Ent. Fr., p. 293, (1852).

Similar to the European *Phlæas*, of which it is the American form, the principal and only difference is in the ground colour of under side of secondaries, which in our form is paler and brighter. In California, examples have been taken that accord perfectly with the European type. Larva green. One of the commonest of our diurnals, occurring from May to October throughout the United States and Canada.

THOE, Gray, (*Polyommatus T.*) Griff. An. King. t. 56, (1832). *Boisduval & Leconte*, Lep. Am., Sept., p. 125, t. 38, (1833). *Guerin*, Icon. Reg. An. Ins., t. 81, (1844). *Morris*, Cat. Lep. N. Am., p. 12, (1860): Syn. Lep. N. Am., p. 84, (1862).
   Lycæna Thoe, Kirby, Cat. Diurnal Lep., p. 343, (1871).
   Chrysophanus Hyllus, Edwards, Syn. N. Am Butt., p. 33, (1872).

Expands 1¼ to 1½ inches. Male, upper surface fore wings brown with purplish reflections; hind wings blackish with orange margin. Female much the same colour and markings as *Phlæas*, but lacks the brilliancy of that species. Canada, New England States, New York and Michigan.

(Pl. X, F. 19, ♂, 20 ♀.)

HELLOIDES, Boisduval, (*Polyommatus H.*) Ann. Soc. Ent. Fr., p. 292, (1852). *Morris*, Cat. Lep. N. Am., p. 12, (1860); Syn. N. Am. Lep., p. 86, (1862).
   Lycæna Helloides, Kirby, Cat. Diurnal Lep., p. 342, (1871).
   Chrysophanus Helloides, Edwards, Syn. N. Am. Butt., p. 32, (1872).
   ? *Polyommatus Castro, Reakirt*, Proc. Ent. Soc. Phila., Vol. VI, p. 148, (1866).
   Chrysophanus Castro, Edwards, Syn. N. Am. Butt., p. 32, (1872).
   ? *Polyommatus Zeroe, Boisduval*, Lep. Cal., p. 45, (1869).

Common in California, Oregon, Colorado, etc.

(Pl. X, F. 23 ♂, 24 ♀.)

IANTHE, Edwards, (*Chrysophanus I.*) Trans. Am. Ent. Soc., Vol. III, p. 211, (1871); Syn. N. Am. Butt., p. 32, (1872).

Very close to *Helloides*, mainly differing from it in the absence of the black spots of upper surface of male. Colorado, Nevada.

I should remark that in the males of *Helloides*, *Ianthe*, *Gorgon*, and allies, the brown colour of upper surface is beautifully glossed with violet, the effect of which it is impossible to imitate by the colourist's art.

(Pl. X, F. 25 ♂, 26 ♀.)

‡MARIPOSA, Reakirt, (*Polyommatus M.*) Proc. Ent. Soc. Phila., Vol. VI, p. 149, foot note, (1866).
   Lycæna Mariposa, Kirby, Cat. Diurnal Lep., p. 342, (1871).
   Chrysophanus Mariposa, Edwards, Syn. N. Am. Butt., p. 32, (1872).
   Polyommatus Niralis, Boisduval, Lep. Cal., p. 44, (1869).

Lower California. I have seen no examples of this except the original types now in my cabinet.

(Pl. X, F. 27 ♂, 28 ♀.)

AROTA, Boisduval, (*Polyommatus A.*) Ann. Soc. Ent. Fr., p. 293, (1852). *Morris*, Cat. Lep. N. Am., p. 12, (1860); Syn. Lep. N. Am., p. 86, (1862).
   Lycæna Arota, Kirby, Cat. Lep. N. Am., p. 343, (1871).
   Chrysophanus Arota, Edwards, Syn. N. Am. Butt., p. 32, (1872).

California, not uncommon.

(Pl. X, F. 21 ♂, 22 ♀.)

VIRGINIENSIS, Edwards, (*Chrysophanus V.*) Trans. Am. Ent. Soc., Vol. III, p. 21, (1870); Syn. N. Am. Butt., p. 32, (1872).
   Lycæna Virginiensis, Kirby, Cat. Diurnal Lep., p. 345, (1871).

Nevada. Is larger than the preceding, and spots of under surface much larger and better defined.

*†HERMES, Edwards, (*Chrysophanus H.*) Trans. Am. Ent. Soc., Vol. III, p. 21, (1870); Syn. N. Am. Butt., p. 33, (1872).
   Lycæna Hermes, Kirby, Cat. Diurnal Lep., p. 345, (1871).

Unknown to me. I give the author's description as follows:

"*Male.* Expands 9-10 inch. Upper side pale fulvous; costal edge and hind margin of primaries brown; base obscured; on disk several brown spots of which the outer ones form an irregular row across the wing; a spot on arc and a second in cell. Secondaries have a long pointed tail; on the margin next anal angle an indistinct row of blackish spots; on the arc a recurved black stripe, surface of wing much obscured. Under side of primaries pale buff, the spots repeated but large and more distinct; margin grayish; secondaries have the basal two-thirds grayish, the margin buff clouded grey; the disk crossed by a row of black spots, those at the extremities crescent; on the arc a black streak; three small spots above in a transverse line and three others near base; at anal angle a black spot and near it others almost obsolete. *Female.* Expands 1 1-10 inch. Similar to male, the markings more distinct."

California.

(Pl. X, F. 12 ♂, 13 ♀.)
XANTHOIDES, Boisduval, (*Polyommatus N.*) Ann. Soc. Ent. Fr., p. 292, (1852); Lep. Cal., p. 45, (1869).
    Morris, Cat. Lep. N. Am., p. 12, (1860); Syn. Lep. N. Am., p. 86, (1862).
    Lycaena Xanthoides, Kirby, Cat. Diurnal Lep., p. 343, (1871).
    Chrysophanus Xanthoides, Edwards, Syn. N. Am. Butt., p. 33, (1872).
    California. The male of this fine species differs remarkably in colour of upper surface from analogous forms. In many examples the pale parts of the upper surface of ♀ is not as red as depicted in fig. 13, (Pl. X), more of a greyish buff.

*†RUBIDUS, Behr, (*Chrysophanus R.*) Proc. Ent. Soc. Phil., Vol. VI, p. 208, (1866). *Edwards*, Syn. N. Am. Butt., p. 33, (1872).
    Lycaena Rubidus, Kirby, Cat. Diurnal Lep., p. 345, (1871).
    Appears from author's following description to be somewhat allied to *Sirius*:

    Male. Expands 1 2.10 inch. Upper side uniform bright copper-red, secondaries having a narrow border along the hind margin of lighter color; both wings edged by a black line: fringes grey, several of the spots of under side of primaries show faintly through the wing; on secondaries a faint discal streak. Under side white, with a faint tinge of orange; no spots on secondaries; primaries have a marginal row of not very distinct brownish spots, wanting on the upper half of the wing; a sinuous row of six clear, black, rounded spots across the disc, the felt spot double; a long spot on the arc; two round spots in the cell and one below. Antennæ black above, ringed with white, whitish below; tips ferruginous. One ♂ received from the interior of Oregon."

*†CUPREUS, Edwards, (*Chrysophanus C.*) Trans. Am. Ent. Soc., Vol. III, p. 20, (1870); Syn. N. Am. Butt., p. 33, (1872).
    Lycaena Cupreus, Kirby, Cat. Diurnal Lep., p. 345, (1871).
    I have as yet had no opportunity of seeing examples of this insect. The author describes it as below:

    "*Male.* Expands 1.1 inch. Upper side bright copper-red, color of *Rubidus*; hind margins edged by black, the secondaries narrowly; both wings crossed by a tortuous extra discal row of small brown spots and points; a spot on arc of primaries and a faint spot in cell; on arc of secondaries a black point. Under side of primaries ochraceous inclining to red; spots as above, larger, edged with white; a spot near base in cell; marginal border fawn colour, on the anterior edge of which is a row of brown points. Secondaries paler, mottled with white, obscured at base; a marginal series of orange crescents, the one next anal angle long and narrow; traces of brown spots on marginal edge; extra discal spots as above, in addition to which are eight others, three on costa, two on arc, two in cell and one in abdominal margin. *Female.* 1.2 inch. Paler red, similarly marked, spots large: under side like male. (Oregon.)"

(Pl. X, F. 29 ♂, 30 ♀.)
SIRIUS, Edwards, (*Chrysophanus S.*) Trans. Am. Ent. Soc., Vol. III, p. 270, (1871); Syn. N. Am. Butt., p. 50, (1872).
    Colorado, rare. The male has much the same fiery colour as in the European *Hippothoe, Virgaureae*, etc.

(Pl. X, F. 6 ♂, 7 ♀.)
HETERONEA, Boisduval, Ann. Soc. Ent. Fr., p. 297, (1852). *Edwards*, Syn. N. Am. Butt., p. 33, (1872).
    Polyommatus Hetronea, Morris, Cat. Lep. N. Am., p. 12, (1860); Syn. Lep. N. Am., p. 89, (1862).
    Cupido Heteronea, Kirby, Cat. Diurnal Lep., p. 363, (1871).
    California. A beautiful species, closely allied to Sirius and Xanthoides, notwithstanding the dissimilarity of colour on upper surface of males.

*†DIONE, Scudder, (*Chrysophanus D.*) Jul. Bost. Soc. Nat. Hist., XI, p. 401, (1868); Trans. Chicago Acad. Nat. Sc., I, p. 330, (1869). *Edwards*, Syn. N. Am. Butt., p. 33, (1872).
    I do not know of the existence of this species in any collection, nor have I access to the works in which it is described, but I do not hesitate to hazard the assertion that I believe it to be nothing more than a synonym of some one or other of those already alluded to, probably *Thoe*.

NAIS, Edwards,————(*Chrysophanus N.*).
    Mr. W. H. Edwards, in his "Synopsis N. Am. Butterflies," has this name cited thus: " 6. Nais, Edwards, Trans. Am. Ent. Soc., 1870. *Hab.*—California, Nevada." I have no knowledge of this insect, and on turning, for my better information, to the index of Vol. III of that work, which was issued in 1870-1871, I could find no *Nais*; I then hunted for Mr. Edwards' articles in that volume, but after turning page by page I became satisfied that no description of *Nais* was to be found there. Kirby, in his Catalogue Diurnal Lep., p. 653, has " 42. L. Nais, Edw., (*Chrys. N.*) Trans. Am. Ent. Soc., 1871. Unio Amer." But a thorough re-examination would produce no better results, thus an hour's time was irrevocably lost because Mr. Edwards

has inserted in his "Synopsis" a name which has no corresponding description. It would be always a great convenience to the student, even if not so much to the compiler, if the latter would favour us always with the No. of the Vol. and page on which the species cited may be found, and the **year of its** publication; it is this that has made the great catalogues of Kirby and Staudinger indispensable to every Lepidopterist. When an author cites his species, but withholds the number of the vol. and page, it causes unwholesome ideas to suggest themselves that the less such species are investigated the better for their stability.

HESPERIA KIOWAH, Reakirt, Proc. Ent. Soc., **Phil., Vol. VI,** p. 150, (1866), is a synonym of *Hesperia Metacomet*, Harris, Ins. Mass., p. 317, (1862).

HESPERIA POWESHEIK, Parker, Am. Ent. II, p. 271, (1870), is identical with *Hesperia Garita*, Reakirt, Proc. Ent. Soc., Phil. VI, p. 150, (1866).

BAPTA VIATICA, Harvey, Bull. Buff. Soc., p. 265, t. 11, f. 6, (1874), is the same as the old *Corycia Semiclarata*, Walker.

AGROTIS DEPRESSUS, Grote, Can. Ent. III, p. 192, (1871), is the *Amphipyra Tragopoginis* of Linnæus in the Fauna Svecica, 316, (1746), and later in his Systema Naturæ, and also in Hubner, Esper, Treitschke, Godart and others. It is a species common to both Europe and America.

NOTICES OF SOME NEW AND RARE SPECIES WHICH I SHALL FIGURE AT AN EARLY DATE.

MACROGLOSSA FUMOSA, nov. sp.
Expands 1¾ inch. Head and body same colour as *Difinis*, but the yellow of head and upper part of thorax slightly more inclined to green. Primaries costa black with the middle third yellow; exterior margin black, 3-16 inch wide on costa at apex, and diminishing to a mere line at the inner angle; as base of wing a triangular patch, inner two-thirds black, outer, which extends along the inner margin and diminishes to a point, is yellow; all the space, which in *Difinis* and allied species is vitreous, is here clothed, on both upper and under surface, with large, heavy, ashen or smoky scales. Secondaries have the exterior margin narrowly edged with black (but still broader than in *Difinis, Tennis* and *Thetis*); a broad black border on abdominal margin; disc of wing covered with same heavy ashen scales as on primaries. Three examples, two in collection of Mr. O. Meske, at Albany, the third—owing to the goodness of that gentleman—is now in my cabinet.

SPHINX CONIFERARUM, Abbot & Smith, Ins. Georgia, p. 81, t. 41, (1797).
Of this species, which has been lost since the time of Abbot, I have had the rare good fortune to obtain two examples, both males; the first one I received about two years since from my old friend, Edward Baumbauer, of Baltimore, who bred it from a chequered caterpillar which he found feeding on pine, and the description of which agreed with Abbot's figure; the second example came from northern New York, and was taken on the wing. Both examples expand a trifle over 2¾ inches, and in all their details agree with Abbot's figure; it is as different from *Ellema Harrisii*, with which it was so long confounded, as it is from *Sphinx Eremitus*. It comes amazingly close to the European *Sphinx Pinastri*, but on the primaries of the latter are two irregular, oblique, transverse, brownish shades which are wanting in Coniferarum, but what most strikingly distinguishes this from all allied forms is the immaculate abdomen which has **not** the shadow of a line or of any mark whatsoever.

SPHINX EREMITOIDES, nov. sp.
Male expands 3 inches. Head and thorax whitish grey; tegulæ edged with a black line; abdomen with a broad, whitish-grey, dorsal band, destitute of any dorsal line, on its sides are seven short black bands, the ones at the base largest, and lessening in size as they approach the extremity where the last two are mere spots. Primaries whitish or silvery-grey, marked much as in *Eremitus* but not come so heavily. Secondaries greyish-white; on exterior margin a broad band, black inwardly and dark greyish nearest the margin; at base of wing a black patch which **does** not extend to the abdominal margin, between this patch and the marginal band or border is a narrower black mesial band, but this is not entire—being broken near the middle by the white ground colour—giving to the latter the appearance of a large white II on a dark back-ground. Female expands 3½ inches; resembles the male with the exception of the mesial band of secondaries which is **not broken** but narrowly continued over the white space that breaks it in the male. Under surface, both sexes; primaries grey; secondaries white, greyish at exterior margin; an irregular, narrow, mesial, brown band which is also continued up the primaries in a double transverse line or shade. Allied to *Eremitus*, but easily to be known by its pale silvery-grey colour, by the almost entire absence of a dorsal stripe, and the peculiar ornamentation of secondaries. Taken in Kansas by Mr. T. B. Ashton, to whose kindness I am indebted for the examples above described.

EUDÆMONIA JEHOVAH, nov. sp.
Male. Expands 4½ inches. Same form as *Semiramis*, Cram., but the exterior margin of primaries, between the veins, is indented a little more deeply, and the tails are not quite so long in proportion. The colour is a dark grey or mouse color ornamented in darker shades almost in the same manner as *Semiramis*. Brazil—in collection of Prof. J. E. Mayer, in New York.
There are known to me three species of these most wonderful moths, all S. American, viz.:
EUDÆMONIA SEMIRAMIS, Cramer, Vol. I, t. XIII, a, ♂, (1775). *Maassen*, Brit. Schnoett., f. 5 ♀, 6 ♂, (1869).
EUDÆMONIA DERCETO, Maassen, Beit. Schmett., f. 13, 14 ♂, (1872).
EUDÆMONIA JEHOVAH, mihi, (1874).

CATOCALA MAGDALENA, nov. sp.
Female expands 2¾ inches. Head and thorax, above, pale ashen; abdomen yellow, body beneath whitish. Primaries pale uniform ash-grey nearly like the paler examples of *Concumbens*; transverse posterior, anterior and sub-basal lines very narrow and inconspicuous; reniform faint and double ringed; sub-reniform caused by a widely open sinus of the transverse posterior line; fringe concolorous with

the rest of wing. Secondaries light yellow; marginal and mesial bands of ordinary width, the latter somewhat elbowed and does not reach to the abdominal margin by over an eighth of an inch; fringe whitish grey. Under surface pale yellow, primaries with a pale brownish or grey marginal band darkest inwardly; mesial band wider in the middle and terminates at both costa and interior margin in a point; no sub-basal band, only a faint grey spot midway between costa and interior margin; fringe white, edged outwardly with pale brown. Secondaries have a pale grey marginal band, palest anteriorly where it is lost in the ground color of wing, a brown mesial band strongly elbowed; fringe white.

For the knowledge of this beautiful species I am indebted to Dr. G. M. Levette, of the Indiana Geological Survey, a most ardent admirer of nature, who captured it in the grove near the State House, in Indianapolis. I have also since seen examples sent from Texas, in the Cambridge Museum.

CATOCALA ASPASIA, nov. sp.
Expands 4 to 4½ inches. Head and body brownish grey. Upper surface, primaries greyish brown, having much the general appearance, in colour and markings, of the darker examples of *Amatrix*. Secondaries scarlet, with black marginal and mesial bands. This species vies in size, though not in beauty, with *Marmorata*, and looks, at a casual glance, not unlike a gigantic edition of *Amatrix*.
Received with a number of other things from Lower California; it also occurs in Texas.

May, 1874.

Dr. Herman Behr lately had the kindness to send me the types of his following unpublished Californian species:

DRYOBOTA CALIFORNICA, Behr, Mss.
Size and general appearance of *D. Protea*, Bkh., but the primaries are more ash-grey and devoid of the greenish cast of that species. Secondaries white, with faint mesial line and discal spot.

TAENIOCAMPA PAVLE, Behr, Mss.
A trifle smaller than *T. Gracilis*, F., which it otherwise closely resembles; the most marked differences are in the sub-marginal line of primaries which is more uneven, in the presence of a pale transverse line running from the reniform to the interior margin, and in most examples the prevalence of a more reddish cast of colour throughout the upper surface of primaries.

COSMIA SAMBUCI, Behr, Mss.
Expands 1¼-1½ inch. Upper surface primaries very pale greenish yellow, in some specimens very pale cinnamon colour, transverse posterior and anterior and discal spot a darker shade of the ground colour, between the two lines a transverse shade of same colour, broadest and darkest at costa, which it reaches by crossing the transverse posterior line, which towards the costa gives an abrupt bend inwards. Secondaries white. Head and body same colour as primaries.

CUCULLIA SOLIDAGINIS, Behr, Mss.
Expands 1¼-2 inches. Head and regular ashen, thorax between the latter dark brown, abdomen dark grey with a brown dorsal line. Primaries grey, the median space suffused with darker colour, which in one example extends to the base; it comes nearer in the markings to *Lactuca*, Esp., but is entirely distinct from that or any other species known of. Secondaries smoky.

CUCULLIA MATRICARIAE, Behr, Mss.
Expands 1¼-1½ inch. Head and body grey, latter with a dark dorsal line. Primaries uniform pale ashen, the fine dark lines are all abbreviated and have the appearance of being heavier than in any other known species, and gives the whole wing a spotted rather than striped appearance, entirely different from its allies. Secondaries white, edged exteriorly with a fine brown line. In both this and the preceding the wings are a little broader in proportion to their length than in the eastern species.

PLUSIA ECHINOCYSTIDIS, Behr, Mss.
Size of *Precationis*. Head and thorax dark greyish brown; abdomen pale brown. Primaries dull dark greyish brown, which on close inspection proves to be intricately varied and shaded; a silver mark much like a U, attached to the lower part of which is a small round or oval silver spot. Secondaries smoky, outer half darkest; fringe white.

PLUSIA GAMMA, Linnaeus, Syst. Nat., X. 513, (1758).
Thirteen examples agreeing in every respect with those from various parts of Europe.

May, 1874.

Of the following species I am anxious to obtain examples, either by exchange or purchase; any Naturalists having duplicates of any of them will confer a great favor by communicating with

HERMAN STRECKER,
Box 111 Reading, P. O.,
Berks Co., Pennsylvania, U. S. of N. America

Ornithoptera Hippolytus, Cram.
Ornithoptera Lydius, Feld.
Ornithoptera Helena, Linn. ♀
Ornithoptera Croesus, Wall.
Ornithoptera Brookiana ♀, Wall.
Papilio Evan, Doubl.
Papilio Pericles, Wall.
Papilio Blumei, Boisd.
Papilio Macedon, Wall.
Papilio Philippus, Wall.
Papilio Arcturus, West.
Papilio Phorbanta, Linn.
Papilio Homerus, Fabr.
Papilio Garamas, Hub.
Papilio Cuiguanabus, Poey.
Papilio Ascanius, Cram.
Papilio Wallacei, Hew.
Papilio Slateri, Hew.
Papilio Endochus, Boisd.
Papilio Icarius, West.
Papilio Dionysos, Dbldy.
Papilio Gundlachianus, Feld.
Papilio Elephenor, Dbldy.
Papilio Disparilis, Herr-Sch.
Argynnis Rudra, Moore.
Argynnis Oscarus, Evers.
Argynnis Cnidia, Feld.
Argynnis Jerdoni, Lang.
Argynnis Desameue, Boisd.
Argynnis Jainadeva, Moore.
Argynnis Ruslana, Motsch.
Argynnis Anna, Blanch.
Argynnis Childreni, Gray.
Argynnis Aruna, Moore.
Parthenos Tigrina, Voll.
Penetes Pamphanis, Dbl., Hew.
Caerois Chorinaeus, Fabr.
Charaxes Epijasius Reiche.
Charaxes Kadenii, Feld.
Charaxes Jupiter, But.
Charaxes Etheocles, Cram.
Catagramma Excelsior, Hew.
Dyctis Binoculatis, Guerin.
Romaleosoma Sophron, Dbldy.
Romaleosoma Pratinas, Dbldy.
Romaleosoma Arcadius, Fabr.
Nymphalis Calydonia, Hew.
Saturnia Epimethea, Dru.
Calinaga Buddha, Moore.
Diadema Boisduvalii, Dbldy.
Pavonia Aorsa, West.
Any species of Phyllodes.

Dynastor Napoleon, Doubl, Hew.
Argynnis Sagana ♂, Doubl, Hew.
Zeuxidia Aurelias, Cram.
Brahmaea Whitei.
Brahmaea Certhia.
Urania Sloanus, Cram.
Eudemonia Semiramis, Cram.
Thaliura Croesus.
Nyctalemon Menurus.
Nyctalemon Cydnus, Feld.
Aetias Maenas
Saturnia Dercoto, Mssn.
Saturnia Argus, Drury.
Any Asiatic species of Parnassius.
Citheronia Phoronea.
Pyrameis Gonerilla, Fabr.
Pyrameis Abyssinica, Feld.
Pyrameis Dejeanii, Godt.
Pyrameis Tameamea, Esch.
Sphinx Cluentius.
Vanessa v. Elymi, Rbr.
Dasyopthalmia Rusina, Godt.
Morpho Phanodemus, Hew.
Any species of Agrias.
Any species of Callithea.
Pandora Chalcothea, Bates.
Pandora Hypochlora, Cates.
Pandora Divalis, Bates.
Colias Viluiensis, Men.
Colias Ponteni, Wallengr.
Bunaea Deroyllei, Thom.
Bunaea Phaedusa, Dru.
Rimaea Zuleica, Hope.
Acraea Perenna, Dbldy.
Opsiphanes Boisduvalii, Dbldy.
Clothilde Jaegeri, Herr-Sch.
Pieris Celestina, Boisd.
Euploea Eurypon, Hew.
Paphia Panariste, Hew.
Limenitis Lymire, Hew.
Io Beckeri Herr-Sch.
Eacles Kadenii, Herr-Sch.
Hepialus Giganteus, H.-S.
Smerinthus Panopus, Cram.
Sphinx Substrigilis, West.
Saturnia Larissa, West.
Eusemia Vietrix, Bellatrix, Amatrix, Dentatrix, West.
Smerinthus Modesta, Fab. (nec. Harris.)
Smerinthus Tartarinovii, Brem.
Smerinthus Dentatus, Cram.

These are a few of the very many of the rarer species that I am eager to procure; of course there are numberless others from all parts of the world, equally desirable and coveted by me.

American Oceanic Express Co.,

NEW YORK,
OFFICE, 48 BROADWAY.

Chartered by Special Act of Incorporation.

CENTRAL EUROPEAN OFFICE:
5 RUE SCRIBE, PARIS.

PRINCIPAL OFFICE IN GREAT BRITAIN:
4 Moorgate St., London, E. C.
B. W. & H. HORNE, AGENTS.

BRANCH OFFICES: Golden Cross, Charing Cross; George & Blue Boar, High Holborn; 108 New Bond Street; 474½ New Oxford Street.

OFFICE IN LIVERPOOL: Old Castle Buildings Tithebarn Street.

CONTINENTAL OFFICES: 5 Rue Scribe, Paris; 82 Rue d'Orleans, Havre; 88 Rodingsmarkt, Hamburg; 29 Bahnhoff Strasse, Bremen.

Merchandise, specie, bullion, stocks, bonds, or other valuables and packages and parcels of every description, personal effects, luggage, etc., forwarded to and from Europe and all parts of the United States, the States and Territories of the Pacific Coast, British Columbia and the Canadas included, at fixed Tariff rates, with no extra charges whatsoever for Custom House brokerage, commission, delivery, etc., etc., the shipper or receiver being under no other costs or expense than the stipulated freight from the point of shipment to place of delivery, and the amount of duties and government fees actually paid at the Custom-houses of the United States or Europe.

For the convenience of shippers, where agencies of the Company are not established, packages or heavy goods may be forwarded to either of the offices or agencies of the Company, by either of the express or transportation companies in the United States, or by post, by railway, through the parcel delivery companies, or forwarding houses in any part of Great Britain or the Continent of Europe.

All packages, trunks, or parcels forwarded by this Company will be landed on arrival simultaneously with the mails, or immediately thereafter and will be entered at the Custom house, duties paid and delivered to the parties to whom addressed in any part of Europe, the United States, the Canadas, or British Columbia, with the greatest possible dispatch. Transportation charges and duties collected on delivery, or may be prepaid, at the option of shipper.

Insurance against marine risk taken by the Company, when desired by the shipper, at the lowest current rates; premium payable in all cases in advance.

Shippers to or from any part of America, and Americans traveling in Europe, will find this the quickest, cheapest and most reliable medium of transportation, the business of this Company being conducted upon the well known prompt American express system, which has become so great a commercial necessity and convenience throughout the United States.

Purchases made, and collections and communications in every part of Europe and the United States promptly executed.

☞ Circulars sent on application to

S. D. JONES,
Manager,
48 BROADWAY, N. Y.

No. 11.   PRICE 50 CENTS.

LEPIDOPTERA,

RHOPALOCERES AND HETEROCERES,

INDIGENOUS AND EXOTIC;

WITH

Descriptions and Colored Illustrations,

BY

HERMAN STRECKER.

Reading, Pa., 1874.

Reading, Pa.:
Owen's Steam Book and Job Printing Office, 515 Court Street,
1874.

CATOCALA AGRIPPINA. Nov. Sp.

(PLATE XI, FIG. 1 ♂, 2 ♀, 3 ♂ variety.)

MALE. Expands 3 inches.
Head and body, above, blackish grey. Beneath, greyish white.
Upper surface; primaries blackish grey powdered with reddish brown; transverse lines black, heaviest towards the costa, less distinct as they near the inner margin; between the transverse posterior and sub-marginal lines the space is brownish; reniform indistinct and brown; sub-reniform almost obsolete.
Secondaries black, with greyish hairs at the base and abdominal margin; fringe white, partly cut with black at the terminations of the veins.
Under surface; primaries white; a broad marginal band, black inwardly and greyish exteriorly; black mesial and sub-basal bands which are connected with the same colour along the inner margin; fringes white, edged outwardly with blackish.
Secondaries white, with black marginal and mesial bands, the space between which is narrow; fringes white.
FEMALE. Expands 3¼ inches.
Ground colour of wings lighter; the reddish brown sub-marginal band more conspicuous; all the lines, as well as the reniform and sub-reniform, distinct and much more sharply defined than in the male.
Described from examples taken in Texas by Mr. J. Boll.
Prominently distinguishable by the reddish brown, which appears to overlay, more or less, the dark grey of primaries and thorax.
FIG. 3 is a variety of the above, having the inner half of primaries whitish; from same locality.

CATOCALA SAPPHO. Nov. Sp.

(PLATE XI, FIG. 4, ♀.)

Expands 3 inches.
Head and thorax, white; abdomen blackish grey. Beneath, white.
Upper surface; primaries milky white; transverse anterior line broad and black at and near costa, faint and brownish on the inner half; transverse posterior fine and partially obsolete, outwardly this is succeeded by a brownish band faint until it nears the costa where it becomes darker and more conspicuous; sub-terminal line, faint; a row of small black sub-marginal points; reniform, dark brown, which color is continued from thence to the costa.
Secondaries, black; fringes white.
Under surface, marked as in *Agrippina*, but the black bands are much heavier, leaving with the exception of base of wings but little white.
Habitat. Texas. Described from one example received from Mr. J. Boll.
I can imagine nothing more lovely than this ermine of the Catocalæ, which in beauty is not even excelled by the queenly *Relicta*. I hesitated a long time ere I could bring myself to describe it as a separate species from *Agrippina*, to which, notwithstanding its white color, it is closely allied, and to which I thought it might bear the same relation as does *Phalanga* to *Pakroyama*. Its smaller size and some differences in the undulations of the lines, however, have led me to the conclusion that it may be a species distinct.

CATOCALA JUDITH. Nov. Sp.

(PLATE XI, FIG. 5, ♂.)

Expands 2 inches.
Head and thorax, above, light grey; abdomen blackish grey.
Upper surface; primaries, pale ashen, the same color as in *Robinsonii*; transverse anterior and posterior lines, black, only moderately conspicuous; sub-terminal line, whitish.

Secondaries black, with black fringes.
Under surface; primaries, a small white basal patch; the black bands are so nearly confluent as to leave but little white.
Secondaries, basal third white, rest black with but faint indications of the pale space between the marginal and mesial bands.
Described from two examples, ♂ ♀, in the collection of Mr. James Angus, of West Farms, New York, near which place they were captured several years since.

CATOCALA AMESTRIS. Nov. Sp.

(PLATE XI, FIG. 6, ♂.)

Expands 2 inches.
Head and thorax, above, ash grey; abdomen yellow. Beneath, dirty white.
Upper surface; primaries ashen, transverse anterior, double and from middle to costa very heavy, forming a diagonal black bar; transverse posterior, fine but distinct; reniform moderate size, the space between this and transverse anterior line is much paler than the rest of the wing; fringes concolourous with the wings.
Secondaries bright yellow; marginal band moderate, broken towards the anal angle where it is replaced by a spot; mesial band rather narrow and does not reach to the abdominal margin; fringes white.
Under surface yellow; sub-basal and marginal bands brownish; mesial bands darker and better defined at their edges.
Habitat. Texas. One example from Mr. J. Boll.
Resembles the Russian *C. Neonympha*[*] more than any American species.

CATOCALA DELILAH. Nov. Sp.

(PLATE XI, FIG. 7, ♀.)

Expands 2½ inches.
Head and thorax brown; abdomen greyish brown. Beneath, yellowish.
Upper surface; primaries rich velvety brown; transverse anterior line very heavy and dark, from the middle of this to the base runs a dark shade; transverse posterior dark brown or black, and distinct; sub-terminal pale fawn-coloured, edged outwardly with dark brown, where it is prolonged into acute parallel teeth which reach to the exterior margin, the three upper ones being the most prominent; reniform and sub-reniform forming, as it were, one spot, the latter entirely disconnected from the transverse posterior line; fringe brown.
Secondaries dark warm yellow; marginal band broad; mesial almost straight from costa to where it elbows towards the lower part of wing; apex yellow; fringe yellow, with some greyish at the terminations of veins.
Under surface dark yellow, even coloured; sub-basal band of primaries not as dark coloured as other two bands, which are more sharply defined and comparatively narrow.
Habitat. Texas; received from Mr. J. Boll.

CATOCALA AHOLAH. Nov. Sp.

(PLATE XI, FIG. 8, ♀.)

Expands 1¼ inches.
Head and thorax ash grey; abdomen yellowish. Beneath, whitish.
Upper surface; primaries ashen, even coloured; transverse anterior and posterior lines fine but distinct, the latter but little dentated; reniform large, greenish white, extending from this to the apex is a dark brown dash or shade, the edge of which towards the costa is sharply defined on the pale ground color of wings, whilst

[*] Esper, Schmett., 198, 1, 2, (1796).

that towards the exterior margin does not so abruptly terminate, but somewhat gradually shades off; sub-reniform entirely disconnected from transverse posterior line; sub-terminal line brownish; in the space between the base and transverse anterior line is a kidney-shaped spot of the same greenish white and same size as the reniform; this spot is not prominent, though easily discernible; fringe concolourous with the ground colour of wing.

Secondaries bright yellow; marginal band moderately broad, interrupted before reaching the anal angle, where it is replaced by a black spot; mesial narrow, somewhat elbowed, and extends to the abdominal margin and thence upwards towards the base; a very small, yellow apical spot; fringe white, greyish inwardly.

Under surface yellow, darkest near base, at costal and apical parts white or nearly so; no sub-basal band; marginal and mesial bands blackish brown, very sharp and distinct; fringes white, cut with black at terminations of veins.

Texas; from Mr. J. Boll.

A most beautiful species, whose nearest eastern ally is *C. Formula*, than which, however, it is larger and much brighter in colour, and presents many points of difference in ornamentation.

CATOCALA MAGDALENA. Strecker.

(PLATE XI, FIG. 9, ♂.)

Since describing this most beautiful species on page 93, from a ♀ taken in Indianapolis, I have received other examples, ♂ ♀ taken by Mr. Boll in Texas.

CATOCALA ATARAH. Nov. Sp.

(PLATE XI, FIG. 10 ♂, 11 ♀.)

MALE. Expands 1¼ inches.
Head and thorax brownish grey; abdomen brown. Beneath, pale greyish.
Upper surface; primaries brown; transverse anterior and posterior lines broad and dark; reniform indistinct; sub-reniform pale and free from the transverse posterior line; the mesial area, between the reniform and transverse anterior line, paler than any other part, the outer half, between this and the transverse posterior line, is the darkest part of the wing.

Secondaries yellow; marginal band deeply indented towards anal angle; mesial narrow, prolonged in an acute angle above the indentation of marginal band, and extending almost to the abdominal margin, and from thence upwards towards the base; a streak of dark colour extends from base to the indentation above the elbow of mesial; fringes yellow and brownish.

Under surface yellow; no sub-basal bands; mesial and marginal bands blackish.

FEMALE. Same size and markings, but the superiors are brighter in colour; the transverse anterior is narrower and lined inwardly with whitish grey; reniform better defined; sub-reniform not so distinct; sub-terminal line white and most conspicuous near and at the costa.

Texas; from Mr. J. Boll.

It is with some slight hesitation that I place these as sexes of the same species. As there were but two examples taken, the material for study and comparison was, consequently, limited; I am inclined to think, however, that future observation will verify my present decision, as the only difference is the narrower transverse lines and brighter colour of the female.

CATOCALA MYRRHA. Nov. Sp.

(PLATE XI, FIG. 12, ♂.)

Expands 2 inches.
Head and thorax, above, grey; abdomen yellowish. Beneath, greyish.
Upper surface; primaries uniform sombre grey, much the same colour as in *Serena*; transverse lines

nearly obsolete, except near costa where they are black and distinct; the reniform is the distinguishing feature in this species, being deep velvety black; sub-reniform scarcely discernible.

Secondaries yellow; marginal band ordinary width; mesial does not extend to within some distance of the abdominal margin.

Under surface; primaries with blackish sub-basal, mesial and marginal bands; secondaries have mesial and marginal bands corresponding to those on upper surface.

Texas; one example from Mr. J. Boll.

CATOCALA CALIFORNICA. W. H. Edwards.
Proc. Ent. Soc., Phila., Vol. II. p. 509, (1864).

(PLATE XI, FIG. 13, ♂.)

Expands 2¾ to 2⅞ inches.

Belongs to the group of which the European *Nupta** and *Elocata*† are types, and which appears to be the prevalent form west of the Rocky Mountains. The primaries resemble very much in colour and ornamentation those of *Elocata*, but the lines are heavier and better defined, with more contrast of light and dark colours. Inferiors same red as in that species. Under surface; primaries white; marginal band black, greyish at apex of wing; mesial and sub-basal as usual in this group. Inferiors; costal third white; inner two-thirds red; bands as on upper surface.

Mr. W. H. Edwards' type, now in Mus. Am. Ent. Soc., Phila., was taken in Yreka, California; one of my examples was brought by Mr. Mead from Colorado. I have also examined an example from collection of Mr. Henry Edwards, which was taken in Arizona.

CATOCALA CARA. Guenée.
Spec. Gen., Vol. VII. p. 87, (1852).

(PLATE XI, FIG. 14, ♂.)

Expands 3 to 3½ inches.

Upper surface of superiors and body rich deep maroon or reddish brown; transverse lines black, narrow, and accompanied with scattered, inconspicuous grey or pale olivaceous scales; reniform indicated by a circle and pupil of same pale colour, but scarcely noticeable except on close inspection.

Secondaries beautiful crimson, with very broad, deep black marginal band, broadest at apex and gradually diminishing until it terminates at the anal angle; mesial nearly even width, slightly elbowed on outer edge, extends to abdominal margin and continued thence up to the base which is clothed with black hairs; fringes dirty white.

This is one of the commonest and, at the same time, the handsomest of all our known red-winged species. It is found from New York to Florida, and as far west as Texas, from which latter state I have received fine examples.

There occur, occasionally, examples in which the primaries are heavily powdered with whitish yellow or olivaceous scales, especially at the apex, on the part surrounding the reniform and more or less on the anterior half of the wing. In this variety the transverse lines are much more heavily marked than in the ordinary form.

CATOCALA AMATRIX. Hübner.
Lamprosia Amatrix, Samml. Exot. Schmett. II. Verz. Bek. Schmett., p. 277, (1816).
Catocala Amatrix, Guenée, Spec. Gen., Vol. VII. p. 86, (1852).

(PLATE XI, FIG. 15 ♂, 16 ♀.)

Expands from 3 to 3½ inches.

Upper surface; primaries and body brownish grey; transverse lines dark brown, but not heavy; inferiors scarlet; black bands extend to inner margin; fringe dirty white.

* Linné, Syst. Nat., XII, (1767).
† Esper, Schmett., 99, 1, 2, (1786).

Of the following species I am anxious to obtain examples, either by exchange or purchase; any Naturalists having duplicates of any of them will confer a great favor by communicating with

HERMAN STRECKER,
Box 111 Reading, P. O.,
Berks Co., Pennsylvania, U. S. of N. America

Ornithoptera Hippolytus, Cram.
Ornithoptera Lydius, Feld.
Ornithoptera Helena, Linn. ♀
Ornithoptera Croesus, Wall.
Ornithoptera Brookiana ♀, Wall.
Papilio Evan, Doubl.
Papilio Pericles, Wall.
Papilio Blumei, Boisd.
Papilio Macedon, Wall.
Papilio Philippus, Wall.
Papilio Arcturus, West.
Papilio Phorbanta, Linn.
Papilio Homerus, Fabr.
Papilio Garamas, Hub.
Papilio Chignanabus, Poey.
Papilio Ascanius, Cram.
Papilio Wallacei, Hew.
Papilio Slateri, Hew.
Papilio Endochus, Boisd.
Papilio Icarius, West.
Papilio Dionysos, Dbldy.
Papilio Gundlachianus, Feld.
Papilio Elephenor, Dbldy.
Papilio Disparilis, Herr-Sch.
Argynnis Rudra, Moore.
Argynnis Oscarus, Evers.
Argynnis Cnidia, Feld.
Argynnis Jerdoni, Lang.
Argynnis Dexamene, Boisd.
Argynnis Jainadeva, Moore.
Argynnis Ruslana, Motsch.
Argynnis Anna, Blanch.
Argynnis Childreni, Gray.
Argynnis Aruna, Moore.
Parthenos Tigrina, Voll.
Penetes Pamphanis, Dbl., Hew.
Cœrous Chorinæus, Fabr.
Charaxes Epijasius Reiche.
Charaxes Kadenii, Feld.
Charaxes Jupiter, But.
Charaxes Etheocles, Cram.
Catagramma Excelsior, Hew.
Dyctis Bisculatis, Guerin.
Romalæosoma Sophron, Dbldy.
Romalæosoma Pratinos, Dbldy.
Romalæosoma Arcadius, Fabr.
Nymphalis Calydonia, Hew.
Saturnia Epimethea, Dru.
Calinaga Buddha, Moore.
Diadema Boisduvalii, Dbldy.
Pavonia Aorsa, West.
Any species of Phyllodes.

Dynastor Napoleon, Doubl, Hew.
Argynnis Sagana ♂, Doubl, Hew.
Zeuxidia Aurelius, Cram.
Brahmæa Whitei.
Brahmæa Certhia.
Urania Sloanus, Cram.
Euchæmonia Semiramis, Cram.
Thaliura Crœsus.
Nyctalemon Meturnus.
Nyctalemon Cydnus, Feld.
Actias Mænas
Saturnia Derceto, Mæn.
Saturnia Argus, Drury.
Any Asiatic species of Parnassius.
Citheronia Phoronea.
Pyrameis Gonerilla, Fabr.
Pyrameis Abyssinica, Feld.
Pyrameis Dejeanii, Godt.
Pyrameis Tameamea, Esch.
Sphinx Cluentius.
Vanessa v. Elymi, Rbr.
Dasyopthalmia Rusina, Godt.
Morpho Phanodemus, Hew.
Any species of Agrias.
Any species of Callithea.
Pandora Chalcothea, Bates.
Pandora Hypochlora, Cates.
Pandora Divalis, Bates.
Colias Viluiensis, Men.
Colias Ponteni, Wallengr.
Bunæa Deroyllei, Thom.
Bunæa Phædusa, Dru.
Rinæa Zuleica, Hope.
Acræa Perenna, Dbldy.
Opsiphanes Boisduvalii, Dbldy.
Clothilde Jægeri, Herr-Sch.
Pieris Celestina, Boisd.
Euplœa Eurypon, Hew.
Paphia Panariste, Hew.
Limenitis Lymire, Hew.
Io Beckeri Herr-Sch.
Eacles Kadenii, Herr-Sch.
Hepialus Gigantens, H.-S.
Smerinthus Panopus, Cram.
Sphinx Substrigilis, West.
Saturnia Larissa, West.
Eusemia Victrix, Amatrix, Dentatrix, West.
Smerinthus Modesta, Fab. (nec. Harris.)
Smerinthus Tartarinovii, Brem.
Smerinthus Dentatus, Cram.
Any Asiatic Catocalae.

These are a few of the very many of the rarer species that I am eager to procure; of course there are numberless others from all parts of the world, equally desirable and coveted by me.

American Oceanic Express Co.,

NEW YORK,
OFFICE, 48 BROADWAY.

Chartered by Special Act of Incorporation.

CENTRAL EUROPEAN OFFICE:
5 RUE SCRIBE, PARIS.

PRINCIPAL OFFICE IN GREAT BRITAIN:
4 Moorgate St., London, E. C.
B. W. & H. HORNE, AGENTS.

BRANCH OFFICES: Golden Cross, Charing Cross; George & Blue Boar, High Holborn; 108 New Bond Street; 474½ New Oxford Street.
OFFICE IN LIVERPOOL: Old Castle Buildings Tithebarn Street.
CONTINENTAL OFFICES: 5 Rue Scribe, Paris; 82 Rue d'Orleans, Havre; 88 Roding-marks, Hamburg; 29 Bahnhofs Strasse, Bremen.

Merchandise, specie, bullion, stocks, bonds, and other valuables and packages and parcels of every description, personal effects, baggage, etc., forwarded to and from Europe and all parts of the United States, the States and Territories of the Pacific Coast, British Columbia and the Canadas included, at *fixed Tariff rates, with no extra charges whatsoever for Custom-House brokerage, commissions, delivery, etc., etc., the shipper or receiver being under no other care or expense than the stipulated freight from the point of shipment to place of delivery, and the amount of duties and government fees actually paid at the Custom-houses of the United States or Europe.*

For the convenience of shippers, where agencies of the Company are not established, packages or heavy goods may be forwarded to either of the offices or agencies of the Company, by either of the express or transportation companies in the United States, or by post, by railway, through the parcel delivery companies, or forwarding houses in any part of Great Britain or the Continent of Europe.

All packages, trunks, or parcels forwarded by this Company will be landed on arrival simultaneously with the mails, or immediately thereafter and will be entered at the Custom-house, duties paid and delivered to the parties to whom addressed in any part of Europe, the United States, the Canadas, or British Columbia, with the greatest possible dispatch. Transportation charges and duties collected on delivery, or may be prepaid, at the option of shipper.

Insurance against marine risk taken by the Company, when desired by the shipper, at the lowest current rates; premium payable in all cases in advance.

Shippers to or from any part of America, and Americans traveling in Europe, will find this the quickest, cheapest and most reliable medium of transportation, the business of this Company being conducted upon the well known prompt American express system, which has become so great a commercial necessity and convenience throughout the United States.

Purchases made, and collections and communications in every part of Europe and the United States promptly executed.

☞ Circulars sent on application to

S. D. JONES,
Manager,
48 BROADWAY, N. Y.

W. V. ANDREWS,
ENTOMOLOGIST, &C.,
Room No. 4, No. 117 Broadway, New York.

☞ Purchasing Agent for Books and Apparatus in connection with Natural History. Also, Cork Pins, &c. Eggs of the different varieties of Silk Worms, to order. Lepidoptera and Coleoptera for sale or exchange. Agent for WALLACE'S SILK REELER, and for KIRBY'S SYNONYMIC CATALOGUE OF DIURNAL LEPIDOPTERA.

No. 12.   Issued Quarterly, at 50 Cents per Part.

LEPIDOPTERA,

RHOPALOCERES AND HETEROCERES,

INDIGENOUS AND EXOTIC;

WITH

Descriptions and Colored Illustrations,

BY

HERMAN STRECKER.

Reading, Pa., 1875.

Reading, Pa.:
Owen's Steam Book and Job Printing Office, 515 Court Street,
1875.

CATOCALA AMATRIX.

Under surface yellowish white; the secondaries suffused with scarlet on the inner half; the usual bands; also a discal lune, more or less prominent in different examples.

There are two common forms of this species; the one, Fig. 15 of Plate XI, has the primaries unicolourous; the other, Fig. 16, has a broad dash of dark brown on the primaries, extending the length of the wing, from the base to the apex interrupted only by the sub-reniform. Neither of these are sexual varieties, as plenty of both form occur in either sex.

This species is found over the same great extent of country as the preceding (Cara). The Texan examples are the largest, averaging 3½ inches in expanse.

The first Catocala that I ever saw in nature was of this species. "Ah! distinctly I remember," though twenty-five years have passed since then with their dreary cortege of woes, how Christian Sproesser, a stout German apprentice of my father, returned home one Sunday—full of beer—with a specimen of Amatrix carefully impaled on a board with a big common pin. I sat for hours feasting my eyes on the splendor of its scarlet wings, and hunting through an old German illustrated book, without a title page, which then constituted my entomological library, to find out what it was. After profound deliberation, I arrived at the erroneous conclusion that it was *C. Nupta*, and labeled it accordingly. I then pictured it in three positions, upper and under surface, and with the wings closed. I remember, also, how I manufactured lemonade to sumptuously regale my Teutonic friend and to show my appreciation of his kindness in procuring me this peerless treasure, and, finally, how the facial nerves of the said Sproesser contracted, especially around and about the region of the nose, when I proffered him the mild beverage. But that example of Amatrix, and the solid youth who captured it, have long passed out of sight; the former to dust, and the latter, whom I still hold—on account of that Catocala—in kindly remembrance, if living, I hope is well and more prosperous than the writer of these lines, or, if dead, has gone to where he belongs.

"For all have their day, the grave and the gay,
Then blow to the devil and vanish away."

NOTICES OF SOME NEW SPECIES OF WESTERN CATOCALA.

From Mr. Henry Edwards, of San Francisco, I have lately received the types of his following unpublished species, all of which I intend to figure at an early day. These, in common with most of the species from the Pacific Slope, belong to the *Nupta* group. And whilst on this subject I would here mention that *Nupta* occurs, and unfrequently, in the Atlantic States. Mr. J. Hooper captured an example near Brooklyn, N. Y.; Mr. Jos. Chos, of Holyoke, Mass., has a specimen which was taken near that place; I have myself taken three, at different times, on willows near Reading, and I have seen several other American examples in various collections.

CATOCALA MARIANA, Henry Edwards, MSS.
Expands 2½ inches. Thorax and head squamose, dark grey; primaries very dark grey or blackish, sparsely powdered with white; transverse anterior and posterior lines blackish, accompanied with white; sub-terminal white; reniform black and indistinct; inferior to the reniform a whitish space; in the single specimen received the sub-reniform is open. Secondaries scarlet; marginal band of moderate width; mesial does not extend to abdominal margin; fringe white. Under surface white, with usual bands; inner half of secondaries scarlet. Closely allied to *Briseis*, but the transverse posterior line is entirely different, having the teeth much longer and more acute, and presenting many other points of difference. In all the examples of *Briseis* that I have, the mesial band of secondaries extends to the abdominal margin, whilst in this example of *Mariana* it does not reach to it by 3-16 of an inch. Taken on Vancouver's Island.

CATOCALA HIPPOLYTA, Henry Edwards, MSS.
Expands 2½ inches. Head and body above pale grey mixed with white. Upper surface primaries very pale powdery grey, transverse anterior and posterior lines brown, not dark, the latter with teeth nearly in a line with each other, of almost equal length; reniform dull brown, indistinct; sub-reniform whitish, not connected with the transverse posterior line; sub-terminal line pale and indistinct. Secondaries, colour as in *Parta*, marginal band narrow and has two indentations on inner edge on half of wing nearest to abdomen; mesial band exceedingly narrow, widest in the middle, extends to within 2-10 of the abdominal margin, fringe white. Under surface white, inner half of secondaries red; usual bands. This is a beautiful insect. The grey of the upper surface of body and primaries is paler than in any red-winged species I have ever seen, and in certain lights is almost silvery in appearance. The extreme narrowness of the mesial on upper surface of secondaries is also most remarkable. Taken in San Mateo County, California.

CATOCALA CLEOPATRA, Henry Edwards, MSS.
Expands 2½ inches. Head and thorax dark bluish grey; abdomen brownish grey. Upper surface; primaries rather uniform dark bluish grey; transverse lines, reniform and sub-reniform, not very distinct and accompanied by yellowish brown shades; sub-terminal line paler, not prominent; secondaries deep scarlet, bands not broad; mesial censes 2-10 of an inch from the inner margin; fringe white on exterior margin, grey on abdominal. Under surface white, inner two-thirds of secondaries red; usual bands. Taken in Contra Costa County, California.

The peculiar blue-grey tinge of superior will serve to easily separate this from allied forms.

CATOCALA LUCIANA, Henry Edwards, MSS.
Expands 3 inches. Body and primaries above colour of *Amatrix*; transverse anterior and posterior lines dark brown and very heavy and prominent, the latter have the principal tooth more prolonged than in any of the others above alluded to; reniform large, double lined; sub-reniform open. Secondaries, colour of *Parta*; bands narrow; mesial same distance from inner margin as the preceding species; fringe dirty white. Under surface yellowish white; inner half of primaries red; bands all rather narrow. Habitat, Colorado.

In colour and general appearance, though not in size and detail, it strongly reminds one of the European *Pacrpera.**

CATOCALA PERDITA, Henry Edwards, MSS.
Expands 2½ inches. Very close to *Faustina*, but the primaries are of a more bluish tinge, having none of the yellowish cast of that species; the transverse lines are heavier, and there is generally more sprinkling of black atoms throughout. The apices of the secondaries of *Faustina* are partly rosy, in this they are entirely white. San Mateo County, California.

There is also another example, much mutilated, which is so close to *Irene*, Behr, that I should have considered it identical had it not been for the circumstance that Dr. Behr's species has a good sized white apical spot on tip of secondaries, whilst in the example in question the black marginal band fills out the whole apex to the fringe. Also the mesial band in this example is narrower, although that is not necessarily specific, as in some of my examples of *C. Nupta* this band is only one-half the width that it is in other examples. I have returned this example to Mr. Edwards for his further examination, considering that as he discovered it, and is, moreover, working up the Heterocera of the Far West, it is only simple justice that he should name it. Well knowing, at the same time, that it will not in his hands, at least, be degraded with such associations as Browniana, Snuggiensis or Tompkinsii. Shades of the mighty! with what names do some of the American Entomologists associate Humboldt, Cuvier and Latreille. It remindeth one not of the lamb and lion lying down together, but of an illustration I once saw, where a small poodle, with closely-shaved hind quarters, was complacently gazing on the caged monarch of the forest.

Among a large quantity of material captured in Texas by Mr. J. Boll, and lately received by me, were two examples of Colias Chrysothene ♂ ♀, which after the most careful comparison I found to agree exactly with the large suite of European specimens in my cabinet. There is the same suffusion of greyish atoms on upper surface of secondaries, and the same heavy greenish on under surface; they are the same size as the European examples, and agree with them throughout, to the utmost minutiae of shade and marking, and are as distinct from *Eurytheme*† and its var. *Keewaydin*‡ as they are from *Aurora*§ or *Pyrrothoe*.‖ Dr. Boisduval long since credited this species to N. America, but the American Lepidopterists have united in erroneously maintaining that *Eurytheme* was the insect he had in view, and that *Chrysothene* was not found here at all.

* Gioena, Cat. Ent., Torino, 1791.
† Boisduval, Ann. Soc. Ent. Fr., p. 286, (1852).
‡ Edwards, Butt. N. Am., Colias 4, (1869).
§ Esper, Schmett. 1, 2, t. 83, (1783).
‖ Hubner, Samml. Ex. Schmett., (1816-1836).

August, 1874.

EUDÆMONIA JEHOVAH.

Nov. sp. on page 93 of this work.

(PLATE XII, FIG. 1, ♂.)

MALE. Expands 4½ inches.
Head and collar grey, with a **pinkish** tinge. Thorax and abdomen **dark brownish** grey.
Upper surface; primaries grey, paler and with a somewhat pinkish **tinge towards the costa**; a dark brown **basal** patch; a triangular, transparent discal mark which is prolonged **in a fine line upwards** to the costal nervure, and is surrounded with a dark brown cloud; beyond this, two irregular transverse lines cross the wing from inner to costal margin, the innermost one broken near the middle, from which inwards it is double or accompanied by a narrow shade; these two lines are close to each other from the inner margin half way, whence they begin to diverge and become widely separated as they near the costa; midway, and joining on the outer edge of the outermost of these two lines, is a conspicuous white spot; further up, joining, or rather emanating from the same line, is a small white spot; and a little below the large spot, and joining the same line, is a small black, triangular spot; the middle third of the outer area is clouded with dark brown. Secondaries with long narrow tails; **colour somewhat darker than** primaries; several narrow, submarginal lines, and a brownish, not very wide, **transverse band**; a small patch of same colour on abdominal margin, and **an obscure transparent discal spot.**

Under surface greyish brown, darkest towards **exterior margin**, mottled with indistinct dots or points of a darker shade; **a transverse brown line crosses all wings** from costa to inner margin; discal spots transparent; the large white spot of upper surface repeated.

I know of but one example of this grand insect, the original of the accompanying figure, which was sent to Prof. Meyer of New York, from Brazil by his son, and to the courtesy of the former I am indebted for the opportunity of publishing the species. In what precise locality it was captured is not known, as it was, I believe, purchased from a dealer in Rio Janiero.

It is a hundred years since Cramer, in his voluminous work, figured the first species of Eudæmonia, the wonderful *Semiramis*; and **the three species** now composing that genus are in appearance the most remarkable Lepidopterous insects yet known.

In the "Verzeichniss"* Hübner has placed two very dissimilar insects in his genus Eudæmonia; the *Semiramis*, Cramer, from Surinam, and the West African *Argus* of Stoll; the latter, Duncan placed in his genus Eustero, (in the Nat. Library, Vol. VII, p. 125), and it certainly ought not to be retained in the same genus with *Semiramis*, as it comes much nearer the true Saturnidæ, especially those composing the genus Tropœa, Hüb. (*Actias*, Leach), whilst *Semiramis*, *Derceto* and *Jehovah* are near to Rhescyntis, and still nearer to Dysdæmonia, to which latter I think they are very closely allied.

The present species, though not so far removed from *Semiramis* as is Maassen's *Derceto*, still differs from it very considerably in most particulars; in Cramer's species the wings are narrower, especially the secondaries, the outer margins are scarcely dentated, the tails are nearly a third longer in proportion, the ground colours are various shades of buff and reddish, the discal mark is not triangular, and the white spot emanating from the outer edge of the transverse line is triangular, instead of having the peculiar six-sided shape of that in the **present** species. Maassen's species, in the plain edges of outer margins, is closer to *Semiramis*, but in the **much** greater width of the tails, which give it a heavy appearance, as well as in ornamentation, it is equally aberrant from both Cramer's and the species here figured. It was scarcely necessary, perhaps, to go into these comparative details; but, unfortunately, so little attention is paid to Exotic Lepidoptera in this country, and in consequence **so** little is known, that I may not, perhaps, be entirely amiss.

The above remarks are entirely in connection with the males of the species **alluded to**, as of *Derceto* and *Jehovah* the females are, I believe, entirely unknown, and that of *Semiramis*, according to Maassen's figure, is widely different in **appearance** from the male, having much broader wings, shorter, broader tails, and differing considerably in **the ornamentation.** I here cite Cramer's and Maassen's species more fully than on page 93:

SEMIRAMIS, Cramer, (*Phalæna Bombyx Attacus*) Papillons Exotiques, Vol. I, T. XIII, A, ♂ (1775). *Gmelin*, Ed. Systema Naturæ, I, 5, 2404, 470 (1788).
Bombyx Semiramis, Fabricius, Genera Insectorum, 277 (1777); Species Insect. II, 170, 13 (1781); Mantissa Insect. II, 109, 15 (1787); Ent. Syst. III, 1, 413, 20. *Olivier* Enc. Méth. Ins., V, 28, 18, pl. 69, f. 8.

* Verzeichniss bekannter Schmetterlinge, p. 151 (1816).

EUDÆMONIA JEHOVAH.

Eudæmonia Semiramis, *Hübner*, Verzeichnisz Bekannter Schmetterlinge, 151, 1585 (1816). *Walker*, Cat. B. M., VI, p. 1265 (1855).
Copiopteryx Semiramis, *Duncan*, Naturalists' Library, VII, 125 (1841).
Eudæmonia Phœnix, *Deyrolle*, in Maassen's Beitrage zur Schmetterlingskunde, Heft I, fig. 5, ♀ figs. 6, 7, ♂ (1869). Ann. de la. Soc. Ent. Belg., T. XII, 1 (1869).
Aricia Phœnix, *Felder*, Novara, T. 92, 1 ♀.
DERCETO, Maassen, Beitrage Schmett., figs. 13, 14, ♂ (1872).

Since the publication of this species there have been objections urged to the specific name by which I have designated it, and in some few instances from sources for which I have every consideration, but in the major part from those whose good or ill opinion weighs alike with me: to the former only are my present words addressed.

One friend, in objecting, writes thus: "The name brings up to serious and contemplative minds everything that is sacred;" if such be the case, then indeed am I happy in my selection, for methinks anything that would lead us to think of the Creator, and would take our thoughts away from the contemplation of the minnes with which he has peopled the earth, cannot but be well; and what better than to reflect on sacred things,—on the evidences of the majesty and power of the Supreme Being? Even as I write, thoughts arise of Nineveh in her magnificence, of her utter destruction, of the writing by an unearthly hand on the wall in the Babylonic palace; and visions of old Egypt's temples rise up—of the first-born smitten, in that awful night by the death-Angel's hand, in every home throughout the land—of the children of Israel guided by the fiery pillar—of the destruction of Pharoah's army—of the Pyramids of Cheops, of Karnak, and of their desolation—of the researches of Champollion and Gliddon, of the quiet resting-place of the latter in his vault in Laurel Hill Cemetery, where he has most company in Drs. Morton, Wilson and Kane.

Why may not the Hebrew word, used to designate the Supreme Being, be bestowed on a species as well as the Hindostanee, which latter has several times been applied to species" and once to a genus?† Is it because our conventional ears are more familiar with the former than with the latter? Had I used the term Par-ob-naw'yah, it is scarcely likely that much objection would have been raised, and yet this is the word used by Dr. Judson to express "Jehovah" in his translation of the sacred writings into the Burman language.

It certainly cannot be more offensive to apply sacred names to animals than to persons, and in Spanish countries almost every tenth person is surnamed Jesus, pronounced by them, Hessoe; this may sound shockingly irreverent to the fastidious ears of Americans, but I doubt much if the Hidalgoes who bear the name of the second person of the Godhead would feel at all elated to know that their spear-ors had acted with irreverence, and that they themselves, on account of their names, were living offences against decency and good men,—and they, too, the most devout and punctilious people on the face of the earth.

That there should be any reasonable objection to the bestowal of the Creator's name on one of the most interesting of His works, I cannot possibly surmise, and the only cause of offence, in scientific nomenclature, is given where the terms are orthographically or etymologically incorrect, or where nature's noble works are degraded to the vile purpose of chariotting into notice the names of obscure individuals whose only merit in the case was in giving the describer a dinner or landing him money; or what is worse still, to attach to scientific objects the names of political demagogues; this is, without doubt, the vilest of all, especially in our own country where political eminence is now solely attained by the most corrupt means, and success ensured only by the sacrifice of every principle of honour and honesty.

There are names enough still left in the grand annals of past history and science, without having recourse to those of persons who, however estimable, have no claims for scientific honours.

There have been hints that, notwithstanding the claims of priority, the specific name I have used will not be accepted; should such be the case, and the name employed by me ignored, it matters nothing, as the species is now known and my work as regards it completed in the figuring and description, as far, at least, as at present possible to me, and whether it be known hereafter by the name I have used, or by another, can be of no possible moment.

As genera are each year becoming more and more divided and sub-divided, some aspirant for scientific fame may make of each species, comprised at present in Eudæmonia, a new genus, as there is, without doubt, the difference of a brush-hair in the details of the several species; of course, the author of the new arrangement would place his own name to the combination, and in that case the present species would be the only instance in which the name of the compiler would be secondary to that of the object, though, of course, no one could doubt that even then the great synonymist would make a mental reservation in favour of his own superiority.

In a recent interesting paper on Ent. nomenclature by Dr. J. Leconte, in that influential organ, the *Canadian Entomologist*, the author concludes with the following: "Unfortunately, under the influence of personal peculiarities, the excitement of political struggles," etc., "names are sometimes proposed which are in the highest degree offensive."‡ After the word "offensive" is an asterisk (*) which directs us to a foot-note by Mr. Wm. Saunders, the able editor of the periodical, who doubtless fearing that the reader might suppose the author alluded to such names as Houjopteryx Saundersii, Lemonthinia Saundersella, and others of the same nature, gives as the benefit of his conjecture, that "the author here evidently alludes to such names as Pleocoma Staffi, Eudæmonia Jehovah, and others of the same nature."

Professor Huxley somewhere says: "Happily, the reputation and real success of a votary of the physical sciences are now wholly independent of the periodicals which are pleased to call themselves "influential organs of public opinion;" the only opinion he need care about—if he care for any, *and he is all the wiser and happier if he care for none*—is that of about a dozen men; two or three in these islands, as many in America, and half a dozen on the Continent. If these think well of his work, his reputation is secure from all the attacks of all the "able editors" of all the "influential organs" put together."

SAMIA EURYALUS. BOISDUVAL.

Ann. Soc. Ent. Fr. III, 2me Ser. xxxii (1855).
Morris, Cat. Lep. N. Am., p. 21 (1860).
Walker, C. B. M., part XXXII, p. 525 (1865).
Samia Euryalus, *Packard*, Proceedings Ent. Soc. Phil., Vol. III, p. 380 (1864).
Saturnia Cynothia, *Behr*, Proc. California Academy of Sciences, Vol. III, part IV (previous to September, 1867).
Platysamia Californica, *Grote*.

* *Ilerda Brahma*, Moore, Cat. Lep., E. I. C., I, p. 29, T. 1 (1857).
Papilio Brama, Guerin, Rev. Zool., p. 48, T. 1 (1840).
† Brahmæa, Walker, C. B. M., VI, p. 1315.
‡ I have only quoted portions of the two concluding paragraphs of the article.

SAMIA EURYALUS.

(PLATE XII, FIG. 2, ♂.)

MALE AND FEMALE. Expand 4 to 5 inches.
Head red; thorax red, with white collar; abdomen annulated with red and white.
Upper surface dull carmine red, except at the outer margins, which are clay-coloured (ochrey); ornamentation as in allied species, but the discal lune of primaries is in some instances prolonged in a line until it almost touches the sub-basal band, and that of secondaries until it joins the transverse white line. In this species the ground colour beyond the transverse white line is of the same even hue as that interior to it, differing in this respect from *Cecropia*, *Columbia* and *Gloveri*, in which the ground color beyond the transverse line, especially on primaries, is composed of atoms of black or grey, and red, more or less segregated.
Under surface much as above, but the ground colour more brownish and dotted in part with white atoms.
The larva has been described by Mr. Henry Edwards,* in Proceedings Cal. Acad. Sciences, who says:

"Length, 2.30 inches. Pale apple-green, of a very vivid tint throughout, with a slight whitish bloom over the whole surface. Head, with some purplish-black streaks in front and at the sides. Mouth parts, pale-green, pitchy internally. Second segment with four minute black dots, edged with white anteriorly, and two very small white mammiform tubercles on the sides. Third, fourth, and fifth segments, with long raised protuberances, pale yellow, with a black, swollen band in the middle, and each surmounted by six blackish spines. The third segment has also four lateral raised white spots. The fourth and fifth segments have two mammiform white spots, the lateral ones on these segments becoming merely black points. On the sixth segment is a faint white raised spot, in the same position as the white swollen tubercles on the preceding segments. Seventh and eighth, with only black points laterally. Ninth, tenth, and eleventh, without any trace of spots. Twelfth segment bears in the middle a long, raised protuberance, yellow, banded with black, exactly similar to those of three, four, and five. On this segment there are also two lateral points, white, tipped with black. Anal segment with four black dots arranged in a square, and two white and black points as in twelve. Stigmata white, edged anteriorly with black. Below the stigmata, and parallel with them, is a row of very minute black dots, edged with greenish-white. Feet, yellowish-green, with the tips purplish-black. Abdominal legs, greenish-yellow, with the edges purplish-black. Viewed from behind, the anal segment is yellowish-green.

"Food plants, *Ceanothus thyrsiflorus*, Esch.; *Frangula Californica*, Gray; *Rhamnus croceus*, Nutt; *Alnus viridis*, D. C.

"When about to undergo its change, the caterpillar attaches itself usually to the under side of a twig, and spins a rather coarse and very compact outer case, with which no leaves or other extraneous substances are incorporated, and within this a reddish-brown cocoon, the filaments of which are strong, rather coarse, but glossy. The cocoon and its outer case are oval, produced into a cone at the end, by which the insect escapes.

"*Chrysalis.* Pitchy, almost black, very short, rounded in front, and much swollen about the abdominal region. Segments rough, and transversely wrinkled.

"Length, 1.15 inch.

"The caterpillar changes to a chrysalis in September, and the imago appears in the following May or June."

Without giving the matter sufficient attention, I adopted Grote's generic term, Platysamia, in connection with the species first described in this work, but have discarded it in the species here described. Kirby says, in a paper read before the Dublin Society, March, 1872, "Grote has changed the generic name, Samia, (used by Walker) without sufficient reason, applying it to *Attacus Cynthia*;" which latter, if breadth of wing be a foundation for generic distinction, would be in the highest sense a Darwinian species, as in its fatherland, China, it has, in a slightly modified degree, much the same form as the great *Atlas*; narrow, falcate primaries, and secondaries prolonged towards the anal angle, and would be placed, consequently, in the genus Attacus, but after a several years' acclimatisation in the United States, a curious change takes place; the fore wings become less falcate, (being now not more so than in *Cecropia*, *Angulifera*, etc.), the secondaries much less elongated, and all wings increased much in breadth; the discal lines also become shorter and broader, more like those of *Cecropia*, and we can now place the insect in *Platysamia*.

The Chinese examples in my cabinet average 5 inches in expanse, those raised from eggs brought from China the same; the first brood, raised from ova deposited by the latter, were all small, averaging only 4 inches, but preserved the typical Asiatic form; afterwards I let all fly as they emerged, and those that issued from cocoons collected in the woods near Reading, four or five years later, (doubtless the descendants of those that I let loose), averaged 5½ inches and were as broad-winged and un-Asiatic in appearance as Cecropia and allies.

In 1856 or 1857, I saw, in the collection of Mr. J. P. Wild of Baltimore, an example of this species labeled "Euryale, California," which he had received from Mr. Becker in Europe. Dr. Boisduval, it appears, however, only published the species by that name, omitting the formula of a description, hence, though well known to most scientists for many years as Euryale or Euryalus, that name had to give way to the later one of *Cecrosthi*, Behr, or *Californica*, Grote; which of these has precedence I cannot say, as I do not know where or when Grote described it, though he claims priority over the other authors.

SAMIA COLUMBIA. S. I. SMITH.

Proceedings Boston Society Natural History, Vol. IX, p. 343 (1863).
Packard, Proc. Ent. Soc. Phila., Vol. III, p. 380 (1864).
Walker, Cat. B. M., Supplement Vol. V, p. 1943 (1866).
G. J. Bowles, Canadian Entomologist, Vol. III, p. 201 (1871).
Hagen, Bulletin Buffalo Soc. Nat. Sc., Vol. II, p. 201 (1875).

(PLATE XII, FIG. 3, ♂.)

The accurate original description of both sexes of this species by Prof. Smith, which was also republished in Mr. Bowles' paper in *Canadian Entomologist*, as well as the illustrations on annexed plate, make it unnecessary for me to go through the same routine.

* In order to avoid misapprehension, it might be well to caution students against confounding the work of two authors of the same name, Henry Edwards, of California, and W. H. Edwards, of Kanawha, West Virginia.

SAMIA COLUMBIA.

The claims of this fine moth to be a distinct species have been considered very doubtful, and its history is, in fact, yet to be completed, as we have thus far no exact description or any figure of the larva. Mr. S. I. Smith, who first discovered it, obtained his examples, 2 ♂, 1 ♀, from cocoons, one of which was spun upon a maple twig and the others on *Rhodora Canadensis*, among which latter the maple was growing, also many other cocoons from which the imago failed to emerge, owing to the presence of the parasites, *Cryptus Smithi*, Pack., and *C. Nasica*, Pack., as shown by the careful examinations of Dr. H. Hagen, of the Cambridge Museum, where all Mr. Smith's types of both pupa and mothe are preserved.

The first notice we have of the larva is from Mr. G. J. Bowles. He says: "In August, 1864, I captured a full-grown larva of this moth (*Columbia*) crawling along a fence in search of some place to make its cocoon. It closely resembled a *Cecropia* larva in size and appearance; thinking it, therefore, to be a larva of that species, I did not take notes at the time, though, on a close examination, I could not quite reconcile the colour and arrangement of the tubercles with the description of *Cecropia* given by Morris. The principal difference (as far as I can remember), was in the number of red warts with which the larva was ornamented. *S. Columbia* possessing more than the other species;" and further, "the larva above mentioned duly spun its cocoon, which was at first of a whitish colour, but in a few days gradually turned to dark brown; the moth died in the chrysalis state, owing, perhaps, to the presence of parasites. Two years afterwards I found another cocoon attached to a twig of thorn (*Cratægus*), but it was full of large parasites, all dead in the pupa."

He further states that he found, in fall of 1867, yet another cocoon spun on a gate-post, which in the following May produced a ♀ *Columbia*, of which he gives a lithographic figure, not differing materially from the figure of the ♂ on the present Plate XII.

The above contains all that at present is known of the larva.

The ♂ example, the original of fig. 3, was sent me from Montreal, Canada, by Messrs. C. W. & G. B. Pearson, most ardent students of Lepidoptera, to whom I am indebted for many favours; these gentlemen wrote me, October 13, 1874, on the occasion of their sending the example, "concerning *Columbia* we cannot say anything further than that we found the cocoon on a maple tree in the vast suburbs of the city, which produced the moth on the 13th of May last; we also send the cocoon, which you will easily distinguish from *Cecropia* by its smaller size and different colour."

The above cited are all the examples that I knew of; i. e. the three types in the Cambridge Museum, found on maple and *Rhodora Canadensis* in Norway, Maine, the one found by Mr. Bowles near Quebec, Canada, and, lastly, the example found on maple near Montreal, Canada, by the Messrs. Pearson, and now in my possession. I have examined those in the Cambridge Museum; the ♀ does not differ in appearance from the male I received from Montreal, except that the discal spots of primaries are not so plainly defined; the scales are both smaller, being not over 4 inches in expanse.

The cocoons, which are attached longitudinally to the twig, are double and not much more than half the size of *Cecropia*; the outer surface is somewhat uneven, of a dark greyish-brown, with little shining spots caused by the crowding together, here and there, of the silk woven around it. The inner cocoon is paler in colour and woven closely to the outer. My cocoon is not as dark as some of those in the Cambridge Museum, though still much darker than any *Cecropia* I have ever seen.

Dr. Hagen in his valuable paper, cited at head of this article, says: " I confess frankly that only the peculiar features of the cocoons support the opinion that *Columbia* is a different species." The differences in the imago are, it is true, slight; when taken, however, in detail, they are the following: The average smaller size of *Columbia*; the almost entire absence of red on the wings, which gives the whole insect a sooty appearance; the white transverse lines much further removed inward from the exterior margin of both wings, making the space interior to the transverse lines much less in comparison than in *Cecropia* or *Ceanothi*, but assimilating in this respect to *Gloveri*. As regards the smallness or almost total obsolescence of the discal lunes, I have seen the same thing often in *Cecropia*, and in my own material of that species are four large males in which the discal lunes of primaries are as small as in *Columbia*, and so dark in colour that only by close inspection are they to be defined from the ground colour of the wing. I have also two examples of *Cecropia* which have the abdomen annulated with blackish-brown and red instead of red and white, but the lateral ornamentation is in some style as in the normal red form and in *Ceanothi*, whereas in *Columbia* it is entirely different, as will be seen by comparing the figures 2 and 3 on Plate XII; but, notwithstanding the apparent similarity, it does not take more than a glance to discern that *Columbia* is different; it can easily be picked out amidst a hundred *Cecropias* through the prevalence of the sooty hue and the absence of red before alluded to, and it looks exactly as we might suppose a hybrid of *Cecropia* and *Promethea* would look—a possibility suggested by Dr. Hagen in his paper, where he cites various instances of hybrids; and, in connection with which I would state that in my cabinet are examples of hybrids from *Anthera Jamamai* and *Pernyi*, *Sericataria Cecthera* and *Pegali*, *Ostacala Despecta* and *Emeros*, *Chiva Evata* and *Ednea*, and others; but one fact militates strongly against the hypothesis in this case, which is that *Promethea* does not occur in Canada, or at least not in those parts where *Columbia* is found, though *Cecropia* does, I believe, abundantly. Were *Promethea* and *Cecropia* both found in the same locality with *Columbia*, I should certainly believe that the supposition of its being a hybrid of these would be the correct one, as the whole appearance of both cocoons and imago would seem to substantiate such a belief.

In my assertion that *Promethea* does not occur in Canada I may, perhaps, be incorrect; my principal connections there have been in Montreal, in the neighborhood of which, my valued correspondents inform me, *Cecropia* and *Polyphemus* are found, but neither *Luna*, *Angulifera* or *Promethea*, and it is only by the non-occurrence of the latter that I am led to doubt that *Columbia* is the result of bastardy; but we must wait until further observation and larger material will solve the riddle.

SATURNIA GALBINA. Clemens.

Proceedings Acad. Nat. Sc. Phila., p. 156 (1860).
Morris, Synopsis Lep. N. America, p. 222 (1862).
Packard, Proc. Ent. Phila., Vol. III, p. 383 (1864).
Walker, Cat. B. M. (Supplement) Vol. XXXII, p. 530 (1865).

(PLATE XII, FIG. 4 ♂, 5 ♀.)

MALE. Expands 2¾ inches.
Head and body brown.
Upper surface white; primaries, a sub-basal band formed by two brown parallel elbowed lines; a discal ocellus consisting of a black spot crossed in the middle by a vitreous line, and surrounded with a narrow yellow circle, to which is added on the inner side a fine blue crescent; directly beyond this a narrow brown band crosses the wing from inner margin to costa; midway between this and the exterior margin is another much broader brown band, which is traversed by an indistinct paler line; a black sub-apical spot, connected at its lower side with the exterior margin by a crimson line.

SATURNIA GALBINA.

Secondaries, discal ocellus same as on primaries, but a little smaller; sub-marginal brown band narrowest in the middle.

Under surface much as above, but the sub-marginal bands have a white, undulate line running through them from inner margin to costa.

The female expands 2¾ inches, and was evidently the only sex known to Dr. Clemens, as his description, which I here append, applies to it and not the male.

"Antennæ luteous. Body and head rather dark brown. Fore wings yellowish-brown, with a rather faint, whitish angulated band at the base. On the discal nervure is a round, black ocellus having a central subvitreous streak containing a yellow circle, and towards the base of the wing a slender blue crescent. A whitish band crosses the middle of the nervules, with a faint wavy one between it and the hind margin. In the apical interspace is a black spot, with a crimson streak to the tip of the wing. The marginal portion of the wing is whitish and is tinged in the terminal edge with pale yellowish-brown. Hind wings similar in color and ornamentation to the fore wings, the ocelli being somewhat smaller. On the under surface, which is similar in hue to the upper, the faint wavy bands of the fore and hind wings are very distinct.

"TEXAS. Collection of Capt. Pope."

Though in the above, which is all the author says, there is nothing to indicate which sex was described, nor is the important item of size mentioned, still I have no doubt that this is the species alluded to.

I received six examples, five ♂, one ♀, from south-western Texas, on the border of the Rio Grande, but most of them before coming into my possession had suffered to such an extent, from the ravages of mites or other insect depredators, as to be utterly worthless. In the Museum of the Department of Agriculture, at Washington, are also examples which, I believe, are from the same locality.

This species is exceedingly rare, the examples cited being, so far as I am aware, the only ones extant in any collection. As far as my knowledge extends, this is the only true *Saturnia*, Schrank, known to occur in the western hemisphere; and, though of a somewhat slighter form, closely resembles the common European species, especially *Spini*, Schiff.

HEPIALUS THULE, Nov Sp.

(PLATE XII, FIG. 6 ♀.)

FEMALE. Expands 3 inches.

Upper surface yellowish-white. Primaries, costa from base to two-thirds its length reddish-brown, within which colour, about equidistant from each other, are three ⊓ shaped yellowish-white marks; the brown colour of costa extends into the discal space, at its base there enclosing two small silver spots, the one nearest the costa being the largest; directly beyond the outer extremity of the discal space, the brownish colour also extends the same distance, likewise enclosing two closely-connected silver spots; two sub-marginal bands composed of rather small, reddish-brown spots, the innermost extending from near inner angle to costa near the apex, the other not reaching to the costa, but connected with the first at the last sub-costal nervule; half way between the innermost sub-marginal band and the brownish costal space is a brown line extending from costa inwards as far as where the sub-marginal bands connect; a marginal row of small brown spots; on inner margin, near the termination of the sub-median nervure, is a small brown spot.

Secondaries tinged with reddish on the costa, and very slightly so on cilliæ of exterior and abdominal margins; otherwise immaculate.

Under surface same colour as above, with the markings faintly repeated.

From a single female sent me from Montreal, Canada, by my excellent entomological friend, Mr. F. B. Caulfield, who stated that it was captured in a park in that city.

It is the only example of this grand insect I have ever seen or heard of, and is so far probably unique.

NOTES ON VARIOUS SPECIES AND VARIETIES.

LYCÆNA REGIA, Boisduval, Lep. Cal., p. 46 (1869), is *Lycæna Sonorensis*, Felder, Reise Nov., Lep. II, p. 281, T. 35, f. 3, 4 (1865).

LYCÆNA RHÆA, Boisduval, Lep. Cal., p. 51 (1869), is a synonym of *Lycæna Catalina*, Reakirt, Proc. Acad. Nat. Sc. Phila., p. 244 (1866).

CATOCALA LEVETTEI, Grote, in advance sheets of Trans. Am. Ent. Soc. (under date of September, 1874), is identical with *C. Judith*, p. 96, T. XI, f. 5, in this work (printed August, 1874).

CATOCALA ANNA, Grote, l. c., is *Catocala Amestris*, l. c.

CATOCALA ADOPTIVA, Grote, l. c., is *Catocala Delilah*, l. c.

GORGOPIS QUDRIGUTTATUS, Grote, Proc. Ent. Soc. Phila., Vol. III, p. 73, T. I, f. 6 (1864), appears to be, and I am of the opinion undoubtedly is, the same as Harris' species, *Argentro-maculatus*, Cat. Ins. Mass., p. 72 (1835), to which species, Grote says in his description of the former, it is allied, and he further adds, "the disposition of the median bands on the anterior wings is somewhat different, and they are not so largely tinged with ochraceous, the two white spots are much smaller and the apex apparently not so falcate, while the coloration of abdomen, metathorax and posterior wings, readily distinguishes the present from Harris' species."

106   NOTES ON VARIOUS SPECIES AND VARIETIES.

Among my examples of *Argenteo-maculatus*, six in number, is one from Lake Superior which agrees almost exactly with Grote's figure, excepting that the white spot nearest the inner margin is much larger, in fact, larger than in any other example I have ever seen, being fully one-tenth of an inch at its greatest diameter; the posterior wings also are darker than Grote's description would lead us to infer those in his type of Quadriguttatus were. Another example, captured in Hunter Co., N. Y., is entirely devoid of all the silver spots; there is in this example considerable difference in the arrangement of the bands, etc., of primaries from the one from Lake Superior. Several others from Hunter County agree in the general markings with the one just alluded to, but have the usual silver spots the same size as and some larger than in Grote's figure; the posterior wings of the different examples vary in depth of colour from pale salmon or fawn to smoky-grey, and the sub-apical and apical marks are either very distinct, half obsolete or entirely wanting, and I much fear that, if Grote's assumed species be distinct from Harris', I possess at least three more new and undescribed allied species!

ARCTIA ANNA, Grote, Proc. Ent. Soc. Phila., II, p. 335, T. 8 (1863), is, without doubt, a melanotic variety of *Arctia Persephone*, l.c., p. 60; varieties with black wings are of not uncommon occurrence among the Arctians; I have seen them of *Caja*, *Figurata*, *Lena*, *Virginalis* and *Plantaginis*.

ARGYNNIS LETO, Behr, Proc. Cal. Acad. Sc. (1862), I hold to be a western form of *A. Cybele*, Fabr., as also may be *A. Nokomis*, W. H. Edwards, but this latter I have not yet had sufficient opportunity to examine in nature, to speak of with any certainty. An analogous case is presented in *Argynnis Alexandra*, Men., which is an aberrant Asiatic form of *A. Aglaia*, Linn.

CATOCALA CONCUMBENS, Walker. I have previously remarked the close affinity of this species with the Russian *C. Pacta*, L. (on p. 40); this latter, except that it is smaller and has the abdomen rosy, resembles very closely our species. My friend, Mr. Paul Knetzing, sent me this winter, from Montreal, Canada, an example of *Concumbens* which, to my unbounded astonishment, has the abdomen coloured, precisely like the hind wings, thus making its resemblance to *Pacta* almost perfect. I received also, about the same time, from Canada, another example with the abdomen rosy, but mixed with grey; and my friends, the Messrs. Pearson, write me from Montreal that they have likewise an example "with the body red, just like the *Pacta* which you sent us." The red-bodied *Concumbens* which I received differ in no other point whatever from the ordinary form, which is seldom subject to any variation.

CATOCALA SIMULATILIS, Grote, is, as its author evidently mistrusted, *Obscura* ♀. I here insert his description, published in advance sheets of Trans. Am. Ent. Soc.
"This species is intimately related with *C. obscura*, somewhat as *C. residua* with *C. insolabilis*. *C. residua* has blackish fringes, the general color of the primaries is dusky ashen, without the limner deepening in color above internal margin of *C. insolabilis*, while the whitish gray subterminal shade contrasts with the dusky tone of the wing. This species has also a black oblique subapical shade beyond the subterminal line, more or less distinctly following the teeth of the line and apparent sometimes within the line, following the two prominent teeth of the t. p. line. This black shading is wanting in *C. obscura* and *C. simulatilis*, which agree in the general smoky ashen primaries and the white fringes to the hind wings, but may be separated by the course of the t. p. line. This, in *C. simulatilis*, is much as in *C. residua* and the other species, with two very prominent teeth and wide open subreniform, whereas, in *C. obscura* the line is more perpendicular and presents a series of fine teeth. The resemblance is otherwise so great between the two that other comparison or description seems unnecessary. Since I only know males of *C. obscura* and females of *C. simulatilis*, I thought that my specimens of the latter might belong, as the opposite sex, to *C. obscura*. Such a sexual difference would be quite new and unusual, and I cannot now be blamed for not adopting such a determination."

And under the circumstances I wonder the author described his species at all; he worked on entirely too scant material, and lays by far too much stress on the teeth-structure of his "t. p. lines." At present are before me thirty examples of *C. Obscura*, ♂ ♀, including the original type; one of these, a male, has the sub-reniform almost closed, all the others have it open, some widely so, others more moderately; the "t. p. line" is, in 14 examples, 5 ♂ 9 ♀, with two very prominent teeth, in 16 examples, ♂♀, with only one prominent tooth, like in the figure on Pl. III; so this ceases, at any rate, to be "a sexual difference." As to the perpendicularity of the lines, in different examples they appear to be like in a battalion of country militia—each one varying in position; in some instances the t. p. line and t. s. line are almost confluent at the inner margin, and in one instance quite so. As to general colour of primaries, it varies much, in some instances being almost black, especially on the area interior to the sub-terminal line.

C. Residua, Grote, is only a common form of *C. Insolabilis*, Gueneé, which is a species that varies much in the depth of ground colour of primaries, and in the dark shading of their inner margin.

In describing Catocalæ, too much weight has been attached by Mr. Grote to whether the sub-reniforms be open or closed, also to the breadth of the secondaries, and to whether the mesial extends to the abdominal margin or not; the following will show how utterly valueless would be diagnoses founded on these points:

In twelve examples of *C. Subnata* I find two males and three females with the sub-reniform open, and four males and three females with it completely closed, and in two it is entirely isolated from the transverse posterior line; yet Grote made the open sub-reniform one of the specific characteristics of this species. Out of six examples of *C. Piatrix*, two have the subreniform open and four closed. The same also in *C. Crataegi*. In four *C. Unijuga*, one has the sub-reniform open, a second has it closed but connected with the transverse posterior line, the other two have it entirely isolated. I have also examples of *C. Fraxini*, *C. Viduata*, *C. Desperata* and *C. Agrippina*, in some of which the subreniform is closed, and in others open. Twenty-two examples of *C. Polygama* display great variation in the width of the black bands, in some instances they being twice as wide as in others. In *C. Parta*, *C. Nupta* and *C. Electa*, the same wide difference in width is displayed, and in these species in some examples the mesial band extends to the abdominal margin; in others it does not reach it within almost one-fourth of an inch.

MEAGRE DESCRIPTIONS OF SOME NEW SPECIES; TO BE FOLLOWED IN A SUBSEQUENT PART BY WHAT IS INFINITELY BETTER—GOOD REPRESENTATIONS.

SPHINX PLOTA, Nov. Sp.
Male. Expands 3¼ inches.
Head and thorax rather light brownish-grey. Tegulæ edged outwardly with white, inwardly by a dark-brown line, also a brown line in middle, parallel to that of edge. Abdomen same light brownish-grey as thorax, with a narrow black dorsal line, seven short black bands on each side, the space between which are dirty white; beneath: pale brownish-grey, almost white.

Primaries same colour as thorax, with a rather short longitudinal line in each cell; an apical line; a submarginal black line accompanied inwardly by another, broader but not so dark in colour, a black spot at base of wing; all these lines, etc., are accompanied more or less with whitish streaks or patches. Secondaries greyish, with dark-brown submarginal and mesial bands.

Under surface greyish-brown.
I might have made shorter work with the above by simply saying that this species was between *Chersis* and *Kalmiæ*; it is neither blue as the one nor reddish as the other, and it has more markings on the primaries than either, giving them an appearance somewhat like *Ellena Harrisii*, but on a larger scale.

MEAGRE DESCRIPTIONS OF SOME NEW SPECIES, &c.   107

Several males, taken by Mr. P. Kneezing near Montreal, Canada, were, through his goodness, added to my collection.

It is evident that we have as yet but a limited acquaintance with the fauna of Canada and British Columbia, as is proven by the many new and undescribed species lately received therefrom; conspicuous in the East are the above described *Sphinx* and *Hepialus Thule*, and in the West, *Euleucophaea tri-color*, Packard, and Henry Edwards' *Saturniidus Occidentalis*, a monstrous form of *Modestus*, expanding nearly 6 inches, and of a very pale yellowish fawn-colour, much like in the European *S. Quercus*.

MACROGLOSSA ÆTHRA, Nov. Sp. or Var.
Female. Expand—1¼ inches.
Above, head and body olivaceous of a paler yellow shade towards the sides; caudal brush yellow and black, beneath same as *Diffinis*. Primaries, margin much broader than in *Diffinis* and serrated on inner edge; a large carmine apical spot; base and interior margin reddish with olivaceous hairs on the former. Inferiors, narrow brown exterior margin; abdominal margin carmine; beneath, costa of both wings red. One example from Montreal, Canada; from Mr. P. Kneezing.

If this be not a new species, it is certainly a most remarkable aberrant form of *Diffinis*; the total absence of the broad black transverse band of upper side of abdomen is a most noticeable feature, as well as the entirely red costa of all wings beneath.

HEPIALUS DESOLATUS, Nov. Sp.
Expands 2 inches. Brown, same shade as *H. Sylvinus*, L., to which the whole insect bears a tolerable resemblance, but the lines, etc., are better defined in the European species, than which ours is much more obscure; on primaries the principal markings are a narrow sub-basal band accompanied with a darker shade, midway between this and the outer margin, running from apex to interior margin, is another paler line with its darker shade, from which at the inner margin emanates another short line which runs somewhat diagonally towards the sub-basal line. Secondaries brown, with a few barely distinguishable paler spots on costa. One example taken at Owen's Lake, Nevada, by one of the naturalists of Lieut. Wheeler's Expedition in 1871.

CATOCALA JOCASTA, Nov. Sp.
Female. 1¼ inches.
Head and thorax grey, abdomen yellowish-grey; beneath, dirty white. Primaries grey, on costa signs of a transverse anterior line; reniform indicated by a few darker scales; transverse posterior is not a line, but a broad shade; beyond this, except a small intervening space, the wing is darker; fringes same colour.

Secondaries yellow, with a broad black marginal band which is deeply indented on inner edge, towards the abdominal angle, where it becomes much narrower; fringes white. Under surface pale yellow, primaries with a broad marginal band, narrowest at inner angle; a rather narrow median band which does not reach to inner margin; no traces of a sub-basal band whatever; fringes grey. Secondaries, marginal band as above, no medial; a few scattered scales on costa, and a few more where the discal lune ought to be; fringes white.

A very curious and interesting species, evidently allied, notwithstanding the absence of the median band of secondaries, to *Whitneyi*, Dodge, and *Myrrha*, figured on Plate XI; there are no distinct markings on the primaries, which are only clouded and have a powdery appearance. One example received from Dr. W. B. Carpenter, Kansas.

PSEUDOHAZIS NUTTALLI, Nov. Sp.
Male. Expands three inches. Head and body ochrey yellow. Abdomen with very faint indications of a chain or row of confluent dark rings, reaching from thorax to anal segment on each side. Primaries pale flesh-colour. Secondaries same yellow as body; all wings with a submarginal black band, narrower than in *Eglanterina* or *Pica*; veins, from outer margin to this band, accompanied with black, broad at margin, narrowing to a point as they near the transverse band; a large black discal spot on all wings.

Female larger than male. Primaries whitish-yellow, very pale. Secondaries ochrey yellow; the same black ornamentation as in male, but not near so heavy.

One ♂ and one ♀ taken by Mr. Nuttall in 1836, at the Rocky Mountains, head of Snake River, and now in possession of Mr. Titian R. Peale, who dedicated the species to its discoverer, in his MSS. description and unpublished plates.

Differs from all allied forms in the immaculate abdomen, which in all the other species is heavily annulated with black.

Harris's species, *Hera*, described in Report of Insects of Massachusetts, 1841, and figured in Audubon's "Birds of America," T. 359) is nothing more than *Eglanterina*, Boisduval (Lep. Cal., Ann. Soc. Ent. Fr., p. 51, 1852). The examples of *Hera* were taken by Mr. Nuttall at the Rocky Mountains; three of them were in collection of Mr. T. R. Peale, who received them from Mr. Nuttall himself; one of these three, a ♂, Mr. Peale still has in excellent preservation, the other two were destroyed through accident. Another specimen was in collection of Mr. Doubleday, England; this one was the original of Dr. Harris' description; and two more are in my cabinet. All these specimens cited were taken in 1836 by Mr. Nuttall. The species is also common in California; but, as a general thing, these are not quite as heavily marked with black as those found in the Rocky Mountains.

There is also much variation in the position of the black discal spot; in none examples there connect with the transverse band, in others are very close but disconnected, and in still others are far removed.

Walker's species, *Pica*, (British Mus., Cat. 6, p. 1318, 1855) of which I possess one ♂ example, taken by Mr. Drexeler in the Rocky Mountains, is a somewhat narrower-winged species, easily known by the uniform white ground colour of all wings.

Dr. Leconte, in his paper on nomenclature already alluded to on page 102, in speaking of the binomial system, says: "The arguments in favour of the original describer of the species on the one hand, and of the author of the binomial combination adopted on the other hand, are equally strong, perhaps, as regards the convenience of science, and each side has been argued with the utmost ability; practically, I do not regard it as a matter of any consequence, if each person will distinctly declare in his work *which system* he used. The number of instances in which any confusion can result are few, and the synonymy in catalogues which are always at hand will at once resolve the doubt."

There can absolutely be nothing said in favour of the author of the combination; the specific name is the one by which we know the insect. No one speaks of *Vanessa Antiopa* = *Vanessa*, but every one knows what insect is meant when we say *Antiopa*. If Grote & Robinson's Last Lep. N. Am., Boisduval's *Adelocephala*, is coupled with Harris' species, "*bicolor*," and Mr. Grote's name placed behind the combination, thus, "*Adelocephala bicolor*, Grote;" again, Hubner's genus, *Aniota*, and Abbot & Smith's species, *Pellucida*, are made *Anisota Pellucida*, Grote. The *Cecropia* of Linnæus, Mr. Grote has placed in his genus, Platysamia, and transmogrified it into *Platysamia Cecropia*, Grote, and so on in this manner have the names of Linnæus, Fabricius, Abbot & Smith, Harris, Walker, etc., been put aside to make place for the greater one of Grote.

Were this method generally followed, the confusion consequent would be truly astounding; the dragon's teeth of Cadmus, or the fecundity of the louse, would be as nothing to the multiplicity of synonyms that would issue from each species. For instance, take *Catopsilia Argante*: Fabricius first noticed this species in his Syst. Ent., and we cite him, in consequence, as the author; gentlemen of the above school would say *Catopsilia Argante*, Hubner, though that author did not even place *Argante* in his genus Catopsilia, but in his contemporary genus, Colias; so we have also *Colias Argante*, Hubner, and as his genera Phoebis and Murtia are also contemporary with the above, and embrace insects structurally the same as in his **Catopsilia** and Colias, we then also have *Phoebis Argante*, Hubner, and *Murtia Argante*, Hubner, and later Dr. Boisduval placed *Argante* in his genus Callidryas, where we have *Callidryas Argante* of that author. Again, of *Pararge Moera*, Linnæus' species: During the time that has elapsed since Linnæus first described it as *Papilio Moera*, it has been *Dira Moera*, *Pararge Moera*, *Satyrus Moera*, *Lasiommata Moera*, *Hipparchia Moera* and *Amecera Moera*. Is this not enough to condemn a system which could only have had its foundation on personal vanity? It is a very convenient thing for the author of a new genus founded, in most instances, on some indefinite-ical point, to place in it the species of Linnæus, Fabricius, Hubner, etc., etc., and then to attach his own name to each species so pirated, or else to resurrect some obsolete or forgotten genus and to crowd into it the species of various authors, living and dead, and behind each such combination to place the name of the industrious researcher who exhumed from the dust on the top shelf of some library the doubtful genus. This procedure is precisely analogous to that of a signpainter placing a picture of Rembrandt's in a frame of somebody or other's make, and erasing the artist's name from the picture and the maker's from the back of the frame, and then putting his own more important name across the face of both picture and frame, and of course rendering both valueless by the hideous defacement.

The specific name is and always will be the abiding one, always standing intact, the one by which we designate the object, though bandied from genus to genus; the generic name is ephemeral,—a thing, as it were, of to-day—therefore it is of the utmost consequence that the authority for the species be given, doubly necessary on the account of the hosts of synonyms which, with frightful recklessness, ambitious aspirants are continually overloading science.

As regards the "catalogues which are always at hand," that may be so in large cities blessed with such Entomological Libraries as that of the Acad. Nat. Sc. of Philadelphia, or the Peabody Institute of Baltimore, etc., or where the student fortunately possesses ample means to enable him to obtain all the requisite literature; but to the less fortunate, but perhaps equally zealous student, who neither lives in a large city nor is blessed (or cursed, as demagogues preach) with wealth, it would be in the highest degree inconvenient, for when we see the species' name we want to know something about it, why so named, where found, etc.,—facts which generally are only fully recorded in the original description, and which we like to see ourselves and not depend entirely on others, however reliable.

As the learned Dr. says, "whether the author's name remains connected permanently with his observation, or not, is a matter of small importance." Unfortunately, were that same name not to the species many and many an error now rectified would be still undetected; the ill with the good we must take, and tolerate the pitiful vanity that influences some to consider that the name placed behind their species should be printed in golden (brazen) letters, in order to eventually arrive at the truth. Finally, I would add that not only should the author of the specific name be added, but also the work, vol. and page in which his species was first described should be cited; this would save many precious hours to those who, too often, are obliged to encroach on time that should be devoted to lucrative pursuits, in order to pursue their unremunerative but beloved and fascinating studies.

Feb. 22, 1875.

Of the following species I am anxious to obtain examples, either by exchange or purchase; any Naturalists having duplicates of any of them will confer a great favor by communicating with

HERMAN STRECKER,
Box 111 Reading, P. O.,
Berks Co., Pennsylvania, U. S. of N. America

Ornithoptera Hippolytus, Cram.
Ornithoptera Lydius, Feld.
Ornithoptera Helena, Linn. ♀
Ornithoptera Croesus, Wall.
Ornithoptera Brookiana ♀, Wall.
Papilio Evan, Doubl.
Papilio Pericles, Wall.
Papilio Blumei, Boisd.
Papilio Macedon, Wall.
Papilio Philippus, Wall.
Papilio Arcturus, West.
Papilio Pherbanta, Linn.
Papilio Homerus, Fabr.
Papilio Garamas, Hub.
Papilio Caiguanabus, Poey.
Papilio Ascanius, Cram.
Papilio Wallacei, Hew.
Papilio Slateri, Hew.
Papilio Endochus Boisd.
Papilio Icarius, West.
Papilio Dionysos, Dbldy.
Papilio Gundlachianus, Feld.
Papilio Elephenor, Dbldy.
Papilio Disparilis, Herr-Sch.
Argynnis Rudra, Moore.
Argynnis Oscarus, Evers.
Argynnis Cnidia, Feld.
Argynnis Jerdoni, Lang.
Argynnis Dexamene, Boisd.
Argynnis Jainedeva, Moore.
Argynnis Ruslana, Motsch.
Argynnis Anna, Blanch.
Argynnis Childreni, Gray.
Argynnis Aruna, Moore.
Parthenos Tigrina, Voll.
Penetes Pamphanis, Dbl., Hew.
Cœrous Chorinæus, Fabr.
Charaxes Epijasius Reiche.
Charaxes Kadenii, Feld.
Charaxes Jupiter, But.
Charaxes Etheocles, Cram.
Catagramma Excelsior, Hew.
Dyctis Bioculatis, Guerin.
Romalæosoma Sophron, Dbldy.
Romalæosoma Pratinos, Dbldy.
Romalæosoma Arcadius, Fabr.
Nymphalis Calydonia, Hew.
Saturnia Epimethea, Dru.
Calinaga Buddha, Moore.
Diadema Boisduvalii, Dbldy.
Pavonia Aorsa, West.
Any species of Phyllodes.

Dynastor Napoleon, Doubl., Hew.
Argynnis Sagana ♂, Doubl, Hew.
Zeuxidia Aurelius, Cram.
Brahmæa Whitei.
Brahmæa Certhia.
Urania Sloanus, Cram.
Eudæmonia Semiramis, Cram.
Thalium Crœsus.
Nyctalemon Metaurus.
Nyctalemon Cydnus, Feld.
Actias Mænas
Saturnia Derceto, Mssn.
Saturnia Argus, Drury.
Any Asiatic species of Parnassius.
Citheronia Phoronea.
Pyrameis Gonerilla, Fabr.
Pyrameis Abyssinica, Feld.
Pyrameis Dejeanii, Godt.
Pyrameis Tameamea, Esch.
Sphinx Cluentius.
Vanessa v. Elymi, Rbr.
Dasyopthalmia Rusina, Godt.
Morpho Phanodemus, Hew.
Any species of Agrias.
Any species of Callithea.
Pandora Chalcothea, Bates.
Pandora Hypochlora, Cates.
Pandora Divalis, Bates.
Colias Viluiensis, Men.
Colias Ponteni, Wallengr.
Bunæa Deroyllei, Thom.
Bunæa Phædusa, Dru.
Rinæa Zuleica, Hope.
Acræa Perenna, Dbldy.
Opsiphanes Boisduvalii, Dbldy.
Clothilde Jægeri, Herr-Sch.
Pieris Celestina, Boisd.
Euplœa Eurypon, Hew.
Paphia Panariste, Hew.
Limenitis Lymire, Hew.
Io Beckeri Herr-Sch.
Eacles Kadenii, Herr-Sch.
Hepialus Giganteus, H.-S.
Smerinthus Panopus, Cram.
Sphinx Substrigilis, West.
Saturnia Larissa, West.
Eusemia Victrix, Amatrix, Dentatrix, West.
Smerinthus Modesta, Fab. (nec. Harris.)
Smerinthus Tartarinovii, Brem.
Smerinthus Dentatus, Cram.
Any Asiatic Catocalae.

These are a few of the very many of the rarer species that I am eager to procure; of course there are numberless others from all parts of the world, equally desirable and coveted by me.

S. D. JONES,

SUCCESSOR TO THE

North Atlantic Forwarding and Express Co.

OFFICE, 48 BROADWAY.

CENTRAL EUROPEAN OFFICE:
5 RUE SCRIBE, PARIS.

PRINCIPAL OFFICES IN GREAT BRITAIN:
ATLAS PARCEL EXPRESS.

LONDON, 55 Aldermanbury.
LIVERPOOL, 1 Brunswick St.
MANCHESTER, 109 Market St.
BRADFORD, 26 Charles St.
BIRMINGHAM, 2 Bull St.
NEWCASTLE ON TYNE, 39 Pilgrim St.
BRISTOL, 80 Radcliffe St.
LEEDS, 20 Trinity St.

GLASGOW, 18 So. Hanover St.
" 16 Renfield St.
EDINBURGH, 1 Frederick St.
ABERDEEN, 13 St. Nicholas St.
DUNDEE, 24 Union St.
DUBLIN, Bachelors Walk.
BELFAST, 46 Waring St.
LONDONDERRY, 15 Sackville St.

General Agent for France, E. SCHLOSSER, 32 Rue d'Orleans, Havre.
IGNATZ ROSENTHAL'S WWE. & CO., No. 14 Poggenmuhle Strasse, Hamburg.
KARESCH & STOTZKY, 29 Bonhoff Strasse, Bremen.

Receives, forwards and delivers heavy merchandise, case goods, baggage, etc., to and from any part of the United States and Europe, at the lowest possible charges. Interior importers will find this the cheapest and most expeditious medium through which they can obtain their goods.

TRANSPORTATION IN BOND.

Merchandise and packages intended to be transported in bond to an interior port of entry in the United States, must be accompanied by a *certified invoice and regular bill of lading*. The fact that the goods are to be so transported in bond must be expressed on the invoice and bill of lading. Shippers and agents will make special note of these instructions.

FAST EXPRESS FREIGHT.

Merchandise, specie, bullion, stocks, bonds, or other valuables, and packages and parcels of every description, personal effects, baggage, etc., forwarded to and from Europe and all parts of the United States, the States or Territories of the Pacific Coast, British Columbia and the Canadas included, at *fixed Tariff rates*.

For the convenience of shippers, where agencies are not established, packages or heavy goods may be forwarded to either of the offices or agencies, be either of the express or transportation companies in the United States, or by post, by railway, through the parcel delivery companies, or forwarding houses in any part of Great Britain or the Continent of Europe.

All packages, trunks, or parcels forwarded through us will be landed on arrival simultaneously with the mails, or immediately thereafter and will be entered at the Custom-house, duties paid and delivered to the parties to whom addressed in any part of Europe, the United States, the Canadas, or British Columbia, with the greatest possible dispatch. Transportation charges and duties collected on delivery, or may be prepaid, at the option of the shipper.

Insurance against marine risk taken when desired by the shipper, at the lowest current rates; premiums payable in all cases in advance.

Shippers to or from any part of America, and Americans traveling in Europe, will find this the quickest, cheapest and most reliable medium of transportation, our business being conducted upon the well-known prompt American express system, which has become so great a commercial necessity and convenience throughout the United States.

Purchases made, and collections and commissions in every part of Europe and the United States promptly executed.

W. V. ANDREWS,
ENTOMOLOGIST, &C.,
Room No. 4, No. 117 Broadway, New York.

☞ Purchasing Agent for Books and Apparatus in connection with Natural History. Also, Cork Pins, &c. Eggs of the different varieties of Silk Worms, to order. Lepidoptera and Coleoptera for sale or exchange. Agent for WALLACE'S SILK REELER, and for KIRBY'S SYNONYMIC CATALOGUE OF DIURNAL LEPIDOPTERA.

No. 13.   Issued Quarterly, at 50 Cents per Part in U. S.
In Europe, 2 Shillings.

LEPIDOPTERA,

RHOPALOCERES AND HETEROCERES,

INDIGENOUS AND EXOTIC;

WITH

Descriptions and Colored Illustrations,

BY

HERMAN STRECKER.

Reading, Pa., 1876.

Reading, Pa.:
Owen's Steam Book and Job Printing Office, 515 Court Street,
1876.

MACROGLOSSA RUFICAUDIS. Kirby.

(*Sesia R.*) Fauna Boreali Americana, Vol. IV, p. 303 (1837). *Walker*, C. B. M., Vol. VIII, p. 82 (1856). *Morris*, Cat. Lep. N. Am., p. 17 (1860 ;) Synopsis Lep. N. Am., p. 149 (1862). *Clemens*, Jour. Acad. Nat. Sci. Phila., Vol. IV, p. 205 (1872).
Hæmorrhagia Ruficaudis, Grote & Robinson, Proc. Ent. Soc. Phil., Vol. V, pp. 149, 175 (1865).
Hæmorrhagia Buffaloensis, Grote & Robinson, Ann. Lyc. Nat. Hist. N. Y., Vol. VIII, (1867); List Lep. N. Am., p. 3 (1868). *Grote*, Bull. Buff. Soc. Nat. Sc., Vol. I, p. 18 (1873). Vol. II, p. 224 (1875).
Sesia Uniformis, Grote & Robinson, Trans. Am. Ent. Soc., Vol. II, p. 181 (1868). *Lintner*, 23d Report N. Y. State Cabinet Nat. Hist., p. 172 (1872).
Hæmorrhagia Uniformis, Grote & Robinson, List. Lep. N. Am., p. 3 (1868). *Grote*, Bull. Buff. Soc. Nat. Sc., Vol. I, p. 18 (1873); Vol. II, p. 224 (1875).

(PLATE XIII, FIG. 1, ♂.)

"Body yellow-olive, underneath pale yellow. Antennæ black; primaries reddish-brown, hyaline in the disk, with the hyaline part half divided towards the base, with a costal bar, covered with yellow olive hairs at the base; underneath the costa, the posterior margin and the nervures are dark ferruginous; there is also a yellow stripe on the inner side of the base; secondaries hyaline in the disk; base externally and costa yellow; internally the base is ferruginous; underneath the dark part of the wing is ferruginous, and the base pale yellow; two first segments of the body yellow-olive, two next black, the rest ferruginous with pale yellow lateral spots. This species appears to be the American representative of *Sesia fuciformis* which it greatly resembles, but differs in the colour of the tail and the base of the secondaries."

No figure accompanied the above description of Kirby's, but there can be little doubt that a species allied to *Thysbe* was intended.

Walker, in C. B. M., says: "This is probably a mere variety of *S. Thysbe*," and states that specimens were received from United States, Trenton Falls, New York, and Orilla, West Canada.

Dr. Clemens, in his monograph of the Sphingidæ, published in the Journal Academy Natural Sciences, Philadelphia, 1859, also cites it as a synonym of *Thysbe*.

Grote and Robinson first stated it to be distinct from *Thysbe* in Proc. Ent. Soc., Phila., Vol. V, p. 149, and placed it in their genus *Hæmorrhagia*; on page 175, i. e., they give Kirby's description above cited, and remark "were we satisfied as to the species Kirby intended by *S. Fuciformis*, the present species might be regarded as related to *S. Diffinis*, Boisd. sp. As it is, we think that a species of *Hæmorrhagia* is meant, while the species has not been since identified," and further on "a mutilated specimen from the most northern parts of Canada West is before us, which evidently forms a distinct species from *H. Thysbe*. In this species, which is altogether slenderer than its congener, the inner margin of the terminal band of anterior wings is nowhere denticulate in the interspaces, but is toothed ally, somewhat inwardly, produced. We are not indisposed to regard this as Kirby's species, but the inferior condition of the specimen prevents all conclusions. The discal cell is crossed by a longitudinal scale line, the species belonging to the more typical group of the genus *Hæmorrhagia*." Three years later they re-described the species as *Sesia Thysbe*, variety *uniformis*, thus: "As *Sesia thysbe, a uniformis* sub., we will record the *Sesia ruficaudis* of Mr. Walker. This is not Kirby's species, to judge from the description of that author. This is a form of *S. thysbe*, occurring in both sexes, in which the external border of the primaries is not dentate inwardly on the interspaces."*

Although another specific synonym was here created on the assumption that Grote & Robinson knew more about Kirby's species than himself, still their fictitious genus *Hæmorrhagia* was for the time, sensibly enough, suppressed by them, for after a rhodomontade of thinly-veiled and confused excuses in reference to *Hæmorrhagia*, they say "which latter we can, therefore, no longer consider sufficiently distinct from *Sesia* to be retained as a genus." And it was only after Mr. Robinson's death that Grote again attempted to restore it in one of his innumerable and ever-changing systematic Lists of N. Am. Sphingidæ, etc., which, like mushrooms, spring up in every issue of the Buffalo Bull. and kindred publications.

There can be little doubt that the species I have figured, which was the one redescribed by Grote & Robinson as *Uniformis*, is the one meant by Kirby in his description of *Ruficaudis*. The older authors did not lay the same stress on elaborately decorated descriptions as do some of the present day, hence there are frequently trifling omissions or vague sentences in their descriptions, and in some instances, as in *Sm. Ophthalmica*, Bd., a line or two sufficed to describe the insect, and although said description would apply to almost any of the eyed Smerinthi having rosy hind wings, no one would endeavor on this account to question or ignore Boisduval's species.

Ruficaudis occurs in various parts of the Middle and New England States, and more plentifully in Canada and the neighboring island of Anticosti, as also in S. Labrador.

The most prominent point of distinction between this and *Thysbe* is the inner edge of marginal band of primaries which is toothed in the latter, whilst plain in *Ruficaudis*, though increased inwardly in the middle, as in *Thysbe*.

Between *Ruficaudis*, Kirby, *Uniformis*, G. & R., and *Buffaloensis*, G & R., I cannot find any specific difference by which to separate them into distinct species. In concluding the description of *Buffaloensis* the authors say "This species is closely allied to H. thysbe G. & R. from which it may at once be separated by its smaller size and the non-dentate inner margin of the terminal band of the primaries in the male. We have elsewhere drawn attention to the character afforded by the inner margin of the terminal band in H. thysbe; it is, however, in the males alone that it is prominently dentate on the interspaces.†") Consequently there would be nothing to separate it from *Ruficaudis*, (their *Uniformis*), which is also without indentations on inner edge of marginal band, excepting in "smaller size," which also comes to be a distinction, as an example which I received from Mr. Grote himself in May, 1875, is quite as large as that of *Ruficaudis* figured in the accompanying plate.

The authors finally state in connection with their published figures: "We figure a variety of the female, in which the usually wholly vitreous fields of the wings are sparsely and evenly clothed with scales. We have observed a similar variation in specimens of H. thysbe."

This makes the attempt of placing *Buffaloensis*, G & R., as a distinct species further objectionable, as the authors were ignorant that *Thysbe* and all allied species have, on emerging from the pups, the transparent space of the wing lightly covered with scales, which soon disappear under the action of flight or by exposure.

*Trans. Am. Ent. Soc., Vol. II, p. 181.
†About two years later the authors discovered their error, and became aware that the females of *Thysbe*, (the commonest of all the N. Am. species,) had the inner edge to the marginal band of primaries dentate as well as the males; and then it was that *Ruficaudis*, Kirby, was bleezed and again synonymised as *Uniformis*, G. & R.

109

I am not disinclined to believe that *Ruficaudis* is but a form of *Thysbe*, as one of my examples of the former is plainly though not deeply dentate on the inner edge of marginal band of primaries. This marginal band can scarcely be of much value specifically as in another example of *Ruficaudis* it is not widened perceptibly in the middle, but its inner edge runs in a regular line nearly as in *Gracilis*, but in the white anterior and median legs, and all other particulars, it agrees with the many other examples of *Ruficaudis* before me.

In another example of *Thysbe* the interspacal points of inner edge of marginal band are exaggerated to such an extent that one of them reaches to the discoidal cell; in this example the marginal band is of great width, as are also the brown basal parts, leaving comparatively little vitreous space.

MACROGLOSSA ÆTHRA. Strecker.

Described on p. 107.

(PLATE XIII, FIG. 2.)

MACROGLOSSA FUMOSA. Strecker.

Described on p. 93.

(PLATE XIII, FIG. 3.)

I have nothing to add to the original description of these two insects, save the figures on the present plate.

MACROGLOSSA FLAVOFASCIATA. Barnston.

Walker, C. B. M., Vol. VIII, p. 87 (1856).
Clemens, N. Am. Sph., Jnl. Acad. Nat. Sc., Phila., p. 151 (1859). *Morris*, Cat. Lep. N. Am., p. 17 (1860); Synopsis Lep. N. Am., p. 151 (1862).
Lepisesia Flavofasciata, *Grote*, Proc. Ent. Soc., Phila., Vol. V, p. 39 (1865); Bull. Buff. Soc. Nat. Sc., Vol. I, p. 17 (1873); Vol. II, p. 225 (1875). *Grote & Robinson*, Proc. Ent. Soc., Phila., Vol. V, pp. 149, 171 (1865); List Lep. N. Am., p. iii (1868).

(PLATE XIII, FIG. 4 ♀.)

Expands nearly 1⅞ inches.
Head and thorax above yellow; palpi black at sides, yellowish beneath; abdomen black, yellow on basal segment, and yellow lateral tufts on last segment; anal brush black; legs and under surface of body black, a yellow spot on middle of last segment.
Upper surface; primaries blackish, with a broad paler sub-terminal band and black discal spot. Secondaries bright yellow, black at base, and with an even, not broad, black margin.
Under surface: submarginal band sparingly scaled, space interior to this ochraceous, marginal band blackish; costa edged with black. Secondaries same as above, but yellow median space much paler, and inclined to ochraceous, also a yellow spot at base.
Habitat. Canada, Mus. Am. Ent. Soc.; Holyoke, Mass., Mus. Strecker.

The ornamentation of the wings is the only point worthy of note in which this species differs from others of the genus *Macroglossa*; superficially, the wings have more the appearance of *Pterogon*.
Grote, in erecting his genus *Lepisesia*, speaks of it as: "A genus hitherto confounded with *Macroglossa*, but more nearly allied to *Sesia*, from which, however, it is quite distinct." Were he to use the term *Sesia* in its correct sense, the alliance would be exceedingly slight; indeed would extend only to the fact that in many species of both genera the wings are hyaline; but in using the term *Sesia*, he alludes to the clear-winged species of *Macroglossa*, such as *Diffinis*, Bdl., etc., to which *Flavofasciata* is certainly allied, as it belongs to the same genus. The grounds for separating it therefrom, as designated by Grote, are entirely too weak to be of any value."
"Head smaller and more obtuse than in *Macroglossum*." Smaller than *M. Stellatarum* it certainly is, and so is the whole insect, but than *M. Croatica* it is just as certainly not smaller, but the same in size; as are also *Bombyliformis*, *Fuciformis*, and the other clear-winged species. Neither can I see that the head is more obtuse than in those mentioned, though it is more so than in *M. Thysbe*. Nor can I see, after careful measurement, that "the eyes are smaller, compared with *Sesia*," (as he calls the clear-winged species of *Macroglossa*,) compared with *Thysbe* they are larger, taking the relative size of the two insects into consideration.
"The anterior wings are relatively much longer, narrower, external margin more oblique than in *Macroglossum*;" he should have added *Stellatarum*, but agreeing in this with *M. Croatica*.
"The costa is medially depressed;" so is it, more or less, in over half of the examples of *Macroglossa* that I possess, both opaque and clear-winged.
"The sub-costal nervure is curved upward, beyond the discal cell;" so it is in *Croatica* and some examples of *Thysbe*, *Axillaris* and others.
"The posterior wings are small;" no smaller than in *Croatica*, *Axillaris* and *Fuciformis*; larger than in *Bombyliformis*.
"First, second and third median nervules less propinquitous than in *Macroglossum*; more curved;" I cannot see that these nervules are further apart than in *Croatica*, *Bombyliformis* and others.
"The abdomen is more smoothly scaled and less obtusely terminated than in *Sesia*;" not more smoothly scaled than in *Thysbe*, *Axillaris*, *Bombyliformis* and *Croatica*; more smoothly scaled than *Fuciformis* and *Ruficaudis*.

MACROGLOSSA FLAVOFASCIATA.

"The nervulation has undergone important modifications, while the pterogostic characters in their entirety are very distinctive, and, without any sudden change, show the position of this genus as intermediate between *Sesia* and *Macroglossum*, while considerably modified from either." I doubt if any one beside the author of *Lepisesia* would have acuteness of vision sufficient to perceive either the "important modifications," or the "very distinctive" "pterogostic characters," and in the sense that he uses *Sesia* and *Macroglossum* they are but synonyms; thus the genus *Lepisesia* is intermediate between *Sesia* and *Sesia*, or between *Macroglossum* and *Macroglossum*! Superficially, the insect looks a good deal more as if it were between *Macroglossa* and *Pterogon*, (*Proserpinus*, Hb.,) the body favoring the former, and the wings the latter.

There is as much propriety in Grote's separating this species from *Macroglossa* as there was in his making the genus *Calasymbolus* for *Smerinthus Astylus*, or *Cressonia* for *Sm. Juglandis*; although Hubner had first designated the latter as *Amorpha Juglandis* in Samml. Exot. Schmett., and afterwards taken it from that fictitious genus and placed it in *Polyptychus* in Verz. bek. Schmett. (1816), all of which Mr. Grote, with his usual sagacity, has been pleased to ignore in favour of his own genus *Cressonia*, which, of course, is a synonym of Hubner's *Polyptychus*, which latter, we may as well add, is but a synonym of *Smerinthus*, Lat.

Smerinthus Dyras Wlk., *Maassen Bell.*, and others are much more aberrant in appearance than either *Juglandis* or *Astylus*; yet none have had the temerity to create new genera for them, and doubtless none will unless those species should by some mischance come under the observation of the author of *Calasymbolus*, *Lepisesia*, etc.

Lepisesia was first placed by Grote at the head of the N. A. Sphingidæ in 1865; in a later effusion *Arctonotus Lucidus* preceded it; in his latest effort he has placed it behind *Macroglossa* (which he has even cut up into several genera), and wedged it in between *Arctonotus* and *Proserpinus*; and in his next spawn we confidently expect to see it jerked down to *Smerinthus* and placed between *Astylus* and *Juglandis*. Rochefoucauld, I believe it was, who said that the only thing that still ought to be capable of causing us astonishment is that we have the power of being still astonished at anything.

PTEROGON CLARKIÆ. Boisduval.

Ann. Soc. Ent. Fr., 2 me ser. X, p. 319 (1852).
Thyreus! Clarkiæ, Walker, C. B. M., Vol. VIII, p. 202 (1856).
Proserpinus Clarkiæ, Clemens, Jnl. Acad. Nat. Sc., Phila., Vol. IV, p. 134 (1859). *Morris*, Cat. Lep. N. Am., p. 18 (1860); Syn. Lep. N. Am., p. 134 (1862). *Grote & Robinson*, Proc. Ent. Soc., Phila., Vol. V, p. 149 (1865); List Lep. N. Am., p. iii (1868).
Grote, Bull. Buff. Soc. Nat. Sc., Vol. I, p. 20 (1873), Vol. II, p. 225 (1875).
Lepisesia Victoria, Grote, l. c., p. 147 (1874).

(PLATE XIII, FIG. 5 ♀.)

Expands 1½ inches.

Head and body above olivaceous; antennæ brownish, darkest above, tips whitish yellow.

Primaries pale olivaceous with a darker median band and discal spot after the manner of *Proserpina*, Pall.,* a sub-terminal dark line which widens to a large triangular at apex. Inferior wings bright yellow, with narrow black marginal band; fringe white.

Under surface nearly same as *Proserpina* in colour and ornamentation. Primaries olivaceous, darker parts of upper surface faintly defined. Secondaries olivaceous, with a paler broad median band or space.

Habitat. Oregon, Northern California. Mus. Am. Ent. Soc., Hy. Edwards, Strecker.

The American representative of the European *P. Proserpina*, which it strikingly resembles in both colour and markings; but it is smaller, and the wings are not angulated.

Grote, when he discovered that his *Lepisesia Victoria*, lately described in Buff. Bull., Vol. II, p. 147, was a redescription of this species, made the correction in a foot-note in this wise: "From a fresh specimen received from Hy. Edwards I find that my description is based on a faded specimen of this species." How astonishingly powerful must have been the action of the light, to have not only changed the colour of the insect, but also to have actually changed the generic characters of his "faded" *Pterogon* (*Proserpinus*,) into those of his own genus *Lepisesia*. Really, Mr. Grote ought to see that his types are not exposed to this malicious light; but, after all, there is no evil without its accompanying good, for if the action of light in fading is powerful enough to change the genus of a dead insect, why may not the same agent be employed, for purposes of utility, on the higher animals. For instance: Why not place all the half-starved, worthless curs, which range at large through our streets, under the action of powerful Grotesque light, and transform them into porkers, ready-roasted? Would not Mr. Grote thus be immortalized with but a tithe of the labor necessary to create synonyms and combinations to precede his name? besides, look at the reward. For the former, millions yet unborn would bless his name as one of the great benefactors of their race; for the latter he would only receive the maledictions of ungrateful Lepidopterists for the amusement he will have bequeathed them in trying to study what "Proserpinus Hub., Clarkiæ Boisd. Clem. Lepisesia Victoria, Grote," and the like, could possibly or impossibly mean. It appears that Mr. Grote has been lately paying considerable attention to Optics, for another important discovery of his, in that branch of science, is that darkness bleaches specimens of moths, etc., for in one of his numerous redescriptions of Catocolæ† he says in allusion to the pale colour of secondaries: "The condition of the specimen does not allow of the suggestion that this change of colour is owing to etiolation?" we sincerely hope not, for if that be the case we tremble for all the thousands of examples that we so carefully exclude from the light. What if some evening we were to take an inspiring look at our treasures, and found nothing but blanched ghosts in place of our gorgeous children of the tropics! no, we pray Mr. Grote may be mistaken; we don't want our species Darwinized into Pierides through disease contracted by exclusion from light. We know that celery and cabbage are white when kept in the dark whilst living, and that fish in the mammoth cave are white also, probably from same cause, but were it not for Mr. Grote's words we would doubt that deprivation of light would disease or whiten a dead insect. Will not Mr. Grote speedily give this important matter further attention, and see if the pallor of the hind wings of his Catocolæ was not owing to some other cause? Dare we suggest, homœopathically, perhaps, exposure to the light; *Similia similibus curantur*.

Sphinx Proserpina, Pallas, Spicilegia Zoologica 9, p. 26, T. II, 7 (1772).
Sphinx Œnotheræ, Schiffermuller & Denis, Syst. Verz., p. 43 (1776).
†*Catocala Innubens* var. *Flavidalis*, Trans. Am. Ent. Soc., Vol. V, advance sheets of Grote's paper, printed and issued second week of Nov., 1874. The work itself just issued. (Dec., 1876.)

PTEROGON JUANITA. Nov. Sp

(PLATE XIII, FIG. 6 ♂.)

MALE. Expands 2 inches.
Head and body olivaceous.
Upper surface; primaries, colour and markings much as in *P. Clarkiæ*. Secondaries bright yellow, a reddish marginal band deepening into brown nearest the outer edge, broader than in *P. Clarkiæ*; a reddish spot on abdominal margin towards anal angle.
Under surface; inner half of wings reddish-brown, outer third pale olivaceous, darker towards margin; the basal part of primaries is more reddish than on secondaries.
Habitat. Mexico or S. W. Texas on borders of the Rio Grande. One example, Mus. Strecker.

A larger species than *P. Clarkiæ*, and with primaries much narrower and more prolonged apically, resembling more, in this respect, *P. Guara*, Ab. & S., from which it differs, however, in the colour of the hind wings, which are red, margined with black, in the latter species. I have seen but the one example which I have portrayed, I trust faithfully, on the annexed plate.

PTEROGON INSCRIPTUM. HARRIS.

Sill. Am. Jnl. Sc. & Art, Vol. 36, p. 305 (1838).
Thyreus? inscriptus, Walker, C. B. M., Vol. VIII, p. 100 (1856).
Proserpinus et Pterogon Inscriptum, Morris, Cat. Lep. N. Am., p. 18 (1860).
Deidamia Inscripta, Clemens, Jnl. Acad. Nat. Sc. Phila., p. 137 (1859). *Morris*, Syn. Lep. N. Am., p. 150 (1862). *Grote & Robinson*, Proc. Ent. Soc., Phila., Vol. V, p. 151 (1865); List Lep. N. Am., p. 3 (1868). *Grote*, Bull. Buff. Soc., Vol. I, p. 20 (1873), Vol. II, p. 225 (1875).

(PLATE XIII, FIG. 8 ♂.)

Expands 1⅔—2 inches.
Head and body above ashen; thorax shaded with brown; two rows of dark brown spots on abdomen; anal segment trifurcated; antennæ serrated in male, plain in female.
Upper surface. Primaries same colour as body, with bands and marks of rich brown of various shades; a pale discal spot and a small white triangular spot near exterior margin. Secondaries reddish; outer margin greyish.
Under surface ashen, all wings outwardly, with darker colours; the small triangular white spot on upper surface near outer edge of primaries is repeated.
Habitat. Middle and New England States, Maryland, Virginia, Ohio, Indiana and doubtless other States east of the Mississippi, but nowhere common.
Mr. John Akhurst, of Brooklyn, N. Y., who raised a number of examples of this species from the larvæ, describes it as being, when full grown, two inches in length, of a fine green colour, caudal horn whitish at the tip, head small, body from third segment tapers towards the head. It feeds on the leaves of the grape, and of the Virginia creeper, (*Ampelopsis quinquefolia*); to undergo its transformation it enters the ground, but not very deep; it is frequently found near the side of a wall or the bottom of a fence post, and even under a board or flat stone; it is full grown about the last of June, or beginning of July, and is single brooded, the perfect insect appearing about the middle of May. Mr. Akhurst made neither notes nor drawings at the time, but the above, though brief and lacking in details, in consequence of his having to depend entirely on memory, he is sure is substantially correct.

Harris provisionally placed this species in *Pterogon*; Dr. Clemens afterwards made for its reception the genus *Deidamia*. If, however, the Russian *Gorgoniodes** is to be retained in the same genus with *Œnotheræ*, then certainly *Inscriptum* belongs there likewise, as there can be no doubt that *Inscriptum* and *Gorgoniodes* are generically the same, at least as far as comparisons between the males extend. Whether the ♀ of *Gorgoniodes* has simple antennæ like that sex in *Inscriptum* I can not now say, as I have seen only males of the former; but in that sex both species have the antennæ serrated, the eyes sunken, the head produced in a crest, the shape of thorax and abdomen the same, the anal segment trifurcated, and the same style of ornamentation on wings and body; and in whatever genus systematists may place *Gorgoniodes*, there also *Inscriptum* belongs.

Proserpinus Gorgoniodes, Hub. Verz. bek. Schmett., p. 132 (1816).

ARCTONOTUS LUCIDUS. Boisduval.

Ann. Soc. Ent. Fr., 2 me Ser. X, p. 319 (1852).
Walker, C. B. M., Vol. VIII, p. 265 (1856). Clemens, Jnl. Acad. Nat. Sc., Phil., p. 183 (1859). Morris, Syn. Lep. N. Am., p. 217 (1862). Grote & Robinson, Proc. Ent. Soc., Phil., Vol. V, p. 169 (1865); List Lep. N. Am., p. 3 (1868). Grote, Bull. Buff. Soc., Vol. I, p. 17 (1873), Vol. II. p. 225 (1875).

(PLATE XIII, FIG. 7.)

Expands 1¾—1⅞ inches.
Body olive green, tegulæ edged with whitish, antennæ stout and heavily serrated.

Upper surface, primaries same colour as body, crossed by two irregular, not very conspicuous, flesh-coloured bands, which connect at the inner margin; the middle of these bands is dull purplish; the space between these bands, and also the basal space, is darker than the marginal part of wing.

Secondaries pinkish, a sub-marginal wine-red band, a purplish-black anal mark; fringes pale yellowish-grey.

Under surface olivaceous, inclining a little to reddish on inner half of primaries; devoid of ornamentation.

Hab. Oregon; Mus. Behr., Hy. Edwards, Strecker.

To my friend of many years, Henry Edwards, am I beyond measure indebted for two examples of this rarest of N. Am. Sphingidæ. Of its larva, food-plant or habits I know nothing; but Mr. Edwards, in a paper he is about to publish in the Proceedings of the California Acad. Nat. Sc., will doubtless be able to give further particulars.

This insect is much in the same position, or rather no position, as the curious Exotic Diurnal *Calinaga Buddha*; no one seems to know rightly where to place it. Clemens and Walker have put it at the last end of the Sphingidæ; Grote & Robinson, in their List N. Am. Sph., put it at the other end, and commenced the Sphingidæ with it. Grote, in his Cat. in Buff. Bull., Vol. I, still retained it there, but in his latest effort, in Vol. II of same work, he has changed its position and placed it between *Macroglossa Erato*, Bdl., (*Euproserpinus Phaeton*, G. & R.*) and *Macroglossa Florofaciata* (*Lepisesia F.*, Grote,) where it most certainly does not belong; its short tongue, the antennæ and other characters denote its close alliance to *Sueriathus*, near which it should doubtless be placed. Walker says, "this genus appears to connect *Smerinthus* with the *Bombycidæ*."

DARAPSA VERSICOLOR. Harris.

(*Chaerocampa*,) Harris, Sill. Am. Jnl. Sc. & Art, XXXVI, p. 303 (1839); Ins. Inj. Veg., Fline's Ed., p. 328 (1862). Walker, C. B. M. Vol. VIII, p. 171 (1856).
Darapsa Versicolor, Clemens, Jnl. Acad. Nat. Sc., Phila., p. 148 (1859); Morris, Cat. Lep. N. Am., p. 19 (1860); Syn. Lep. N. Am., p. 169 (1862).
Otus Versicolor, Grote & Robinson, Proc. Ent. Soc., Phila., Vol. V, p. 154 (1865).
Darapsa Versicolor, G. & R., List Lep. N. Am., p. 4 (1868). Edwards, Can. Ent., Vol. II, p. 134 (1870). Grote, Buff. Bull. Soc., Vol. I, p. 22 (1873), Vol. II, p. 226 (1875).

(PLATE XIII, FIG. 9 ♂.)

Expands 3 inches.
Body beautiful bright green; tegulæ edged with white; a white central dorsal line runs the whole length from the head to the end of abdomen; tegulæ and prothorax, and some of the last segments of abdomen, edged with white, also white lateral lines on the head. Beneath green and yellow; edges of abdomen white.

Upper surface, primaries with alternate white and green curved bands of varying width; broad green marginal band, a white apical line, the white space on disc tinged with purple, a green discal dash. Secondaries rust-red, white at costa and abdominal margin, exterior edged with an irregular, narrow, greyish and greenish margin.

Under surface, primaries yellowish, basal half suffused with reddish; margin green; white apical line; some white marks at costa not very far from apex. Secondaries green and yellow; three white bands very broad at costa and abdominal margin, almost obsolete on disc of wing.

Habitat. New England and Middle States, and probably others.

*This "name cannot obtain," as Grote & Robinson's description of both the genus and species was based on a picture. "We erect this genus for a small California species of the present family, which, while allied to *Proserpinus*, differs by the small, reduced sec-ondaries, longer antennæ and infixed abdomen. We are indebted to Mr. J. W. Weidemeyer for the information respecting this singular little species, which, we believe, has not been hitherto described, while an excellent figure, shown us by Mr. S. Calverley, enables us to present the present description, and to fix the species. It appears that Dr. Boisduval has etiquetted a specimen in his cabinet as *Proserpinus Phaeton*." All of which we think refreshingly cool. In after years they saw the real insect in Boisduval's collection, the actual example "etiquetted" by that great savant, and then, with impudence unparalleled, from this they made another description in Trans. Am. Ent. Soc., Vol. II, p. 182, (1868,) where they say: "The present description should supersede that given by us as noted above, and which was made from a colored drawing of the species, and is necessarily inaccurate in detail." Language fails!

Prof. Meyer, of Brooklyn, some years since was successful in breeding this splendid insect; the larvæ, he says, resembled those of *D. Myron*, but were larger. They feed on the *Cephalanthus Occidentalis*, a plant which grows on margins of creeks, in swamps, &c; has lanceolate leaves in twos and threes, and white flowers in clusters, and is better known as the Buttonbush. It is a common weed, occurring in various parts of Long Island and New Jersey, and I have little doubt that careful search by collectors who have the plant within reach, would be rewarded by the finding of some of the larvæ. Doubtless owing to the plants growing in and near water, very many of these larvæ are drowned, which may in some measure account for the amazing rarity of this insect, which without exception is the most lovely of all our N. Am. species, and second only to the peerless *Chœrocampa Nerii*, among those of other countries.

CHŒROCAMPA PROCNE. Clemens.

Jnl. Acad. Nat. Sc., Phila., Vol. IV, p. 151 (1859). *Morris*, Cat. Lep. N. Am., p. 20 (1860). Synopsis Lep. N. Am., p. 173 (1862). *Walker*, C. B. M., Supplement Vol. XXXI, p. 30 (1864). *Grote & Robinson*, Proc. Ent. Soc., Phila., Vol. V, p. 155 (1865); List Lep. N. Am., p. 4 (1868).
Metopsilus Procne, *Grote*, Bull. Buff. Soc. Nat. Sc., Vol. I, p. 22 (1873); Vol. II, p. 236 (1875.)

(PLATE XIII, FIG. 10.)

"Head and thorax dull brown, (if not faded,) with a broad whitish stripe on the sides, extended to the lower edge of tegulæ. Abdomen brownish testaceous, with faint dark-brown dorsal marks in atoms. Anterior wings rather pale brownish, punctated with dark atoms and with obscure dark brown lines extending from the base to the tip; discal spot dark brown and small. Posterior wings uniform blackish brown. Under surface of the wings brownish, somewhat tinged with rufous, and with two rows of brown spots in middle of the posterior. California."

Dr. Clemens' description above quoted agrees exactly with the example I have figured, which passed into my keeping along with the Lepidopterous collection of the Rev. Dr. John G. Morris some years since. It had no name attached, but merely the locality, "S. California;" its former possessor could give me no further particulars concerning it, but I have no doubt it is the species described as *Procne*, though this name may perhaps eventually prove a synonym of some species common to Mexico and Tropical America.

SPHINX LUSCITIOSA. Clemens.

Jnl. Acad. Nat. Sc., Phila., Vol. IV, p. 172 (1859).
Morris, Cat. Lep. N. Am., p. 19 (1860); Syn. Lep. N. Am., p. 197 (1862). *Walker*, C. B. M., Supplement Vol. XXXI, p. 36 (1864). *Grote & Robinson*, Proc. Ent. Soc., Phila., Vol. V, p. 165 (1865); List Lep. N. Am., p. 5 (1868).
Lethia Luscitiosa, *Grote*, Bull. Buff. Soc. Nat. Sc., Vol. I, p. 26 (1873); Vol. II, p. 228 (1875.)

(PLATE XIII, FIG. 11 ♀.)

Male. Expands 2¼ inches.
Head and thorax very dark brown above, whitish-grey on sides; abdomen light brown above, with a black dorsal line; on each side a row of black spots; beneath grey.
Superior wings narrow and prolonged, less in length from base to inner angle than from the latter to apex; exterior margin almost straight. Upper surface light brownish, shaded with darker brown at costa and inner margin; a rather broad dark brown marginal band; a very small white discal spot; a narrow black apical line, and a few abbreviated, almost obsolete, black lines in the cells. Fringe brownish.
Posterior wings ochrey, with broad, black marginal band, and very faint evidences of a mesial band; fringe white.
Under surface ochrey, with a rather broad brown marginal band on all wings.
Female. Expands 3¼ inches.
Head and body as in male.
Primaries much broader and not so much prolonged, being less from apex to inner angle than from latter to base. Upper surface clouded with light grey; black marginal band broadest at inner angle, and diminishes to a point before reaching the apex; black apical and other streaks better defined than in the male; exceedingly small white discal spot. Fringe dark brown and white, former colour predominating.
Secondaries as in male.
Under surface, primaries greyish, with a slight yellow tinge; marginal band not well defined, whitish and brown, former colour gaining at and towards apex, the latter at inner angle. Fringes brown and white.

Secondaries yellowish grey; brown marginal band, paler towards outer angle, and not reaching to abdominal margin. Fringe white.

Habitat. Canada, New England and Middle States, also Ohio, Maryland, Virginia, Indiana, Wisconsin, and probably other States. Rare.

Dr. Clemens' description applies only to the male, which differs from the female much more than is generally the case in this genus. The types were captured by Mr. T. B. Ashton, who directed Dr. Clemens' attention to them in the collection of George Newman, Philadelphia, where they were represented, I believe, in both sexes.

Of the larvæ I believe nothing is known.

SPHINX LUGENS. Walker.

C. B. M., Vol. VIII, p. 219 (1856).
Grote & Robinson, List Lep. N. Am., p. 5 (1868).
Agrius Lugens, Grote, Bull. Buff. Soc. Nat. Sc., Vol. I, p. 26 (1873), Vol. II, p. 228 (1875). *Gaumer*, Observer of Nature, Vol. II, No. 5 (1875).
Sphinx Eremitoides, on page 93 of this work.

(PLATE XIII, FIG. 12 ♀.)

Since redescribing this species from examples received from Mr. Ashton, I have become indebted for others to Prof. Snow, of Kansas, who was the first to breed it and who describes the larva, when full-grown and ready to enter the ground, as follows:—

"Length 3¼ inches, greatest thickness .56 in., head greenish brown with a distinct white stripe on each side; general colour of body pale green, with seven oblique lateral white bands; caudal horn black and in length .37 in. It becomes full-grown from 21st of September to 15th of October; imago appears from May 20th to June 10th. Food plants *Salvia Pitcherii*, Torrey, and *Salvia Trichstemnoides*, Parsh. The larvæ were first observed by me in October, 1873, in great abundance, and several imagines were obtained from them in the following May and June; the species is double brooded." Mr. Gaumer, in the "Observer," also states that "two broods of these caterpillars appear during the year, the first in June, the second very late in autumn; the last brood hibernate in the chrysalis state under ground and are much more numerous than the first."

Walker's types in the British Museum were from Mexico, and no N. American collection, I believe, possessed it, nor was it known to occur in the United States until taken by Mr. Ashton and Prof. Snow, in Kansas.

Dr. Clemens, in Jnl. Acad. Nat. Sc., Phila., Vol. IV, p. 169 (1859), gives *Lugens* as a synonym of *Sordida*, Hub. (which latter is, by the way, a synonym of *Eremitus*, Hub.).

The example described by me on page 93 as having the black mesial band of secondaries broken in the middle by the whitish ground is merely an aberration, as I have not observed this peculiarity in any of the examples since received.

SPHINX PLOTA.

Described on p. 106.

(PLATE XIII, FIG. 13 ♂.)

Since describing the males taken in Canada I have received a fine female of this species from Mr. Dury, who captured it near Cincinnati; I have also heard of several other examples that were taken in various localities, so that this is evidently as widespread a species as most of its congeners.

The female resembles the male, save that the primaries are more even coloured, there being fewer dark marks than on the other sex. Of the larva, as yet, nothing has come to my knowledge.

SPHINX JASMINEARUM. Boisduval.

Griffith's Cuvier's Animal King., XV, T. 83, f. 1 (1832?). *Leconte &c.*, Wilson Treat. Ent. in Enc. Brit., p. 236, f. 5, 6 (1835). *Clemens*, Jnl. Acad. Nat. Sc., Phila, Vol. IV, p. 173 (1859). *Morris*, Cat. Lep. N. Am., p. 19 (1860); Syn. Lep. N. Am., p. 198 (1862). *Walker*, C. B. M., Sup. Vol. XXXI, p. 36 (1864). *Grote & Robinson*, Proc. Ent. Soc., Phila., Vol. V, p. 165 (1865).
Diludia Jasminearum, Grote & Robinson, List. Lep. N. Am., p. 4 (1868?). *Grote*, Bull. Buff. Soc. Nat. Sc., Vol. I, p. 26 (1873), Vol. II, p. 227 (1875).

(PLATE XIII, FIG. 14 ♂.)

Expands 3¾ to 5 inches.

Head and body above whitish grey; a black stripe on sides of the thorax; on abdomen lateral white spots edged with black; beneath white.

116   SPHINX JASMINEARUM.

Upper surface; primaries same colour as body, with zig-zag transverse brown lines; a dark shade extends from a little below the middle of exterior margin to the costa interior to the discal spot, this latter white and inconspicuous; fringes white, brown at terminations of veins.

Secondaries dark brown, with obsolete traces of marginal and mesial bands; greyish at abdominal and inner half of exterior margins; fringes as in superiors.

Under surface light brown; basal half of secondaries paler and greyish.

Habitat. New York, New Jersey, Pennsylvania and others of the Atlantic States. Rare.

Larva pale yellowish-green, dorsal lines of darker colour on the sixth, seventh, eighth, ninth and tenth segments; transverse narrow white lateral stripes on all the segments save the two last, where these stripes which extend to the caudal horn are red and green, the latter colour uppermost; caudal horn green, with dull red serrations. Feeds on ash, and probably elm.

The genus Diludis in which Grote places this species he constructed with *Sphinx Brontes*, Drury, for his type. Here are his words, from which it will be seen that the weighty reasons for erecting the genus Diludis are these, that: "From *Macrosila* the species differ by the straighter external margin of the primaries, and by the exserted internal angle, in those characters resembling *Amphonyx*, while the normal palpal conformation, with a number of other characters, amply separate them from Prof. Poey's genus. We do not give further characters here of a genus which we are satisfied should be erected, since we have insufficient material upon which to amplify from needed dissections." And of course, as the authors were "satisfied," it was the duty of the scientific world to humbly submit without putting G. & R. to the trouble of giving "further characters."

SPHINX CONIFERARUM.   ABBOT & SMITH.

Lep. Insects of Georgia, p. 81, T. 42 (1797). *Harris*, Sill. Jnl. Art. & Sc., XXXVI, p. 296 (1839); Ins. Inj. Veg. (Flint's Ed.), p. 328 (1862). *Morris*, Cat. Lep. N. Am., p. 18 (1860); Syn. Lep. N. Am., p. 199 (1862).
Hyloicus Coniferarum, Hubner, Verz. bek. Schmett., p. 139 (1816). *Grote & Robinson*, Proc. Ent. Soc., Phila., Vol. V, p. 166 (1865); List Lep. N. Am., p. 5 (1868).
Anceryx Coniferarum, Walker, C. B. M., Vol. VIII, p. 224 (1856).
Ellema Coniferarum, Gr-te, Bull. Buff. Soc. Nat. Sc., Vol. I, p. 27 (1873).
Lapara Coniferarum, Grote, l. c., Vol. II, p. 228 (1875).

(PLATE XIII, FIG. 15 ♂.)

Expands about 2¼ inches.
Head and collar umber; thorax and abdomen ash-grey and immaculate.
Upper surface, superiors ash-grey with two short black streaks in the cells between the median nervules, and an inconspicuous dentated transverse line succeeded inwardly by a slightly paler shade; fringes white, with brown at terminations of venation. Inferior wings brownish-grey, paler at base; fringes as on superiors.
Under surface pale brownish-grey.

Habitat. New York, Maryland, Georgia, and doubtless others of the Southern and Middle States. Exceedingly rare.

The only examples I know of are in the collection of Titian Peale, Esq., and two in my own possession, one of which was taken in New York State, and the other was raised from a larva found feeding on pine, near Baltimore, Md.†

Abbot has figured the larva, which he says fed on *Pinus Palustris*; it has a yellow head, and the body chequered with light and dark grey squares.

SPHINX HARRISII.   CLEMENS.

Ellema Harrisii, Clemens, Jnl. Acad. Nat. Sc., Phila., Vol. IV, p. 188 (1859). *Morris*, Cat. Lep. N. Am., p. 20 (1860). *Grote & Robinson*, Proc. Ent. Soc., Phila., p. 166 (1865). *Lintner*, 23d Report N. Y. State Cab. Nat. Hist., p. 170, T. 8, f. 10 ♂, 11 ♀ (1869). *Grote*, Bull. Buff. Soc. Nat. Sc., Vol. I, p. 27 (1873).
Ellema Harrisii, Morris, Syn. Lep. N. Am., p. 216 (1862).
Hyloicus Harrisii, Grote & Robinson, List Lep. N. Am., p. 5 (1868).
Ellema Harrisii, Walker, C. B. M., Supplement Vol. XXXI, p. 37 (1864).
Sphinx Coniferarum, Harris, Sill. Am. Jnl. Sc., Vol. XXXVI, p. 297 (1859).
Lapara Bombycoides, Grote, Bull. Buff. Soc., Vol. II, p. 228 (1875).

(PLATE XIII, FIG. 16 ♀.)

Expands 1½ inches.
Head and body ash grey; upper edge of tegulæ edged with brown; abdomen immaculate. Beneath brownish.

*Proc. Ent. Soc., Phila., Vol. V, p. 168 (1865).
†See page 93.

Upper surface; primaries ashen, with transverse undulate lines and shades; two black streaks in the interspaces between the median nervules; fringes white, brown at termination of veins.

Secondaries brownish, basal half pale, nearly white; faint evidences of a mesial band; fringe as on primaries.

Under surface brownish. Secondaries paler at abdominal margin. Fringe as above.

Habitat. New England, Middle, and others of the Atlantic States. Rare.

Larva is green, with lateral pale stripes, destitute of the caudal horn, and feeds on *Pinus Strobus* and doubtless other species of *Coniferae*.

It has long been surmised that this species might be identical with *Lapara Bombycoides*, Walker,* which exists at present, as far as is known, in the single type example in the Hopeian Collection at the University of Oxford; originally it was contained in the collection of Mr. Saunders, of London, England, where it was described by Walker;† afterwards the whole of the Heterocera of Mr. Saunders' collection were added to the Oxford Museum. From this unique, which is in perfect condition, Prof. Westwood had the kindness to make for me an accurate coloured drawing which represents an insect indeed allied to *Sphinx Harrisii*, but separated from it by the following differences, which Prof. Westwood, to whom I submitted a proof of the accompanying figure of *Harrisii*, has intimated in a recent letter: *Bombycoides* is much more brownish in tint, not near so fascien grey; the thorax is destitute of all black lines; the dark dashes on middle, near inner margin of primaries, are more central and nearer to base of the wing; the secondaries are without any appearance of dark central fascia; the fringe on anal margin is of same pale dull brown as the rest of wing; under surface of all wings is uniform pale dull brown.

The figure of Prof. Westwood is also larger than any example of *Harrisii* I have yet seen, being about 2¾ inches in expanse.

In Grote's last List of N. Am. Sphingidae he has given *Ellema Harrisii* as a synonym of *Bombycoides*, but without stating any reason for so doing.

With *Ellema Harrisii* has long been confounded the previously described *Sph. Coniferarum*, which latter was long looked on as a myth, owing doubtless to its great rarity.

Dr. Clemens, in his Monograph in Jnl. Acad. Nat. Sc., Phila., states in connection with *Sph. Coniferarum*: "The specimen Dr. Harris described under this name, as I have ascertained from a photograph, was *E. harrisii*. This is probably likewise identical with *S. coniferarum*. The discovery of the larva of *harrisii* will remove any doubt respecting the identity of the insects."

It is not safe to depend entirely on pictures, be they ever so accurate or even photographic. Whether Dr. Clemens' opinion above cited be correct or not, it is at least evident that Dr. Harris was acquainted with the larva of *Coniferarum*, Abbot & Smith, as he says on page 328 (Ins. Inj. Veg. Ed. 1862): "the curiously checkered caterpillar of *Sphinx Coniferarum* on pines;" the larva of *Harrisii* is not "curiously checkered," but is green with lateral and sublateral stripes of yellow and white.

SPHINX SEQUOIAE. Boisduval.

Lep. de la Californie (1869).
Henry Edwards, Proc. California Acad. Nat. Sc. (1873).
Hyloicus Sequoiae, Grote, Bull. Buff. Soc., Vol. I, p. 27 (1873), Vol. II, p. 228 (1875).

(PLATE XIII, FIG. 17 ♂.)

Expands 2 inches.

Head and body grey, two black lines on head extending thence along upper edge of tegulae; abdomen with a black dorsal line, sides with alternate black and white bands.

Upper surface, primaries grey, with short black streaks in the cells, and one transverse one, accompanied by a shade near the exterior margin; fringes brown and white alternately.

Secondaries brownish, without marks; fringes white on abdominal and inner half of exterior margins, rest brown and white alternately.

Under surface brownish grey.

Habitat. California. Mus. Boisd., Hy. Edwards.

The original of Fig. 17 was lent me by Mr. Henry Edwards, of California, and is, probably, with the exception of Dr. Boisduval's type, the only example extant in any collection. In Mr. Edwards' Memoir on Pacific Coast Lep., above cited, he says: "I had the good fortune to take a fine ♂ of this rare species in Bear Valley, in June, 1872. It was hovering at mid-day over a pool of water, darting down occasionally to drink. The specimen from which Dr. Boisduval made his description was captured by the late M. Lorquin, at Grass Valley, resting on the bark of a Redwood tree, (*Sequoia sempervirens*—Lamb)." To Mr. Edwards I am indebted for the opportunity of presenting the figure of this rare species.

*C. B. M., Vol. VIII, p. 233 (1856).
†"Cinereous. Fore wings with a zigzag oblique black line, and with several lanceolate black marks. Hind wings brownish, paler towards the base; cilia white. Length of the body 10 lines; of the wings 24 lines.
Canada. In Mr. Saunders' collection." Walker, C. B. M., Vol. VIII, p. 233.

SPHINX SANIPTRI.

(PLATE XIII, FIG. 18 ♂.)

MALE. Expands 3 inches.
Upper surface in colour and ornamentation same as the European *S. Pinastri*, L., with this exception—that the latter has two broad transverse brown bands on primaries, the outermost of which is entirely wanting in the present insect, and the innermost is quite narrow and darker in colour than in *Pinastri*.
Under surface uniform brownish-grey, faint traces of a mesial band on secondaries. In *Pinastri* are the marginal part of primaries a little paler and more ashen than the rest of wing; in this species there is no perceptible change in the colouration.

FEMALE. Expands 3¼ inches.
Head and body same as male.
Upper surface, primaries same colour as male, destitute of all markings save a faint apical line and the obscure streaks in cells between the median nervules near the median nervure.
Under surface uniform dull greyish-brown.
Described from one ♂ and one ♀ example. The former was captured in Canada and was received by me from Mr. Reakirt; the female I took sitting on a fence near some pine woods a mile from Reading, Pa. I have never seen any others. Both examples are in good condition, though the female is a little worn; they seem to me to be an intermediate form between *Sequoia* and *Pinastri*, though very close to the latter.

Some years ago, in the month of October, crawling on the ground among the dead pine leaves in this same piece of woods, I found two larvæ which belonged to some insect of this group, perhaps to this species. My notes say: "Not quite three inches long, rather slender, head yellow striped with red; body reddish, surrounded with many transverse fine black lines; a brown stripe on back from head to anal horn, this stripe lined with white on both sides; on sides alternate bands or lines of green and yellow, green predominating from head to last segment (save one); caudal horn dark reddish-brown; first few spiracles white, the others ringed with red and black; from base of anal horn to end of anal segment, a reddish brown dorsal line."

Unfortunately, with all my care, these larvæ, though they entered the ground, failed to produce perfect insects; nor did I ever after see but one other, which was mutilated by some bird, but I have little doubt but that they were the larvæ of the species I have figured.

If this species be a form of *Pinastri*, I know not, as I have never seen an example of the latter destitute of the broad brown transverse shades of primaries; but should this be the case, it is an easy matter to re-anagramize the name back to its original spelling, and alter its wider gut.

ON THE GENERIC PHANTASIES OF S. H. SCUDDER.

We fear Mr. Scudder's terrene existence is in considerable jeopardy, for has it not been said "whom the gods wish to destroy they first make mad," and to recapitulate all the entomological vagaries that gentleman has indulged in would take volumes; at the magnitude of such a synopsis even Holmes's ghost would stand dumfounded.

The Hesperidæ have been separated by him into myriads of genera, and the genera into counties species. *Nisoniades Juvenalis* has been forced to evolve *N. Virgilius*, *Horatius*, *Ennius*, *Ovidius*, *Tibullus*, *Plautus*, *Propertius*, *Funeralis*, *Terentius*"—these separated from the old species and each other only by a twist or two in the shape, or the millionth of an inch difference in the size of the organ of generation! From the genus *Pamphila* he has educed, on what grounds it would be a wise man indeed who could tell, genera without end: *Prenes*, *Limochores*, *Oclodes*, *Anthomaster*, *Polites* and *Hedone* are a few of the many that at the moment occur to me.

In an evil hour, by some mischance, he came into possession of the old obsolete Hubnerian tract, beginning "Tentamen determinationis," etc., printed (without date) sometime about 1806; he must needs get a reprint of the precious document for distribution at ten cents, or thereabouts, per copy, and, lo! broadcast, like seed of thistle, or fibre pestilence, the thing spread, bringing forth no good fruits. As the leading sheep blindly jumps headlong into a ditch, and the flock as blindly follow, so Grote, ever ready, and mad for any means that might bring his name into notice, enrolled himself under the Tentamen banner, and others of still lesser note, stricken with Tentamania flocked around the same standard. The pages of the Can. Ent. are filled with its virus, the Cambridge organ and the Buff. Bull. (Grote's organ) teemed with it. Scudder, on pages 233–269 of publication just cited, gives his "synonymic List of N. Am. Nymphales;" were it written in the language of the Zulu, it could be no whit more unintelligible to the mass of students than it is. He says: "The following list has been prepared to exhibit in the briefest possible manner the classification, nomenclature, etc.," and that "it is" Heaven forfend) "the Prodromus of a more extended catalogue in which the writer hopes to include a fuller synonymy * * * * and which, through the co-operation of his colleague, Mr. A. R. Grote, will embrace all the Lepidoptera of North America."

He goes on to say that "the aim has been to eliminate everything unessential to the points in view," to which he might have added, which was to try to cram down our throats head and shoulders the most monstrously absurd and incongruous compilation that ever emanated from the diseased brain of man since the advent of Adam. Here is the way you are to distinguish his genus *Satyrodes*: "Hind wings entire;" now you know all about it the moment you see this insect (there is but one of the genus); you know what it is, notwithstanding that a thousand others have the "hind wings entire;" there is a mysterious affinity between you and the insect that tells you it belongs to Scudder's genus *Satyrodes*. You turn to the list and find *Satyrodes Eurydice*, Linn.–Johnss., Amer. Acad., 6, 406 (*Papilio*); Scudd., Rev. Amer. Butt., 6 (*Argus*);" shades of the mighty! what an exhumation of old dead bones; the insect meant, by referring to the synonymy, in exceedingly small italics, is the common *Pararge Canthus*, Linn., (*Boisduvalli*, Harr.), by which name it has been known and cited for a hundred years, and now at this late date we are called upon to change it at *Scudder's* behest. "*Neominois*" is erected by Scudder for an insect (*Satyrus Ridingsii*) allied to *Satyrus Beroe*, Fr., *S. Tryphyte*, Esp., *Semele*, Linn., etc. As a synonym of *Oreynois Wheeleri*, Edwards, he cites our *Satyrus Hoffmani*, entirely ignoring the figures of the male of that

*Paper on Asymmetry, Proc. Bost. Soc. N. H., 1870, Vol. XIII, pp. 277, 288.
†Figured on Plate IV of this work.

ON THE GENERIC PHANTASIES OF S. H. SCUDDER.

species on Plate VIII (Figs. 12, 12), which is as different from Mr. Edwards' description of his males as from *Satyrus Podarce*, Esp.; he quotes only our female figures against Mr. Edwards' males, whilst Mr. Edwards himself says in his description that the female was "not known," and whilst our male is a brown butterfly, as dark as *Alope* on both surfaces, Mr. Edwards, according to his text, has a much larger butterfly, "light yellowish brown" above and whitish beneath, more like the ♀ of our species, cited by Scudder as a synonym, but Mr. Edwards himself stated that he knew not the female—that his 9 examples were all males.

Heliconia Charitonia is given as *Apostrophia Charitonia*. Under *Chlorippe Herse* and *C. Lycaon* are disguised beyond all recognition *Apatura Clyton* and *A. Celtis*, Bdl. et Lec. Our common *Limenitis Ursula*, Fabr., is designated as *Basilarchia Astyanax*. *Vanessa* is cut into four different genera, viz.: (1.) *Polygonia* for the Grapsas; (2.) *Eugonia* for *V. J. Album* and *Californicus*; (3.) *Aglais* for *Milberti* in solitary grandeur. Language fails us—our hand refuses to go further—even the ink on our pen pales—must we record it, that actually the fourth genus taken from *Vanessa* Mr. Scudder has called *Pupilio* and placed in it the one insect *Vanessa Antiopa*, L. Surely no man, not gone stark mad, would be guilty of such unheard of, aye, undreamt of absurdities, and—but each page of this most puerile affair exposes new and wilder extravagances. A separate genus called *Speyeria* is constructed for *Argynnis Idalia*—on what grounds? Because, says Mr. Scudder, "outer half of upper surface of hind wings with two rows of pale markings on a blackish ground, none of the spots confluent;" were I Dr. Speyer I really would prosecute, but alas, what do I say, are the unfortunate answerable in law for their vagaries? After *Speyeria* comes the genus *Semnopsyche*, Scudder—for what? to receive *Argynnis Diana*; here is the foundation of *Semnopsyche*: "basal half of hind wings un-spotted beneath, or with only one or two faint light spots;" after this come *Argynnis*, showing in his arrangement that Scudder considered *Idalia* further removed from *Cybele*, *Atlantis*, etc., than from *Diana*. *Euphydryas*, Scudder, contains a single species, the common *Melitæa Phaeton*, whilst its close ally, *M. Chalcedona*, is transmogrified into *Lemonias Chalcedona*. *Thessalia* is made for *Melitæa Leanira*, Bdl., *Theona*, Men., and *Thekla*, Edw. For *Melitæa Harrisii* we now have *Cinclidia Harrisii*. *Charidryas* is for *Melitæa Nycteis* and *Carlota*; and *Anthanassa* for *Mel. Texana*, Edw., and *Punctata*, Edw. He ends with *Hypatus* for the Libytheidæ, having made out of the Nymphalidæ of N. Am. 56 genera and 187 species, averaging about 3⅓ species to each genus, and God save us from what is to follow if this be only "the Prodromus?"

In the same volume his colleague, Grote, has even outstripped him; for in a catalogue of the N. Am. Sphingidæ (not including *Sesia* and *Trochilium*) he has 39 genera and 74 species—about two species to the genus! It is scarce worth while to go over these freaks of this vainest of egotists; suffice it to say, that *Deilephila Gallii* is here *Hyles Chamænerii*; *Philampelus* is cut into three genera, i. e., *Dupo*, *Philampelus* and *Argeus*; *Smerinthus* into five, i. e., *Paonias*, *Calasymbolus*, *Smerinthus*, *Amorpha* and *Cressonia*—the latter at the expense, as previously stated by us, of *Polygonius*, Hbn.; *Sphinx* is divided into *Lethia*, *Daihs*, *Dilophonota*, *Hyloicus*, *Lapara*, *Dilodia*, *Macrosila* and several others. Mr. Moeschler has ably criticised this wholesale manufacture of genera, and Grote, in a feeble attempt to vindicate himself, keeps in a rambling way to the subject for a dozen lines or so, then goes wandering off into the realms of Ornithology, quoting from a paper on *Sesia Chiombia*, by Dr. Hagen, which has nothing to do with the subject, and is evidently far above his (Grote's) comprehension, at any rate; from this he gets to a paper by Prof. Riley, which causes him much wonderment, because that author wouldn't put our N. American *Apatura* into a different genus from the allied European ones; he then is not agreed that in proposing a generic name an author is obliged to construct a perfect diagnosis, and, he might have added, when it is impossible to do so—and excuses himself by telling us there must be differences of opinion in Entomology as in other matters, and finally winds up with a covert hit at Morrison, delicately intimating that two of that author's species are synonyms, so in fact everything must be that had not gone through his mill; then comes a modest notice about "my suggestions," "affinities," something about the mountains—I mean "the animals" which formerly may have taken refuge on Mount Washington," to escape the flood, we suppose, and we all wonder what the deuce he has been trying to get at, and come to but one conclusion, that it was to exalt Grote above Moeschler, and all creation besides.

Here is an idea of the great fundamental principle that Scudder and Grote are working on: They take the first mentioned species, if it happens to suit their purpose, in any one of Hubner's innumerable "Coitus," and make that the type of the genus; thus, the present genus *Vanessa* embraces insects placed by Hubner in *Polygonia* and *Eugonia*, the first name mentioned in the former is *Polygonia Triangulum*, Fabr., (*Vanessa Egea*, Cram.), about as aberrant in appearance from the rest enumerated as it can well be, and not resembling any of our known species in the sub-genus *Grapta*: so Scudder avoids this one, passes by the next, *C. Aureum*, L., likewise heeds not Cramer's *Progne*, but seizes on *C. Album* as the type of *Polygonia*; thus he has resurrected *Polygonia*, which must stand as a distinct genus for the reception of those *Vanessas* previously comprised in Kirby's sub-genus *Grapta*. Now we come to *Eugonia*; Hubner's type of this coitus was a rather unfortunate one, as the enumerated species enumerated after the previous coitus *Polygonia* happens to be the first one which he placed in this next coitus *Eugonia*, namely, *Angulosa*, Cram., which is but a synonym of *C. Aureum*, L., a Japanese species which is as close to such species as *Polychloros* and *J. Album* as it is to *Progne* and *C. Album*; but, to make all things square, Hubner, with wonderful sagacity, places it in each of his two coitus, only under a different name in each one, so that neither Linne nor Cramer would have their species in this instance ignored; so extreme must, as the last species (*V. Album*, Esp.) in *Polygonia* is only the same as the first under another name, or rather under two other names. Scudder, in adopting the genus *Eugonia*, took no notice of the first mentioned species, but passing over it and the next (*Rhinocalpa Polynice*, which has no more to do with *Polychloros* and *C. Album* than it has with *Agrerouia Fauna*), and likewise the next *V. Album*, Wien. Verz.,=*J. Album*, Bdl. et Lec.), he comes on *Polychloros* which he makes the type of *Eugonia*, and then the first of the two species he puts in that genus is the identical *J. Album* = *V. Album* = *L. Album*, Esp., which he has ignored as the type in favor of *Polychloros*.

Grote, in his N. Am. Sphingidæ, imitates Scudder, in following Hubner, but altering, of course, from the latter as his occasions may require. For instance: In adopting Hubner's genus *Lethia* he takes the last species (*Sphinx Gordius*) mentioned as his type, instead of the first, (*Sph. Ligustri*;) and *Sph. Drupiferarum*, the nearest American ally to *Ligustri*, he has placed in another genus. Hubner's coitus *Polyptychi*, which contains *Jugulandis*, he ignores, as it would conflict with his own genus *Cressonia*, created for that species. All that is yet wanting to complete Grote's work is to follow Hubner again and to bend his genera with short lucid descriptions, Scudder has already done so, which would enable the student at a glance to know the species included in their countless tribes, stirps, familie, coitus, etc. Here is Hubner's diagnosis of his family *Angulati*, comprising his *Paonia* and *Mimantes* (*Smerinthus Ocellata*, *Myops* and *Excaecatus* in the first, *Sm. Tiliæ* in the second): "The body beautifully coloured; the wings bluntly angulated, lightly shaded."*

If that isn't enough to identify one of the *Smerinthi* a mile off, I don't know what is.

In Hubner's coitus *Acherontiæ* are *Ach. Atropos*, L., *Sph. Chionanthi*, Abb. & S., (*Muc. Rustica*, Fabr.,) and *Ach. Morta* (*Lethe*, West); the first the common European death's-head, the last the African species, and between them is put our *Sphinx Rustica* (*Chionanthi*,) for the one reason, doubtless, that, like the other two, it has a skull-like marking on the back of the thorax; but I much fear Hubner, like his imitator, Grote, did not know half the things he wrote about, for in his next coitus *Chionanthi* we again find *Chionanthi* under its older name of *Rustica*. In the fifth stirps, *Exhibituæ*: "The body small; the wings large, peculiarly ornamented;"† the third family of this stirps, *Communiformes*, is thus described: "the wings of common form; variously ornamented and coloured;"‡ would the sagacit of any man living recognize *Eudus Imperialis* or *Cith. Regalis* as belonging to the above family? Stirps 7, family D, are coitus 1,

---

*"Der Rumpf farbig ausgezeichnet; die Flügel stumpf eckig, sanftschattig angelegt." Verz. p. 142.
†Der Leib klein; die Flügel gross, sonderbar gezeichnet. Verz. p. 151.
‡Die Flügel gemeinförmig; unterschieden gezeichnet und gefärbt. l. c., 153.

Adelpha containing seven species of the insects more familiarly known as, and now embraced in Boisduval's genus *Heterochroa*, and one *Nymphidium*. Coitus 2, *Nepio*, contains two astonishingly dissimilar insects generically, though superficially bearing some resemblance: the *Eresia* (*Pap.*) *Nauplia*, Linn., and *Olina* (*Pap.*) *Emilia*, Cram. Coitus 4, *Metamorpha*, (defined as having "the wings very oddly formed, grandly decorated with green spots,") is composed of *Victorina Sulpitia*, Cram., *V. Steneles*, L., and *Colænis Dido*, L., whilst *Colænis Delila* and *Aleionea* are put not only in another stirps, with some forty coitus between them and *C. Dido*.

But, after all, there is this one most important difference between Hubner and his present imitators, that though his writings were fanciful he gave most accurate and reliable figures of his species, which they do not.

It is unfortunate, most unfortunate, that owing to the existence of the mutual admiration society which embraces so many of the American Lepidopterists, there has been but little protest against the phantasms of the authors alluded to; there is no fear that the scientists of Europe will at their diction adopt such Laputian nomenclature, but there are here many beginners and less advanced students who have, unhappily, partly adopted the style of nomenclature of these Chams of Lepidopterology, as their catalogues, lists, etc., are published in cheap periodicals, easily obtainable, whilst the solid, real work of the older as well as the present standard authors on entomological science are not so easily accessible, which is the more to be regretted, as though Scudder and Grote are actuated by widely different motives in their writings, still both produce the same pernicious results; Scudder's lists, theories, etc., seem to be gotten up to show what amount of time and labor one human being is capable of completely wasting; whilst, were it not for his overweening egotism, it might possibly occur to Mr. Grote that there was some other object in publishing catalogues, etc., than that of the endless repetition of the name of the compiler.

NOTES, NEW SPECIES, ETC.

SAMIA CEANOTHI is the correct name of the *Saturnia* fig. 2, Plate XII. It was fully described by Dr. Hermann Behr in Proceedings California Acad. Nat. Sc., Vol. I, p. 47, April, 30, 1855; the author at the same time presented the Academy with a drawing of the insect, as well as a specimen of the cocoon, remarking that it was found on *Ceanothus thyrsiflorus*, also on a *Rhamnus* and a *Photinus*, and that it was likely to prove valuable. Again, on pages 68–69 (l. c.), Aug. 27, 1855, in recording donations to the Cabinet of the Academy, is the following: "From Dr. Behr, a specimen of the Cal. Silk Worm (*Saturnia Ceanothi*, Behr)."

Ten years later Grote, entirely ignoring the above description, redescribed the species as *Platysamia Californica* in a foot-note in a paper on "Bombycidæ of Cuba" in Proc. Ent. Soc., Phila., Vol. V, p. 229, Dec., 1865. At the conclusion of his description he adds: "It is not impossible that this species may be *Saturnia Euryale*, Boisd.; if so, this latter name cannot obtain, since it has not, as far as I am aware, been sanctioned by any description."

Sancta simplicitas!

LYCÆNA CATALINA. Dr. Behr informs me that his *Lycæna Lorquini* is identical with this species; its citations are thus:
LYCÆNA CATALINA, Reakirt, Proc. Acad. Nat. Sc., Phila., p. 244 (1866).
Lycæna Lorquini, Behr, Proc. Cal. Acad. Nat. Sc., Vol. III, p. 280 (1867).
Lycæna Rhœa, Boisduval, Lep. Cal., p. 51 (1869).
Lycæna Dœunis, W. H. Edwards, Trans. Am. Ent. Soc., Vol. III, p. 272 (1871).

LYCÆNA RAPAHOE, Reakirt. Have compared the types of this species with a large number of examples of *Lyc. Icarioides*, and can come to no other conclusion than that they are the same. Dr. Behr writes me that *Ly. Rapahoe* as figured in this work is identical with *L. Dædalus*, in which event the nomenclature of the species would be
LYCÆNA ICARIOIDES, Boisduval, Ann. Soc. Ent. Fr., p. 287 (1852).
Lycæna Rapahoe, Reakirt, Proc. Ent. Soc., Phila., Vol. VI, p. 146 (1866).
Lycæna Dædalus, Behr, Proc. Cal. Acad. Nat. Sc., Vol. III, p. 280 (1867).

LYCÆNA PHERES, Bdl., and L. EVIUS, Bdl. Dr. Behr suggests maybe local varieties of the same species, as he has intermediate examples; in my own cabinet are also a number of the latter, showing various gradations from one form to the other, and I have no doubt but Dr. Behr's surmise is correct.

LYCÆNA OPTILETE Knoch. (Papilio O.) Beit. Ins. Ges., I, p. 76, t. 5 (1781).—This species may be added to the N. Am. fauna, as through the kindness of mine honored friend, Dr. Behr, I have received several examples that were taken in Alaska; they present no difference whatever from the European examples; it is a species having a wide range, being found in Germany, Russia and Siberia.

SPHINX STRIGII, Bdl., which has been accredited to California, I have received from South Africa. It belongs to the same group as *S. Chærilus*, Cram., *S. Pinastri*, etc.

Through the kindness of Prof. C. V. Riley I have examined the plates of Sphingidæ recently published by Dr. Boisduval, two new ones from N. America are figured in both imago and larva under the names of *Sphinx Catalpæ*, Bdl., and *S. Cupressi*, Bdl., the latter from Georgia evidently belonging to the *Pinastri* group. Prof. Riley has had the larva of *S. Catalpæ*, but has not so far, I believe, been successful in rearing it.

MELINÆA DORA, N. S.

Expands 3½ inches. Head black, a yellow dorsal line and yellow points at the eyes; antennæ yellow, except a short space towards the head where they are black; body above brown, below yellow; on thorax a yellow central dorsal line. Inner half of primaries brownish red; a yellow, somewhat irregular bar extending from below middle of exterior margin diagonally to middle of costa, exterior to this band the wing is black; in the discoidal cell is an irregular black mark, and another at end of the cell joins the yellow band, this latter is also joined inwardly by another irregular black mark reaching from the middle to the exterior margin of wing; between this mark and the one in the discoidal cell is a black spot; a black dash at base of wing; on costa a black streak, and another longer, broader one at inner margin, extending from base to half the length of the wing. Secondaries brownish red, from middle of exterior margin to apex very narrowly margined with black, widest at veins; beneath as above, but the black marks of primaries smaller; base of secondaries yellow, accompanied with a short black streak; a small black spot in discoidal cell. I received this example from Mr. Reakirt, whose MSS. name I have retained for it; its locality was Esmeraldos. Allied closely to the lately described *M. Röberi*, Staudinger, but differs from it in the absence of all spots on the black apical part of primaries, in the absence of the yellow spot at inner angle, in the absence of the broad black margin of outer half of interior margin, and in the presence of the black dash on inner half of same; and on reverse of secondaries, in the position of the black spot, which is within the discoidal cell in ours, and outside of it in Staudinger's species; the latter is from Central Am. To its author I am indebted for my examples.

It has been accurately figured by Mr. Gustav Weymer in the Stett. Entom. Zeit. (1875). Taf. II, fig. 4.

Ageronia Anomala, Nov. Sp.

Male expands 3 inches; outline of wings same as *A. Amphinome*, except that the outer half of costa of primaries is more arched; colour and ornamentation of upper surface also precisely the same as in that species, with the single exception of the small white lunate spot at costa, mid-way between the broad white band and apex, which is wanting in our species; in *all other respects*, even to the smallest minutiæ, it is the same in appearance, and no one, on even the most critical examination, could by this surface separate it from *Amphinome*. Under surface of all wings same dark shining brown as in *A. Arete*, towards base a shade paler; primaries have the white transverse band of upper surface repeated, and a very small white spot on apex, otherwise immaculate; secondaries have two basal scarlet spots, one on costa and the other just anterior to it, this latter is shaded a little exteriorly with black; a row of scarlet submarginal spots situated as in *Arete*, with the exception of the anal one of that species which is here wanting, there being but four, whilst in *Arete* there are five; a small white apical spot. The whole under surface presents almost the exact appearance of *Arete*, ♀; the only point that strikes the eye as at all different is the irregularity of the edges of the white band of primaries; but the two basal spots of secondaries are also not like in *Arete*, round, but irregular, more like splashes or suffusions.

Taken high up the Amazon in several examples, all of which, save one, however, were lost; in same locality were taken with it many *Megistanis*, *Deucalion* and *Brotis*; also *Morpho Cisseis*, Bates,* *Agrias Lucina*, Feld.,† a *Eunica* (allied closely to *Alcmena*, Dbldy., but smaller, and violet where that species is blue,) besides many other species of both butterflies and moths.

Catocala Herodias. Nov. Sp.

Female expands 2¼ inches. Head and thorax dark smoky gray; abdomen grayish brown; beneath light gray. Primaries above same dark smoky gray as thorax, almost evenly coloured, only a little darker towards exterior margin; transverse lines, fine and black, but inconspicuous; veins from exterior margin to transverse posterior line black; a small black basal streak; reniform almost obsolete, subreniform widely open; the transverse posterior line between subreniform and costa in four unusually long sharp teeth; fringe gray. Secondaries same red as *Corcinata*, a little darker; mesial band, very irregular, from costa not quite one-third in broad, then quite narrow, then broad and strongly curved like an ≈, then a mere line, continued suddenly to a triangular patch which does not reach quite to inner margin; marginal band broad at costa, somewhat gradually narrowing towards inner margin, which it does not quite reach; apical line red; fringe white, black at terminations of some of the veins. Under surface primaries pinkish; all the three dark bands connected by a dark shade on inner margin; secondaries, inner two-thirds pink, costal part white, bands as above.

Texas, one ♀ example, taken by Mr. Belfrage. No. 501 of his collection.

This insect has much the appearance of *C. Lupina*, H-S., from Armenia, the colours are the same, and the upper wings of both species are decorated in much the same manner, save that *Lupina* has a bold black longitudinal basal dash, which is replaced in ours with the merest line; the mesial bands of primaries are entirely different in the two species. *Lupina* is smaller than the present species, and the under side of primaries is destitute of pink or reddish tinge.

Catocala (*Corcinata*) Var. Circe. Nov. Var. or Sp.?

Male and female expand 2¼—2½ inches. Head and thorax whitish gray, with black lines, abdomen grayish brown, beneath white. Primaries same whitish gray as thorax; marks as in *Corcinata*, but heavily black and accompanied with black shades; reniform tolerably distinct, subreniform conspicuous; a broad blackish shade extends, more or less interrupted, from base to exterior margin. Secondaries and under surface as in ordinary forms of *Corcinata*.

I have examined six examples taken by Mr. Belfrage in Bosque Co., Texas, and find them all constant to the form above described. I have little doubt but that this is a form of *Corcinata*, but marked and constant enough to perhaps deserve a separate designation.

Catocala Semirelicta, Grote, is undoubtedly a variety of *Briseis*, W. H. Edw., bearing the same relation to the latter as does *Phalanga* to *Polygama*. I have many and varied intermediate examples.

Bunæa Eblis. Nov. Sp.

Male expands 5½ inches. Upper surface rather dark umber brown. Primaries falcate; a narrow white submarginal band, or rather line extending from costa to inner margin; indistinct transverse mesial and sub-basal shades, the latter undulate, the former almost straight; a small transparent discal lune. Secondaries produced in an angle at middle of exterior margin; a white submarginal band or line much farther from exterior margin, between anal margin and middle, than between latter and costa; from the anal margin this line is nearly straight to a little beyond the middle of wing, when it curves rather abruptly upwards to the costa; in the centre of the wing is an ocellus nearly one inch in diameter, this is formed by a large brown spot shaded into jet black outwardly towards its edge, and containing in its middle a small transparent triangle; this spot is surrounded by a vermilion coloured ring, and this latter by a white line. Under surface brownish gray, faint traces of the white submarginal lines of upper surface made by a brown line; this submarginal line on secondaries is not bent at middle of wing as above, but goes almost straight across from anal margin to apex; the space from the submarginal lines to the outer margin darker colour than rest of wing; a transverse narrow brown median shade crosses all wings; transparent discal spots as above.

From Calabar, W. Africa, presented to me as a Christmas box by my ever dear and tried friend, Mr. T. Chapman, of Glasgow, Scotland, who in his letter announcing the gift remarked that he knew it would be to me far more acceptable than a turkey or a keg of whiskey.

Heliothis Regia, Nov. Sp. Expands 1$\frac{1}{10}$ inches. Head and body white. Ground colour of all wings on both surfaces white. Primaries marked much in same style as *Rivulosa*, Guen., but widely different in colouration; the basal part and submarginal band are purplish crimson; the middle of the white central space is yellowish brown, or olivaceous, with an almost golden tint in some lights; the outer part of white marginal band is also of this same colour. Secondaries with faint ill defined, rather broad purplish crimson marginal band. Under surface, primaries, basal part white, marginal band white, rest crimson. Secondaries with a crimson apical spot, and another smaller one on margin, midway between this and basal angle. Texas, taken by Mr. J. Boll.

Heliothis Fastidiosa, Nov Sp. Expands ¾—1 inch. Head and body olive yellow, collar tinged with red. Primaries yellowish olive, median space tinged with pinkish, and separated from basal and marginal parts by very narrow whitish lines, not in all cases, however, perceptible; basal space palest near body, darkest towards median space; marginal space palest outwardly, deepening into brown towards median space; discal mark large, but somewhat obscure. Secondaries pale dull yellowish, marked with black after the manner of *Dipsaceus*, L. Under surface yellowish white; primaries; a black basal patch; a large black discal spot which connects with the submarginal band; inner two-thirds of this latter is black, and costal third is crimson; inferiors marked below as above, but the outer half of marginal band is crimson, and the costa also is of that colour. Upper surface has, on a superficial glance, some resemblance to *Cardui*, Hub. Texas, J. Boll, (No. 31).

*Wien. Ent. Monat., Vol. IV, p. 189, T. 4 (1860).
†l. c., Vol. VI, p. 110 (1862).

HELIOTHIS SIREN, Nov. Sp. Expands 1⅛ inch. Head and thorax yellow, inclining to rust colour; abdomen above black, beneath yellowish. Upper surface; primaries shining greyish yellow, much same tint as in *Lynx*, Guen., also style of decoration much as in that species; the basal part is rust or sienna coloured, mixed with black on or towards edge nearest the median space; the submarginal band is also rust coloured, with exterior and inner edges mixed with black; this band is suddenly narrowed to a mere line about one-third in from the costa; an indefinite rust coloured shade through middle of median space; discal spot rather small; fringes same colour as ground of primaries. Secondaries all black, with pure white fringes. Under surface of primaries black, with a white exterior margin, and a narrow yellowish white edging on costa. Inferiors black, with large white apical space extending along costa, two-thirds in; fringe white. Texas, J. Boll (No. 45).

HELIOTHIS INCLARA, Nov. Sp. Expands 1 in. Head and body yellowish; back of abdomen, except terminal segments, shaded with blackish gray. Upper surface; primaries yellowish gray, uniform in shade; a basal patch of livelier darker yellow, shaded exteriorly with brown, this basal patch does not reach to costa; a rather narrow submarginal band of darker colour than the ground, and traversed through its length by a macular shade of blackish. Secondaries greyish, with a very broad black margin, discal spot almost imperceptible, fringe pure white. Under side whitish, with a slight yellow tinge, especially on inner and costal parts of primaries, on which are a blackish basal patch, discal spot and broad submarginal band, all three more or less merged into each other. Secondaries with a broad blackish submarginal band, which, however, reaches only from anal angle to about half way between the latter and the apex. Texas, J. Boll (No. 46), allied to the preceding *H. Siren*.

HELIOTHIS NUBILA, Nov. Sp. Expands 1⅕ in. Head and body above greyish, of an olivaceous greenish tinge; beneath whitish. Upper surface; primaries same colour as body; the median lines very faintly defined, especially the inner one, which is almost obsolete; the submarginal space of a darker shade than rest of wing; basal part scarcely distinguishable from median space; no discal spot or blotch perceptible; fringe same colour as wings. Secondaries white, of a yellowish cast, marginal band broad, black and straight on inner edge; discal spot large and black; some black scales at base; fringe white. Under surface yellowish or tawny white; on primaries a longitudinal black basal patch reaching neither to costa or inner margin; a round black discal spot; a broad black submarginal band which does not extend to either inner margin or costa; between it and the latter the colour is reddish. Secondaries; an indistinct discal spot, and a submarginal band, inner half of which is black and apical half reddish. Texas, J. Boll (No. 48.)

HELIOTHIS RUBIGINOSA, Nov. Sp. Expands 1¼ inches. Head, thorax and primaries, above, near the colour of *Xanthia ferrugineoides*, Guen., but a trifle more inclined to reddish, not so yellowish; abdomen paler, more whitish; body below sprinkled with cinnamon coloured scales. Median lines of primaries white, faint, and rather irregular; the outer and basal spaces shaded inconspicuously with grey at these lines; otherwise no difference in shade of colour between median and outer and basal spaces; fringe greyish. Inferiors pale tawny; broad marginal band and discal spot, very faintly defined by slightly darker shade of cinnamon, or reddish tawny. Under surface same colour, palest at inner margin; discal spots barely visible; no other definite ornamentation. Texas, J. Boll (No. 50).

HELIOTHIS IMPERSPICUA, Nov. Sp. Expands 1¼ inches. Head and thorax rather pale gray, inclining to olivaceous; abdomen ochrey white. Upper surface; primaries same colour as thorax, and in style of marking approach *Rivulosa*, Guen., and *Thoreaui*, Grote, though far paler in colour than either, resembling in this respect *Schinia trifascia*, Hub., *Oregona*, Morrison, etc. The white line which separates the basal from median space is but very little bent, and the one between the marginal and median space is not bent inwardly below the middle as much as in the allied species above alluded to, discal mark only perceptible by the slightest possible darker shade. Inferiors as in *Rivulosa*, but much paler. Under surface also much as in that species, but far paler. Texas, J. Boll (No. 53). This species has the same relative appearance to *Thoreaui*, Grote, as the latter has to *Rivulosa*, Guen.

HELIOTHIS ULTIMA, Nov. Sp. Expands 1 inch. Head and thorax above sober brown; abdomen yellowish-brown; beneath yellowish; some reddish scales at terminal half of abdomen. Upper surface; primaries same brown as thorax; median space not differing much in tint from rest of wing; median lines narrow, white and indistinct, and much bent like in *Spinosae* and *Rivulosa*, but the median space is very much more contracted than in those; discal shade almost imperceptible; outer space a little paler at margin; fringe whitish. Superior wings yellow with broad black margin and small discal spot; fringe white. Under surface coloured and marked much as in *Spraguei*, but much paler throughout. Texas; J. Boll (No. 49).

HELIOTHIS SPECTANDA, Nov. Sp. Expands 1¼ inches. Head and thorax above pale greenish; abdomen ochraceous white; beneath white; fore legs dull red or maroon, median tarsi also reddish. Upper surface; primaries, pale whitish green, somewhat like the tints in *Schinia Trifascia*, but greener, and crossed by three almost straight transverse paler bands or lines, the one adjoining the submarginal space the widest, within the latter a transverse row of faint spots; reniform and orbicular spots pale, but clearly defined; fringe same colour as rest of wing. Secondaries white with broad brownish marginal border, broadest in the middle; fringe white. Beneath white with greenish tinge and opalescent reflections; a brown discal spot and faint evidences of an abbreviated submarginal band on the primaries. One example, Texas; J. Boll (No. 52). Is apparently near to *Heliothis Armigera* and allies, but possibly may belong to another genus.

ÆNIGMA, Nov. Gen. Head very small, across the eyes at widest place scarce one-third the width of thorax, which latter is of great breadth; palpi long, tapering and close together, giving the appearance of a snout or proboscis; antennæ nearly the length of abdomen, very slender, filiform; median and hind tibiæ spurred; abdomen, which is rather slight compared with the immense thorax, projects beyond hind wings; fore wings much the same shape as in *Hypoena filicornis*, Guen., but more produced on inner margin towards but not at the base; hind wings broad, not strongly curved on outer margin. As the type is a unique, and not my property, I could not denude it, and can, in consequence, say nothing of the nervation, excepting that there appear to be a great number of subcostal veinlets on the superiors. The insect on which I have based this genus seems in some respects to be allied to *H. filicornis*, Guen., and in many more to be distinct therefrom. It is a ♀ example, as denoted by the forked frenulum.

ÆNIGMA MIRIFICUM, Nov. Sp., ♀. Expands 1⅝ inches. Head, palpi and thorax above dark slate-grey; abdomen black with each segment narrowly edged behind with yellow; anal brush yellow; beneath, palpi, head and body pale yellow, becoming darker towards and at extremity of abdomen. Upper surface; primaries small, same slate-grey as thorax, spotted over whole surface with small dark points or dots which have a tendency to form tolerably regular transverse rows from costa to inner margin; fringe concolourous with wing. Inferiors black, an irregular, interrupted yellow mesial band extending from near anal angle somewhat upwards to middle of wing, where it is broken by a line of the black ground colour, and continued by and terminating in a nearly round yellow spot which lies much nearer the exterior margin than does the middle of the yellow band of which it forms a disconnected part; it does not extend to costa or apex, but has considerable of the black ground colour intervening; from a little within the middle of exterior margin to the anal angle extends a conspicuous yellow marginal patch, almost as in *Hypocola filicornis*; fringe light grey. Under surface rather bright yellow; primaries with black submarginal and median bands, neither of which extend to costa or internal margin, but near the latter are connected by a black band which extends nearly to the base of wing; fringe grey. Inferiors with a number of scattered brown dots; a large black spot at anal angle, and another of same size close to it; these two spots, the latter of which is continued towards apex by some brown scales, form an imperfect submarginal band; an abbreviated black dash near abdominal margin. One ♀, Texas; J. Boll (No. 55).

TEN MINUTES' NOTICE OF

"A Check List of North American Noctuidæ, Part I.—BY A. R. GROTE, A. M.,

Containing Notes and Descriptions, Remarks on Structure and Geographical Distribution of the Group, and one Photographic Plate, illustrating the Species from California and the East. The Species (790) are numbered for convenience of students. The list will be mailed free on receipt of price."

This tempting advertisement allured us into remitting the price, and in due time we were the recipient of a small octavo of 28 pages, title page, preface and index included therein, which, on opening, discovered to us that we had paid our dollar for the intense gratification of possessing twenty-eight pages full of Mr. Grote's name, printed in every imaginable variety of type, pica, long primer, brevier, etc., etc.

The names of the species are in double columns; no citation of place of original description of species, or statement of locality are given—the same old story over and over until the heart sickens at the low egotism which displays itself in everything this compiler does; it is *Heliocopsis Scripta, Grote*, not *Thyatira Scripta, Guen.*; it is not *Harrisimemna Sexguttata, Harris*, but *H. Sexguttata, Grote*, not *Acronycta Clarescens, Guenee*, but *A. Clarescens, Grote*, and so on. Such a tearing away of the old landmarks never was seen.

In some instances the name of the true author of the species is placed below Mr. Grote's; in others he appropriates unblushingly the proprietorship of another's species without giving the poor author a chance between brackets, or even in small type, as in the case of *Mamestra Indefera*, boldly it stands out "*M. Indefera, Grote, Agrotis Ind., Grote*," no hint that Guenee was the author of the species (Noct. II, 76, 768), but a bare-faced piece of pirating. To make up for it he has, however, given *Pachnobia Carnea* as Guenee's species, when it really was described by Thunberg (Mus. Nat. Acad. Upsal, p. 72, f. 1, 1788, before M. Guenee was born. "*Hadena Laterilia*, (Herb.)" happens to be Hufnagel's species, published in 1767, when Hubner was about six years of age; of this species *Apameformis, Guen.*, is a synonym, though given as distinct by Mr. Grote. *Glabulus, Morrison*, which he has placed with *Amphipyra Pyramidioides* and *P. Tragopoginis* in the genus *Pyrophila*, is a true *Agrotis*, allied to *A. Baja*.

The synonyms are by no means always given, though it is stated that they are.

The Catocalæ are not embraced at all in the present list.

Some species are entirely omitted, others are so disguised that their own fathers would never suspicion them; and thus, with new names for old genera, and very antique, obsolete genera—long disused, or scarce ever used—resurrected, the compilation can be of no value to the beginner; and to the advanced student can serve but to excite a smile at the wholesale way in which the father of our science have been invited to step to the rear, whilst Mr. Grote and his friends stand forward in boldest relief, all duly Latinized, as Radcliffe-i, Harvey-ana, Harvey-i, Glenny-i, Hayes-i, Chandler-i, Day-i, Stewart-i, etc., etc., including the officers and the whole board of directors of the Buffalo Society, who will doubtless be grandly carried down to posterity on the wings of these unfortunate little moths.

The foot-notes on each page might be condensed into one, which would take in the full meaning of all, to this effect: That none of Mr. Walker's, or Mr. Morrison's, or anybody else's names can be identified from published data, except Mr. Grote's.

"The "remarks on structure" and "geographical distribution" occupy two entire pages: three and one-fourth pages more are devoted to describing eight new noctuids and in showing the author's superiority as a scientist over Mr. Morrison and everybody else. There are also a few more new genera ground out in these pages.

The last 1½ pages are taken up with the index of genera, where we seek in vain for the old familiar names; all here is new; varnish and veneer glare on every page.

The "photographic plate illustrating the species from California and the East," illustrates in a gloomy, shadowy sort of way ten whole species by ten whole figures, (save the antennæ of some,) of which the *Hadena Badistrига* and *Agrotis Funeralis* are recognizable; two others, *Acronycta Lithospila* and *Xylina Thaxteri* may perhaps be; the remaining six, like the rural artist's drawing of a horse, require the names to be written beneath them to insure identification.

The whole thing is scarcely worth the time devoted to this review, but as the advertisement would lead us to expect quite a different production, than that really furnished, we have given this cursory warning because the price demanded is entirely too big to pay for trash paper.

JAN., 1876.

Of the following species I am anxious to obtain examples, either by exchange or purchase; any Naturalists having duplicates of any of them will confer a great favor by communicating with

HERMAN STRECKER,
Box 111 Reading P. O.,
Berks Co., Pennsylvania, U. S. of N. America.

Ornithoptera Hippolytus, Cram.
Ornithoptera Lydius, Feld.
Ornithoptera Helena, Linn. ♀
Ornithoptera Croesus, Wall. ♂
Ornithoptera Brookiana, Wall.
Papilio Antimachus, Dru.
Papilio Evan, Doubl.
Papilio Pericles, Wall.
Papilio Blumei, Boisd.
Papilio Macedon, Wall.
Papilio Philippus, Wall.
Papilio Phorbanta, Linn.
Papilio Homerus, Fabr.
Papilio Garamas, Hub.
Papilio Caiguanabus, Poey.
Papilio Ascanius, Cram.
Papilio Wallacei, Hew.
Papilio Slateri, Hew.
Papilio Icarius, West.
Papilio Dionysos, Dbldy.
Papilio Gundlachianus, Feld.
Papilio Elephenor, Dbldy.
Papilio Disparilis, Herr-Sch.
Euryades Corethrus, Bdl.
Euryades Duponchelii, Luc.
Thais Honoratii, Bdl.
Argynnis Rudra, Moore.
Argynnis Oscarus, Evers.
Argynnis Cnidia, Feld.
Argynnis Jerdoni, Lang.
Argynnis Dexamene, Boisd.
Argynnis Jainedeva, Moore.
Argynnis Ruslana, Motsch.
Argynnis Anna, Blanch.
Argynnis Aruna, Moore.
Parthenos Tigrina, Voll.
Penetes Pamphanis, Dbl., Hew.
Charaxes Epijasius Reiche.
Charaxes Kadenii, Feld.
Charaxes Jupiter, But.
Charaxes Etheocles, Cram.
Catagramma Excelsior, Hew.
Dyctis Biocolatis, Guerin. et var.
Romaleosoma Sophron, Dbldy.
Romaleosoma Pratinoe, Dbldy.
Romaleosoma Arcadius, Fabr.
Nymphalis Calydonia, Hew.
Saturnia Epimethea, Dru.
Calinaga Buddha, Moore.
Diadema Boisduvalii, Dbldy.
Pavonia Aorsa, West.
Any species of Phyllodes.
Any Asiatic Catocalae.
Dynastor Napoleon, Doubl., Hew.
Argynnis Sagana ♂, Doubl. Hew.
Zeuxidia Aurelius, Cram.
Brahmæa Whitei.
Brahmæa Certhia.
Eudæmonia Semiramis, Cram.
Eudæmonia Derceto, Msen.
Eudæmonia Jehovah, Strock.
Thaliura Croesus.
Nyctalemon Metaurus.
Nyctalemon Cydnus, Feld.
Actias Mænas
Saturnia Argus, Drury.
Saturnia (Actias?) Artemis, Brem.
Saturnia Atlantica, Luc.
Any Asiatic species of Parnassius.
Citheronia Phoronea.
Pyrameis Gonerilla, Fabr.
Pyrameis Abyssinica, Feld.
Pyrameis Dejeanii, Godt.
Pyrameis Tameamea, Esch.
Sphinx Cluentius.
Vanessa v. Elymi, Rbr.
Dasyopthalmia Rusina, Godt.
Morpho Phanodemus, Hew.
Any species of Agrias.
Any species of Callithea.
Pandora Chalcothea, Bates.
Pandora Hypochlora, Cates.
Pandora Divalis, Bates.
Colias Viluiensis, Men.
Colias Pouteni, Wallengr.
Bunæa Deroyllei, Thom.
Bunæa Phædusa, Dru.
Rinæa Zuleica, Hope.
Acræa Perenna, Dbldy.
Opsiphanes Boisduvalii, Dbldy.
Clothilde Jægeri, Herr-Sch.
Pieris Celestina, Boisd.
Euplœa Eurypon, Hew.
Paphia Panariste, Hew.
Limenitis Lymire, Hew.
Io Beckeri Herr-Sch.
Eacles Kadenii, Herr-Sch.
Hepialus Giganteus, H.-S.
Smerinthus Panopus, Cram.
Sphinx Substrigilis, West.
Saturnia Larissa, West. ♀
Eusemia Victrix, Amatrix, Dentatrix, West.
Smerinthus Modesta, Fab. (nec. Harris.)
Smerinthus Tartarinovii, Brem.
Smerinthus Dentatus, Cram.
Smerinthus Meander, Bdl.

These are a few of the very many of the rarer species that I am eager to procure; of course there are numberless others from all parts of the world, equally desirable and coveted by me.

Arctic Bird-Skins, Eggs, Plants, Etc., Etc.

I would bring to the notice of those interested that I have on hand for sale fine examples of BIRD-SKINS and EGGS from Greenland, Labrador and other parts of Arctic America, including many very rare and desirable species of Hawks, Ducks, Gulls, Auks, Grouse, etc., etc., all correctly identified. Also, collections of pressed PLANTS from Arctic America in beautiful examples and correctly named.

LEPIDOPTERA, native and exotic, either on hand or can be obtained for clients at short notice.

COLEOPTERA and other Insects occasionally on hand and can always be obtained if ordered—particulars given on application by letter.

I will also sell Insects on commission for persons having such to dispose of.

I am always glad to exchange for any species of Lepidoptera not in my collection, or to obtain such by purchase if exchanging be not desirable.

HERMAN STRECKER,
Box 111 Reading P. O., Pa.

S. D. JONES,
Successor to the North Atlantic Express Co.

CUSTOM HOUSE BROKER,
Solicitor and Forwarding Agent,
48 Pine Street, Room 4, New York.

Difficult cases arising under the Revenue Laws solicited.

W. V. ANDREWS,
ENTOMOLOGIST, &C.,
36 Boerum Place, Brooklyn, New York.

Purchasing Agent for Books and Apparatus in connection with Natural History. Also, Cork Pins, &c. Eggs of the different varieties of Silk Worms, to order. Lepidoptera and Coleoptera for sale or exchange. Agent for WALLACE'S SILK REELER, and for KIRBY'S SYNONYMIC CATALOGUE OF DIURNAL LEPIDOPTERA.

No. 14. ISSUED QUARTERLY, AT 50 CENTS PER PART IN U. S.
IN EUROPE, 2 SHILLINGS.

LEPIDOPTERA,

RHOPALOCERES AND HETEROCERES.

INDIGENOUS AND EXOTIC;

WITH

Descriptions and Colored Illustrations,

BY

HERMAN STRECKER.

Reading, Pa., 1877.

Reading, Pa.:
OWEN'S STEAM BOOK AND JOB PRINTING OFFICE, 515 COURT STREET,
1877.

MACROGLOSSA ERATO. Boisduval.

Lep. Cal., (Ann. Soc. Ent. Bel. XII), p. 65, (1868).
Butler, Trans. Zool. Soc. Lond., IX, p. 529, (1877).
Euproserpinus Phaeton, Grote & Rob., Proc. Ent. Soc. Phil., V, p. 178, (1865). Trans. Am. Ent. Soc., II, p. 181, (1868). *Hy. Edwds.*, Proc. Cal. Acad. Sc., (1875).

(PLATE XIV, FIG. 1.)

    This rare little species having been described at length by Dr. Boisduval, and also by Grote & R., as I give a figure of the upper surface I may be spared from nauseating the student by further repetition except to add that the under side of primaries is white bordered with fuscous at outer margin, and secondaries pale yellow with black border outwardly.
    It occurs in Los Angeles County and probably in other parts of Southern California. I am indebted for the possession of the original of the accompanying figure, which was taken by the late G. R. Crotch, to the enduring goodness of the great savan, Dr. H. Hagen.

    The name *Phaeton* adopted by Grote cannot for a moment be entertained as his original description in Proc. Ent. Soc. V, 178, was made from a picture and not from any real insect, for particulars of which permical attempt see foot-note on page 115 of this work.
    Several years later when Grote accompanied his patron Robinson on a visit to Europe they received the species from Dr. Boisduval and on their return home gave from said example their redescription, which however was not in time to supersede that of Boisduval published in Ann. Soc. Ent. Belge, (XII), p. 65. To these circumstances Dr. Boisduval alludes in the Lep. Rev. (Suites a Buffon), p. 363, (1874), where he says that at the desire of MM. Grote & Robinson he presented them with this rarity as well as many other Heterocerous Lep., as they were anxious to have them to illustrate some articles on the Lep. Het. of the United States, and he further adds that he cannot understand why they substituted the name of *Phaeton* for his name of *Erato*.
    On the appearance of the above Grote delivered himself in the Canadian Ent. (VIII, p. 28, 1876), along with other equally savory and modest matter, of the following: "*Euproserpinus phaeton* G. & R. Dr. Boisduval (Suites a Buffon, 1874, 363) says as to the species which he calls *Macroglossa phaeton*, quoting Grote and Robinson's original description, that he does not know by what chance we changed the name of this species from *erato* to *phaeton*. This remark is based on a misunderstanding." He then goes on to say that when he and his colleague first described this species from a picture and from information received from Mr. S. Calverly, who also stated that the species was described in MSS. by Dr. Boisduval as *Proserpinus phaeton*, that "we preserved Dr. Boisduval's name, giving him in our paper credit for the species."
    Would the reader like to know how Grote gave Dr. Boisduval credit for the species? By referring to the description in question he will see this line: "It appears that Dr. Boisduval has exhibited a specimen in his cabinet as *Proserpinus Phaeton*;" that is the way he gave him credit for the species. Wasn't it a noble way? Who will dare doubt after this that nobility of soul still finds an abiding place in the human breast? Further on in the same article Grote states than Dr. Boisduval lent him and his colleague an example from which for the first time they made their description from the real insect, in Trans. Am. Ent. Soc. 1878, adding that "at about the same time Dr. Boisduval published this species under the name of *erato*." Thus goes on this Sir Arrogandissimos from fool to fouler, bewraying himself with the filth of his own conceit; for of a verity, hath it not been most truly said by the world-renowned Sancho Panza that the higher a monkey climbs the more he————exposes himself to shame and ridicule.

PTEROGON TERLOOII. Hy. Edwards.

(Proserpinus T.) Proc. Cal. Acad. Sc., (1875).

(PLATE XIV. FIG. 2, ♂.)

    Fully described in the work above cited. The figure on plate XIV was drawn from one of the two original types loaned for the purpose by my very dear friend, Dr. H. Behr, in whose coll. are the only two examples so far known to science. The under side of primaries is greenish yellow shaded broadly in the middle with dull red. Secondaries also greenish yellow with a faint median band of a shade darker. It is closely allied to the Eur. *P. Proserpina*, Pall. (*Oenotherae*, Schiff.). Described from two ♂ taken at Mazatlan, Mex., by the late Baron Terloot, to whom the species was dedicated by its author.

SMERINTHUS IMPERATOR. Nov. Sp.

(PLATE XIV, FIG. 3, ♀.)

FEMALE. Expands 4½ inches.
Head above yellow fawn colour, thorax violaceous grey, not dark; abdomen yellowish fawn shaded, some-

---

*Nous avons prete cette rareté a MM. Grote et Robinson, ainsi que plusieurs autres Lepidopteres heteroceres pour qu'ils puissent, selon leur desir, les faire figurer dans un ouvrage qu'ils ont entrepris sur les Lepidopteres Heteroceres des Etats-Unis d'Amerique. Nous ne savons pas par quel hasard ces messieurs ont change notre nom d'*Erato* pour lui substituer celui de *Phaeton*.

what, dorsally and with a faint dorsal line of violaceous extending the whole length. Beneath pale fawn; legs violaceous.

Upper surface; primaries, general style of ornamentation somewhat as in *Modesta*, Harr. Basal third of wing very pale violaceous grey, yellowish at base, and traversed in its middle from costa to inner margin by an irregular darker shade; the outer edge of the basal third is very irregular and produced in a sharp angle at the innermost median nervule, and is narrowly shaded where it joins the median space by darker tint; the inner half of the median space is tinted with brownish, the outer half is same pale violaceous grey as the basal part; the outer edge of the median space is scalloped and shaded with darker grey; a large pale discal mark; the third or terminal space is of the same pale grey as the major part of rest of wing, shaded on costal half with pale yellowish fawn, a darker patch on inner margin not far from inner angle. Secondaries dull crimson, yellowish white at inner margin, and a large pale grey patch covers that part of the wing at and near the anal angle, within which patch is a blackish dash parallel with outer margin, between which latter and said dash is a faint grey abbreviated line extending from the anal angle inwards to where the crimson colour commences. Under surface of all wings very pale yellowish fawn with a broad terminal band but a shade darker; basal half of primaries dull crimson, which colour does not however extend to either costa or inner margin, and the discal mark is designated by the pale fawn of ground colour of wing.

Hab. Arizona. One ♀, Mus. Strecker.

This differs from its nearest ally, *Modesta*, Harr., in the far greater breadth of wing, the great robustness of body, the entirely different colour and in the difference of the undulations of the transverse lines and shades, also in the shape and greater size of the discal mark or bar. In *Modesta* and its Pacific coast var. *Occidentalis*, Hy. Edwds., the colours are even, smooth shades, well defined and separated from each other by demarcation lines, whilst in *Imperator* the colours are blended more or less into each other and have a heavy powdery appearance, the scales being far heavier and rougher as seen through a lens than in Harris' species. *Imperator* approaches the var. *Occidentalis* somewhat, and somewhat only, in the paleness of the ground colour, but beyond this there is no nearer approach than to the stem form typical *Modesta*, to which *Occidentalis* assimilates in every respect except being paler in colour and generally of larger size.

For this species, which I consider one of the grandest acquisitions our Heterocerous fauna has for a long time received, as well as for a large number of other rare and new species from inner Arizona and Utah, I am indebted to the energy and perseverance of Mr. B. Neumoegen of New York. Heretofore, owing to its being mainly in possession of the Indians, as well as to its unfavorable climate and general sterility, the representation of the Lep. fauna of Arizona was of the most meagre description imaginable, comprised in a few examples in the coll. of W. H. Edwds., and hence still in the Mus. of the Agricultural Dep. at Washington, all derived from the same source, the chance collections of government exploring and surveying parties. For years *Arygnnis Nokomis* was known only by a single entered ♂, and later for a long time by a few more, ♂ and ♀. When my friend Neumoegen commenced a few years since to study and collect Lepidoptera, to which he applied himself with an energy seldom equalled, I impressed on him the importance of obtaining examples from Arizona, giving him drawings and descriptions of *Nokomis* and some other prominent species. By indefatigable industry he secured collectors who from inner Arizona, in a remarkably short time, sent a large quantity of the most interesting material, among which were the above described splendid *Smerinthus*, as well as a number of others new to science, which will be described in the present and future parts of this work. In the first lot received of these Arizona Lep. were a number of both sexes of the coveted *Nokomis*, but unfortunately the season was so far advanced when the onslaught commenced that all were torn or too long flown to be desirable; the glory of later sendings made however ample amends for this first quasi disappointment. A large proportion of these insects seem to be remarkably pale aberrant forms or representations of Pacific or Eastern species; prominent in this respect are *Mel. Alma*, n. s., *Arg. Nokomis*, W. H. Edwds., *Pap. Undulosa*, n. s., or v., *Sph. Elsa*, n. s., *Pseud. Nuttalli* and others. *Arg. Nokomis* I have always considered as an extreme variety of *A. Cybele*, as I believe the Amoor and North China *A. Sagana* may be a form of the East Europe *A. Laodice*; but to this subject I will revert in my description of the various new species from this wonderland.

I cannot omit mentioning another still more astonishing thing, in connection with the reception of those Arizona novelties, which, incredible as it may appear, is nevertheless a fact, to the truth of which I am willing at any time to be qualified with proper jurat appended; it is that when Mr. Neumoegen passed them to me for description he did not even hint, let alone make it the condition, that any of the new species should be named after himself, his wife, his aunts or his cousins-german, his grandparents, the stranger within his gates, or even after his rich neighbor. May his skeleton be preserved!

SPHINX ELSA. Nov. Sp.

(PLATE XIV, FIG. 4 ♂, 5 ♀.)

MALE. Expands 2⅞ inches.

Head and thorax above pale rose colour, latter black towards and at base, but with a mark composed of two contiguous rose coloured crescents at its juncture with the abdomen; the latter marked laterally much as in *Drupiferarum*, Ab.-Sm., but owing to the scales being much rubbed from the back in both ♂ and ♀ it is impossible to describe that part with accuracy, though from the general appearance of the insect I should be led to infer that the broad dorsal band was whitish or tinged with rose. Antennae heavy, serrated and black save

---

* ♂ *A. Sagana*, Dbldy.-Hew., Gen. Diur. Lep., t. 24, fig. 1, (1850). ♀ is *Damora Paulina*, Nord., Bull. Mos., II, p. 440, t. 12, f. 1, 2, (1851).

towards base where they are on upper side pale rose. Beneath head and body white tinged with rose, a dark brown line on sides of thorax, legs brownish mixed with rosy white, tarsi black.

Upper surface primaries white tinged on inner two-thirds with rose colour, a black powdery basal patch extending some distance outwards and terminating in scattered black points about the middle of wing; also three somewhat waved lines or narrow bands composed of more or less segregated black points or atoms, these lines run more or less parallel with the exterior margin, but all unite into one at the apex where it is most distinctly defined. Whole wing loosely scattered more or less with minute black points; no indications of a discal spot; fringe white with blackish at terminations of veins. Secondaries white with black mesial and submarginal bands like in *Drupiferarum*, but not as heavy in proportion.

Under surface primaries white tinged very faintly with rose, powdered with fine black points, two parallel submarginal lines composed of loose black atoms and converging into one better defined line at the apex. Secondaries white tinged with pale rose and with the black bands of upper side faintly repeated.

FEMALE. Expands 3 inches.

Head white, antennae black with white tips, thorax much as in ♂ but with much more black on back and less tinged with rosy, the patagiae and part towards head being pure white, abdomen also as in ♂, but the white parts without the rosy tint.

Upper surface of all wings pure white except in that which accompanies the first black band from the base; black bands, etc., much as in ♂, but more distinct. The wings in this sex are broader; the primaries less pointed at the apex and more rounded on the costal margin, and the whole insect is, except in the two points embraced in above description, devoid of the lively roseate hue of the ♂. Its nearest American congener is *Drupiferarum*, and its European of course *Ligustri*, L., from both of which, as well as from all other species of known Sphingidae, as far as I am aware, it differs in its white colour.

One ♂, Mus. Streck.; one ♀, Mus. Neumoegen; both from Arizona.

SPHINX HAGENI. Grote.

(*Ceratomia H.*) Grote, Buff. Bull., II. p. 149, (1874); Butler, Trans. Zool. Soc. Lond., IX, p. 621, (1877).

(PLATE XIV, FIG. 6 ♂.)

This species was originally described from an example in the Mus. Comp. Zool. Cambridge, taken by Boll in Texas. Since then that gentleman has bred it in some numbers from the larva. Grote's superficiality was again made painfully evident in his description of this species by placing it in the genus (or sub-genus) *Ceratomia*, Harris, of which the larva is distinguished from all others of the Sphingidae, as far as I am aware, in the presence of four horns, two on the back of the second segment and two on the third, from which peculiarity Dr. Harris named his genus as well as the only species in it." The fact is, *Hageni* is nearest to *Sphinx* (*Daremma*) *Undulosa*, Wlk., and it is almost incredible that Grote, who even made some comparisons between that species and *Hageni* in his description of the latter, could overlook their affinity. In truth, so close are the two that in a large series of both species, received from Boll, there are some examples about which it is difficult to decide to which species they belong, and the absence of the greenish or olivaceous hue alone marks it a fair probability that they are *Undulosa*, though as a general thing this latter is by far the larger of the two, but it attains a greater size in the New England and Middle States than it does to the far south or west, and the eastern examples are lighter coloured.

The larva of *Sphinx Hageni*, when full grown, is about 2¾ to 3 inches in length. Generally pale apple green, but occasionally it occurs of yellowish green with darker streaks, like most of the larva of the Sphingidae it varies somewhat in colour, but the apple green is the prevailing hue. The body, all over the back and sides, is covered with whitish points arranged transversely in regular rows; head very closely covered with white points not arranged with any regularity; on the sides are diagonal white lines, shaded with rose red on the upper edge, this red shading being darkest in the middle. Spiracles surrounded with brown, which is further encircled with yellow. Caudal horn flesh coloured, thickly studded with small raised points. Feet rose red; prolegs pale reddish. Undergoes its transformation in the ground. For the above description, accompanied by a faithful drawing, I am under obligations to Mr. Boll, who was the discoverer of the species.

LIPARA BOMBYCOIDES. Wlk.

Cat. B. M. Lep. Het., VIII, p. 233, (1856); Clemens, Jrl. Acad. Nat. Sc. Phil., p. 187, (1859); Morris, Syn., p. 215, (1862); Grote, Buff. Bull., I, p. 28, (1873); Streck., Lep. Rhop.-Het., I, p. 117, (1876); Butler, Trans. Zool. Soc. (Lond.) IX, p. 620, (1877).

(PLATE XIV, FIG. 7.)

Ever since its description by the late Mr. Walker from a unique, at that time in the collection of Mr. Saunders of London, Eng., this insect has been a puzzle to American Lepidopterists. In 1875, Prof. Westwood made for me an accurate coloured figure from the type, which latter is now in the Hopeian coll. of Oxford University. The differences between this figure and *Sphinx Harrisii*, of which

---

Ceratomia Quadricornis, Harr., Sill. Jrl., XXXVI, p. 293, (1839), a synonym of *Agrius Amyntor*, Hub., Samm. Ex. Schmett., II, (1806-1824).

LIPARA BOMBYCOIDES.

most of the American Lepidopterists consider it a synonym, I have designated on page 117 of this volume (after the description of *Sph. Harrisii*). The present figure on accompanying plate XIV was carefully made from Prof. Westwood's drawing, above alluded to.

Although the type was taken in Canada, nothing of the kind is known to exist in any American collection; it approaches nearest to *Sph. Harrisii*, but the latter, which I have figured on plate XIII, f. 10, though having much the same general appearance yet differs considerably in detail, though *Bombycoides* may possibly be an aberration of it. That it is however a distinct species that is yet to be re-discovered is after all not unlikely when we consider that such a conspicuous species as *S. Plota* escaped the notice of the many American collectors and students until three years since when I detected it in a small collection sent me from Montreal, Canada. If I have been able to do little towards elucidating the mystery that enshrouds *Bombycoides*, I at least trust I have not done an unacceptable act in presenting the figure of Walker's type that my friends and the others many have some better knowledge of its appearance.

SAMIA GLOVERI. Streck.

Since figuring and describing this species in the commencement of this work I have received through Mr. Neumoegen a number of cocoons from Utah and Arizona which developed in due time the perfect insect. It is subject to the same variations as *Ceropia* in size of discal lune, breadth of white transverse bands, etc. etc.; also varying considerably in size, the largest being six inches in expanse, the smallest but four. Some examples are much paler than others in the red ground colour. Of the larva I have as yet received no description, and have only learned that it is found on gooseberry and currant; but the cocoons are somewhat of the shape of those of *S. Ceropia* though not so large and unlike that species; the outer case tightly adheres to the inner, and is hard woven and gummed, and has the appearance as though made of rough silver, not as in *S. Columbia* with a few silver threads streaked through here and there, but the entire cocoon looks as if woven of coarse large fibres of rough silver, and is very beautiful indeed. Were it not for the wonderful features of the cocoons I should undoubtedly pronounce *Gloveri* but the Arizona or Utah form of *Ceropia*, for the perfect insects differ in nothing but the ground colour of wings. Dr. Hagen is of the opinion that *Gloveri* is a form of *Columbia*, but as just stated I rather think it a variation of *Ceropia* and have little doubt that excessive breedings of it in the Atlantic States would eventually change the crimson ground colour to the black of *Ceropia*. The figure 3 on plate XIV represents a most astonishing semi-albino ♀ aberration, the left wings of which are normal and the right one suffused with white to the complete extinction of the crimson ground colour, with the exception of a small basal patch on primary; the discal lune on primary is surrounded by a black shade, that on secondary is faintly outlined with grey. The under surface of this monstrosity is normal on both sides. There is a difference, as the figure shows, in the outline of the wings, especially of the primaries, that of the abnormal side being much more arched and fuller on the costa. This strange freak was captured in inner Arizona.

The preponderance of pale coloured or albinous species and examples in the salt regions of Arizona and Utah, is truly wonderful and without precedent. What the cause can be, climatic or local, that results in the production of these astonishing forms is a question which opens a field of investigation of unparalleled interest to every student of natural science.

BUNAEA EBLIS. Streck.

(PLATE XIV. FIG. 9, ♂.)

Of this species, described on page 121, from a single example presented to me by my ever dear friend Mr. Chapman, of Glasgow, Scotland, Prof. Westwood informed me that there are examples in the Oxford Museum, and suggested that it might perhaps be a form of *B. Phedusa*, Dru.

NEW SPECIES, VARIETIES, ETC.

PAPILIO (ASTERIAS) var. UTAHENSIS, n. var. ♂, wings somewhat narrower than in the common form; primaries more falcate. Pale yellow stripes on each side of the head and prothorax; inside also pale yellow; usual lateral rows of yellow dots on abdomen; anal valves pale yellow. Macular bands and border pale yellow on both surfaces, without the orange colour so conspicuous on the under surface of the ordinary examples; the mesial band of secondaries does not extend into the discoidal cell on the upper surface. In cell on under surface are a few yellow scales; anal eye as in common form; submarginal row of spots on under surface primaries confluent in some examples, separated by the nervures only in others—principally it differs from *Asterias* in the head and thorax being striped instead of spotted, in the yellow anal valves, in the different shape of the wings, in the yellow spots and bands being very much paler on both surfaces; in the mesial macular band being, especially on secondaries, much narrower, and in the submarginal spots of primaries, on under side, being confluent or almost so. All the examples I have seen are of larger size than the common run of *Asterias*, though not larger than some examples of the latter. The ♀ differs from the ♂ principally in the partial obsolescence of the mesial macular bands. Hab. Utah.

PAP. RUTULUS, var. or ab.? ♂ expands 2½ inches. Upper surface same shade of yellow as in the ordinary form, black bands and margin all very broad as in the heavier marked examples of *P. Eurymedes*; the broad black exterior border of primaries on its inner edge between the fifth subcostal nervule and the second discoidal nervule is extended abruptly in a bow inwards from the regular line until it is almost merged into the abbreviated transverse band; the submarginal row of yellow spots on primaries very small, and the three nearest apex not in a line with the others but best off at the second discoidal nervule; on the secondaries the yellow submarginal lunules or bars rather of diversified style and size; the apical one is a very narrow bar, a mere line; the one between the first and second subcostal nervules is far broader than in most cases; that between the second subcostal and discoidal nervules is a bar of great size, being twice the length of the last or of the one following it, which with the one between the first and second median nervules is crescent shaped and of large size; the next as well as the anal mark are exceedingly small and rust red; above these two latter, but with a considerable interspace between, are crescents of no great size of blue scales.

Under surface bands, etc., not quite so heavy as above; primaries, the inner edge of black margin does not extend inwards between the fifth subcostal and second discoidal nervules nearly so much as on the upper surface; the submarginal yellow spots are confluent, forming a broad unbroken band which covers the outer half of the black marginal band and separated from the exterior margin by little more than a black line. Secondaries, all the submarginal yellow bars very large and almost confluent at their angles, the one between the first and second median nervules, which is largest, is lunate, the others parallelogram in shape; interior to these are the shining blue and greenish as in the normal form.

The body and head above black with but little indication of yellow on the patagia. Several examples from Arizona.

NEW SPECIES, VARIETIES, &c.

Thecla Kali, n. sp. ♂ expands 1 to 1¼ inches. Upper surface, bright shining yellow copperish; primaries with very broad blackish costal margin extending from sexual stigma to exterior margin where it is broadest, occupying all the space between the second discoidal nervule and the costa. Exterior margins of all wings bordered with blackish, but very narrowly, especially on the secondaries; fringe all blackish, except between the second and third median nervules of secondaries, where it is white.

Under surface shining silky grey or slaty, much the same colour as on under side of *Thecla Alcestis*, Edwds., and *Thecla Guerens*, L., but more glossy. Secondaries at and near base powdered with black and white atoms. Exterior margin of all wings edged with a narrow blackish line; this is succeeded inwardly by a white line. On primaries is a submarginal row of fuscous spots edged outwardly with white on disc; a white band edged with black on inner edge, this band is broken in three entirely separate parts; the last of these parts, between the last median and the submedian nervule, is nearly obsolete; a narrow white discal ring edged outwardly with black and broken towards the costa. Secondaries have a submarginal row of black sagittate spots differing much in size, the sixth from apex is the largest, the seventh is much nearer the exterior margin, being only separated therefrom by the white marginal line, and is surmounted by a large red spot which later is edged inwardly by a black crescent; between the eighth black spot from the apex and the exterior margin the space is filled with grey caused by pale blue and black scales; a black streak at margin above anal angle; interior to the row of submarginal spots on the disc is very irregular sinuous white line heavily edged inwardly with black; in the cell are two white lines, and above these another, all edged outwardly with black. From Arizona.

This more beautiful and conspicuous species bears on the upper side somewhat of a resemblance to such exotic species as *T. Apollo*, Fabr., *Zeritis Pierus*, Cram., *Thestor Callimachus*, Ev., etc., its rich yellowish metallic ground colour contrasting strongly with the blackish margins. It comes in or near to the same group as *T. Damon*, Cram. (*Südzens*, Bff.-Lee.), but is widely different from that species. On the three examples examined I can discover no traces of tails to the wings, though these frail appendages may have been broken off.

Thecla Fotis, n. sp. Size and shape of *T. Augustus*, Kirby. Upper surface uniform dark grey. Under surface, primaries rather dark slaty grey, a submarginal row of almost obsolete dark points, a scarcely distinguishable irregular darker line across the disc; interior to this the wing is more or less scattered with pale atoms; fringe grey. Secondaries edged on exterior margin with a white line which is succeeded by a band composed of some black scales, directly interior to which is a row of round white spots or dots, one in each nervul interspace, each of them is surmounted by a small crescent formed of black atoms; interior to these is a not very conspicuous sinuous pale grey or whitish line inwardly edged with black; the part of the wing interior to this line is darker than any other part of the under surface and is scattered loosely, especially towards the base, with whitish atoms; fringe grey. From Arizona.

Augustus, Kirby, is its nearest ally, from which it widely differs in the slaty grey colour of both surfaces and the row of white submarginal spots on under surface of secondaries.

Satyrus Ashtaroth, n. sp. ♀ expands 2 inches. Upper surface of all wings very pale ochraceous, or yellowish white, dusted with pale brownish at basal parts; across the disc of primaries is a very irregular rather pale brownish band which becomes nearly obsolete towards interior margin, the brownish colour of this band extends outwardly along the second discoidal and the first median nervules until it joins the rather narrow exterior border of same colour; within the broad yellow band or space between this latter and the middle band are two large velvety black oval spots with small white centres, between these two is a quite small black spot; fringe pale brownish and white. Secondaries with a pale almost obsolete zigzag submarginal line which is all that would indicate the inner edge of a border, interior to which across the disc on costal half of wing a faint brownish half obsolete band, the outer edge of which in the cells is prolonged into very long teeth; not far from exterior margin between first and second discoidal nervules is a small oval black spot; fringe white.

Under surface almost the same in all respects as ♀ *S. Hippolyte*, Esp., from the Ural regions, to which the present species is very closely allied. It belongs to the same group as *S. Ridingii*, W. H. Edwds., but is a much larger, paler insect with the brown marks fainter, fewer and differing in detail; it certainly assimilates nearer to the Russian species mentioned than to the Colorado one. Described from a single ♀ from Arizona not in very good condition.

Aedophron Grandis, n. sp. Expands 1½ inches. Head, thorax and primaries above same pale whitish citron yellow as in *A. Phlaphora*, Ld.; abdomen and secondaries pale shining silky white. Under side, body and all wings same shining white; tarsi brownish. The wings are narrower and more elongated than in the Syrian species, but I do not think ours is generically different.
Several examples from Arizona.

Cucullia Antipoda, n. sp. Expands 1¼ inches. Above; head, collar and patagia whitish grey, back of thorax brownish; abdomen whitish dusted with brown atoms. Primaries same pale grey as the thorax more or less sprinkled with brown points which have a tendency to form lines in the interspaces; a dark brown marginal line interrupted at the veins; a large kidney shaped double ringed reniform, orbicular also conspicuous; a dark brown line runs outwards along inner margin from near base to three-fourths the length of the inner margin, thence it turns upwards towards the reniform, but with a great bend inwards between the latter and inner margin, from exterior margin one-fourth its length from inner angle extends a short dark brown line which does not reach to the last described zigzag line; interior to the orbicular is a transverse zigzag line forming two great teeth, one of which connects with the orbicular, the other points towards but does not meet the great tooth formed by the sinus of the brown line between the inner margin and reniform. Secondaries white, venation brown; brownish at exterior margin which is edged with a dark brown rather well defined line; fringe white. Under surface; body whitish grey; primaries greyish, paler on costa, apical parts and exterior margin; exterior margin with a dark brown line broken by the veins. Secondaries whitish faintly dusted, but not thickly, with minute brown atoms; interrupted marginal line as on primaries; a minute brown discal point.

To the old world student it will be more to the point than all the above merely to state that this species is very close to *C. Santonici*, Hub., the most noticeable differences being that the Sarepan species is larger and there is a brownish shade in the region of the reniform and thence to costa; also the brown margin of upper side of secondaries is much broader as well as darker than in ours.
Several examples from Arizona.

Catocala (*Faustina*) var. Zillah, n. var., is distinguishable from the common form by the upper surface of primaries being suffused with rather scattered rust red atoms especially about the reniform and subreniform, and along the transverse posterior lines and thence to submarginal lines. Taken in several examples along with a number of the ordinary form of same species in Arizona.

Catocala Perdita, Hy. Edwds. Two examples from Arizona differ from the type first described on p. 100 of this work in being a little larger and in the upper surface of primaries being less thickly covered with black atoms, showing more distinctly the white ground, especially on the part of the median space interior to the reniform and the space between the transverse posterior and submarginal lines.

NEW SPECIES, VARIETIES, &c.

CATOCALA STRETCHII, Behr. The several examples of what I am almost certain is this species I have not been able to compare with the unique type in the coll. of Dr. Behr in San Francisco, but I have received from that savan a beautifully coloured figure made by Stretch from the type; with this figure these Arizona examples agree except that in them the mesial band of secondaries is narrower and some of the red ground colour extends beyond the outer edge of the black submarginal band; there is also some difference in the red colour, that on the figure being a little darker, but I scarce think I am wrong in deciding these Arizona examples as *Stretchii*. Dr. Behr's type was taken by Mr. Stretch in Nevada and was described in Trans. Am. Ent. Soc., III, p. 24, (1870).

Besides the above and the rare species to be described and illustrated in succeeding pages were received *Papilio Daunus*,[*] *Colias Edwardsii*,[†] *Anth. Creusa*,[‡] *Julia*,[§] *Lycaena Oro*,[‖] *Heteronea*,[¶] *Zeroe*,[**] *Thecla Crysalus*,[††] *Niphon var. Eryphon*,[‡‡] *Acadica*,[§§] *Apodemia Mormo*,[‖‖] *Argyn. Nokomis*,[¶¶] *Aphrodite*,[***] *Hesperia*,[†††] *Edwardsii*,[‡‡‡] et var. *Nevadensis*,[§§§] *Coronis*,[‖‖‖] *Melitaea Palla var. Whitneyi*,[****] *Nubigena*,[††††] *Satyrus Nephele var. Ariane*,[‡‡‡‡] *Pamphila Ottoe*,[§§§§] *Lita Sexsignata*,[‖‖‖‖] *Synuda Socia*,[¶¶¶¶] et var. *Adumbrata*,[*****] *Howlandii*,[†††††] *Gonytodes Trilinearia*,[‡‡‡‡‡] and sundry others which time has not yet enabled me to examine fully.

The following descriptions of New Lepidoptera are from examples captured by Mr. J. Boll, mostly in the vicinity of New Braunfels and San Antonio, Southwestern Texas:

MELITAEA IMITATA, n. sp. ♂ expands 1 inch. Wings narrower and more elongate proportionally than in *M. Vesta* which is probably its nearest ally. Upper surface much as in *Vesta* but the blackish markings especially of secondaries much heavier; in the primaries there is no notable difference, but the secondaries have a broad marginal black border comprising fully the outer third of the wing; within this border are two rows of fulvous crescents, those of the outermost one being the smaller. The under surface is peculiar, resembling much that of *M. Harrisii* both in style and colouration. The inner half of primaries is paler fulvous than above; the outer margin has a narrow darker fulvous band, interior to this is a broad black band, irregular on its inner edge and broadest at costa; within this band are two rows of crescents of various sizes; the outermost row is pale yellow, the other fulvous; this black marginal band is succeeded inwardly at some distance by a very irregular narrow band or line of unequal width which extends from costa to inner margin; between this latter band and the base of wing are several black lines of the usual style; fringe black cut with white between the veins. Secondaries, outer margin narrowly fulvous as in primaries, thence one-third of the wing is black; within this latter colour towards outer margin is a row of pale yellow lunules, the ones nearest to apex and anal angles small, the others large and with the exception of the third from anal angle, which is the greatest, much of the same size. This broad black submarginal space or band is succeeded by an irregular pale yellow mesial band divided transversely by a black line and bordered inwardly by another; the rest of wing interior to this is fulvous, having a yellow spot in discoidal cell; between this spot and the base is an irregular yellow band lined on both edges with black also a yellow band lined with black at base. Body black or blackish brown above, below pale yellow; antennae black ringed with white.

♀ a little larger and with the black markings not quite as heavy.

MELITAEA LABUNDA, n. sp. Expands 1¼ inches. Wings even more elongate than in the preceding. Upper surface fulvous, not dark. Fringes whitish, grey at termination of nervules. A fuscous band on exterior margin of all wings; this band is almost entirely occupied with lunules of the same colour as ground of wings; on the primaries three of these lunules nearest to the inner angle are largest and nearly of equal size, the remaining four are smaller and the one nearest to apex is a mere dot. On secondaries the lunules nearest apex and anal angle are the smallest, the others do not differ much in size. Interior to these marginal lunules on primaries are two irregular partly obsolete black lines extending from costa to inner margin, heaviest near costa; in the discoidal cell and at basal part are some more irregular blackish lines. On basal half of secondaries a few faint abbreviated wavy lines. Under surface; primaries much the same colour as above; a narrow darker fulvous band on exterior margin succeeded inwardly by a row of lunules; the first six from costa are pale yellow edged with blackish, the remaining two at inner angle are fulvous and merged into the narrow margin of same colour; at some distance interior to this row of lunules is a blackish line heaviest at costa and not reaching quite to the interior margin; five other abbreviated lines extend from costa inwards to the median nervure. Secondaries fulvous with a marginal row of seven pale yellow lunules, the ones nearest apex and anal angle are the smallest, the third one from anal angle the largest, and the other four are nearly of equal size; an irregular pale yellow mesial band lined inwardly with black and divided through the middle by a black line and further at the costal third by another black line; a pale yellow spot in the discoidal cell, between which and the base an irregular pale yellow band edged with black extends from costa to inner margin; at base of wing also a narrow yellow band edged outwardly with black. Fringe white with grey at termination of nervures.

In spite of its diminutiveness this little species, which is one of the most remarkable yet discovered in this country, brings strongly to mind certain species of Acraea such as *Viola*, Fab., and *Rahira*, Boll.

LIBYTHEA LARVATA, n. sp. ♂ size and shape of *Bachmanni*; the black and fulvous colours of upper surface arranged nearly the same as in that species, but the shape and colour of the subapical spots and bars are different; these are washed with fulvous in the present species whilst in *Bachmanni* they are pure white; in the latter the white discal bar is on both surfaces entirely disconnected and distant from the white spot near the middle of exterior margin, neither is it in a line with it; the present species differs entirely therefrom in the arrangement of these spots, as follows: besides the subapical one, which is quite small, an interrupted band composed of three spots extends from middle of costa nearly to the middle of exterior margin; the first of these spots is small and is on the costal nervure, the second which does not quite join it is large and at its lower point nearest outer margin is joined by innermost angle, nearest costa, of the last spot which is square. Under surface, primaries have the chain of three spots nearly as above but a little larger and

---

[*] *P. Daunus*, Boll, Sp. Gen. I, p. 342, (1836). [†] *C. Edwardsii*, Behr, Edwds.' Butt. N. Am. I, t. 6. Col. (1870). [‡] *Anth. Creusa*, Dbldy.–Hew., Gen. Diur. Lep., p. 56, t. 7, (1847). [§] *Anth. Julia*, W. H. Edwds., Trans. Am. Ent. Soc. IV, p. 61, (1872). [‖] *Lyc. Oro*, Scud., Can. Ent. VIII, p. 23, (1876). [¶] *Lyc. Heteronea*, Bdl., Ann. Soc. Ent. Fr., 2me Ser. X, p. 295, (1852). [**] *Lyc. Zeroe*, Bdl., Lep. Cal., p. 45, (1869). [††] *Thecla Crysalus*, W. H. Edwds., Trans. Am. Ent. Soc. IV, p. 344, (1873). [‡‡] *Th. Eryphon*, Bdl., Ann. Soc. Ent. Fr. 2me Ser. X, p. 289, (1852). [§§] *Th. Acadica*, W. H. Edwds., Proc. Acad. Nat. Sc. Phil., p. 55, (1862). [‖‖] *Ap. Mormo*, Feld., Wien. Ent. Mon. III, p. 271, (1859). [¶¶] *Arg. Nokomis*, W. H. Edwds., Proc. Acad. Nat. Sc. Phil., p. 221, (1862). [***] *Arg. Aphrodite*, Fabr., Mant. Ins. II, p. 62, (1787). [†††] *Arg. Hesperis*, W. H. Edwds., Proc. Ent. Soc. Phil. II, p. 502, (1864). [‡‡‡] *Arg. Edwardsii*, Reak., l.c. VI, p. 137, (1866). [§§§] *Arg. Nevadensis*, W. H. Edwds., Trans. Am. Ent. Soc. III, p. 14, (1870). [‖‖‖] *Arg. Coronis*, Behr, Proc. Cal. Acad. Sc. II, p. 173, (1858–1862). [¶¶¶] *Arg. Myrina*, Cram., Pap. Ex. II, t. 189, (1779). [****] *Mel. Whitneyi*, Behr, Proc. Cal. Acad. Sc. III, p. 88, (1863). [††††] *M. Nubigena*, Behr, l. c. p. 91. [‡‡‡‡] *Satyrus Ariane*, Bdl., Ann. Soc. Ent. Fr., 2me Ser. X, p. 307, (1852). [§§§§] *Pamphila Ottoe*, W. H. Edwds., Proc. Ent. Soc. VI, p. 207, (1867). [‖‖‖‖] *Lita Sexsignata*, Harvey, Buff. Bull. II, p. 280, (1875). [¶¶¶¶] *Synuda Socia*, Behr, Trans. Am. Ent. Soc. III, p. 27, (1870). [*****] *S. Adumbrata*, Behr, l. c. [†††††] *S. Howlandii*, Grote, Proc. Ent. Soc. Phil. III, p. 533, t. VI, (1864). [‡‡‡‡‡] *Gonytodes Trilinearia*, Pack., Hayden's Geo. Survey, X, p. 202, t. IX, f. 53, (1876).

white, and from the lowermost one to the costa extends a narrow white bar, the colour exterior to this on the apical part is white mottled with brown points, rest of outer margin brownish with a few reticulations of darker hue; the costa the same as this latter; space from base of wing to chain of white spots dark ochraceous excepting a blotch on disc of wing towards inner margin which is whitish with slight sprinkling of ochre. Secondaries lustrous white mottled with dark brown points or flecks; across the middle of the wing from costa to inner margin these flecks become more or less confluent forming a broad irregular band heaviest at costal half; another band, but not so wide and more irregular, extends along the outer margin two-thirds of its length from anal angle; half way between the middle band and base of wing, on costa, is a brown patch formed by the confluence of the dark specks. Fringes on both surfaces of secondaries brown; of primaries brown from apex to first angle, rest whitish.

The most remarkable difference between this and the three other American species, *Bachmani*, *Carinenta* and *Terina*, is, as above shown, in the shape, disposition and colour of the discal and accompanying spots. It is nearer to *Bachmani* than to either of the others and may perhaps be a variety of that species.

CHARIS GUADELOUPE, n. sp. Allied to *Borealis*, G.-R., but differs from it, as well as from all other species of the genus that I am acquainted with, most remarkably in the shape of the exterior margin of the primaries, which are undulated, being strongly produced opposite the discoidal cell and again at the last median nervule. The colour of upper surface is not so reddish as in *Borealis*, being somewhat of a greyish or fuscous tinge; two submarginal silvery lines with row of small black spots between them, these as well as the other dark lines are much the same as in *Borealis*, but not quite as heavy. Under surface same colour, reddish yellow, as in *Borealis*, markings nearly similar though not as heavy. Fringes on both surfaces of primaries white near apex then blackish to the middle indentation of margin, then white for a short space succeeded again by blackish, then white at the indentation near inner angle, at angle itself blackish. On secondaries the fringes are blackish from apex to near middle of margin then white, then black, again white and finally near and at the anal angle blackish. ♂ ♀ expand a little over 1 inch. The primaries in ♂ are much more produced apically, and the inferiors are smaller than in ♀, but the remarkable undulate outer margin of primaries is the same in both sexes. The outer margin of secondaries is also undulate but not to such a marked degree as the primaries.

PAMPHILA SIMILIS, n. sp. ♂ ♀ same size and shape as *Vialis* and *Eos*, W. H. Edwds.; colour and markings on upper surface same as in the latter species, but differing considerably on the under surface of secondaries which are dark blackish brown with a large somewhat triangular whitish grey basal patch, another patch of same colour on middle of costa and still another half way between the basal patch and outer margin not very far from inner margin; along the outer margin is also a little grey. Fringes brown from anal angle to middle of wing, thence to the apex alternate brown and white. The primaries on under surface are blackish brown with some whitish grey at exterior margin from middle to apex; the white spot on costa and the smaller one below it are repeated as on upper surface. Fringes brown except at apex and a spot near the inner angle which are white.

SPILOTHYRUS NOTABILIS, n. sp. Size and shape of *Malvarum*, Hffg., and on upper surface somewhat resembles in general appearance that species. Body and head above dark brown, beneath white. Ground colour of upper surface rather pale olive brown, but the spaces between the nervules are so filled with sagittate and other dark brown markings as to exclude in a great measure the paler colour. On the middle of costa of primaries are two small white semi-translucent spots, these are joined by a third in the discoidal cell; half way between these and the apex are three more connected minute white spots, and in the middle of the wing are two more which are almost joined at their inner points; on secondaries are also two minute white spots in middle of wing and about midway between them and the costa is another single one. Fringes are pale olive alternated with dark brown at terminations of veins. Under surface yellowish white with the dark lines, etc., in cells, and the small white spots same as on upper surface, but owing to the very pale ground colour the contrast between the latter and the dark streaks is very marked and pleasing, but this is one of the host of species that it is useless to attempt to describe in a way that it can be with any certainty recognized, and my only excuse for worrying the student with such useless trumpery is that I will as soon as possible add figures of this as well as of all the other species herein described, for to use the words of one of the greatest of living authorities, "it is by a very faithful figure alone that they can be satisfactorily separated," and that "descriptions alone are utterly inadequate," and "unaided by figures more than worthless."

This is the first insect of its genus that has been found to occur in N. America.

ARCTIA OTHONA, n. sp. ♂ expands $1\frac{1}{4}$, ♀ $1\frac{1}{2}$ inches. On upper surface, head and thorax are pale pinkish yellow, the collar with two broad black stripes, the thorax with three; abdomen crimson with a dorsal and two lateral rows of confluent black spots. Primaries black with pale pinkish yellow lines arranged in precisely the same manner as in *A. Speciosa*, Mosch., (which is nearly as in *A. Virguncula*, Kirby,) with the single exception that there is a cross bar of yellow extending from the junction of the discoidal and discocellular nervules to the costa. Secondaries same crimson as the abdomen, with an irregular narrow black border to the outer margin and a broader one to the costa; at anal angle a large black triangular spot joining at its lower edge the outer margin; on the costa towards apex are two more large black spots, between the outermost or apical one of these and the large spot at anal angle is a large black triangular spot. These spots, as is always the case with the Arctians, vary in size, shape and number in different individuals. Fringes of all wings pale pinkish yellow. Under surface; body pale yellow with black at sides. Primaries much as above excepting that the neuration is not denoted by pale yellow lines. Secondaries as above but colours not so intense.

Several examples taken near Dallas, Tex.

DATANA ROBUSTA, n. sp. ♂ expands $1\frac{3}{4}$–2, ♀ $2\frac{1}{4}$–$2\frac{1}{2}$ inches. Tawny yellow or buff, very much the colour of *Nadata Gibbosa*, Wlk. Exterior margin of primaries entire. Head, thorax and primaries unicolourous; outer edges of tegulæ of a more greyish tinge in some examples, in others darker ferruginous or tawny than the thoracic patch. Abdomen and inferiors of same colour as but much paler than the primaries; first segments of abdomen darker, same colour as thorax. Primaries with two principal darker lines answering in position to the transverse anterior and posterior lines of the *Noctuidæ*; in the disc between these two lines are three more transverse lines, all running more or less parallel with the outer one; the third of these three lines is sometimes almost obsolete. A curved line runs from apex to midway between costa and inner angle, sometimes joining the transverse posterior line. A large darker discal spot accompanied interiorly by a smaller spot. In some instances the space between the transverse anterior and posterior lines is suffused from the median nervure to the interior margin with reddish brown or ferruginous which in some instances extends narrowly interior to the transverse anterior line to the base of thorax and to the tegulæ. The veins, except the costal, between the two main transverse lines are all marked with some ferruginous; outer margin a little darker and somewhat greyish. Outer half of secondaries darker than near base. Fringe of primaries ferruginous, of secondaries yellowish white.

This species differs from all the other species and varieties as follows: That the thoracic patch is same colour as ground of wing instead of being darker; in some instances as above stated the tegulæ and base of thorax are darker, leaving the thoracic patch paler instead of darker as is the case in all other species; again, in that the veins between the two principal cross lines are marked in same dark colour as on the transverse lines; it is apparently more robust and its whole appearance is so peculiar that if once seen there is no likelihood of its ever being confounded with any of the other known species. From vicinity of Dallas, Texas.

EUDRYAS WILSONII, Grote, (*Civis W.*) Proc. Ent. Soc. Phil. II, p. 65, t. 3, (1863). This beautiful species has hitherto only been known by two examples, one in Mus. Comp. Zool. at Cambridge and the other, from which the original description and figure were made, in Mus. Am. Ent. Soc. Phila. It differs remarkably from the other species in having pectinated antennæ.

HELIOTHIS LANUL, n. sp. Expands 1½ inches. Head and body white lightly tinged with sienna or rust brown at basal part of thorax and tips of patagia. Upper surface; primaries white and not very dark rust brown; the basal third is at and near base brown, then white with scattered brown points, then comes a line which extends from costa to inner margin as in *Rivulosa*, *Regia*, etc., but not so much bent as in these species; this line is succeeded by the median space which is brown and encloses two conspicuous white spots, the largest in the discoidal cell, the other half way between it and the inner margin; the largest of these white spots is joined exteriorly by a metallic lead coloured discal spot; the outer edge of the median space, which is very much produced opposite the discal spot, is succeeded by a white line edged outwardly with brown, beyond this is again white, then an irregular jagged line of brown edged outwardly with white; this is succeeded by the marginal band of brown with a row of minute black spots. Fringe white with brown points at tips of veins. Secondaries silky white with a very faint brownish submarginal band. Fringe long and white. Under surface white with markings of primaries in a manner somewhat faintly repeated on costal and exterior parts.

HELIOTHIS GLORIOSA, n. sp. Expands 1¼ inches. Head and body above white with pale olivaceous shades; beneath white. Antennæ and legs white. Upper surface; primaries dull purplish red and olivaceous, neither of these colours intense, marked somewhat after the manner of *Rivulosa* and allies. The basal third of wing is purplish and is separated from the median space by a pure white line which widens at the veins thus forming teeth; the median space is olivaceous and encloses a purplish discal spot which latter is prolonged outwardly to and beyond another white toothed line which separates the median from the outer space; the latter is purplish interiorly and olivaceous marginally, the latter colour is more or less at the nervules encroached on by the purple. Fringe light and dark olivaceous. Secondaries dirty white; a faint discal mark; marginal third of wing broadly shaded with brownish. Fringe white with brown at veins. Under surface; primaries shining white; fuscous discal spot; a pale crimson shade near apex; a broad fuscous submarginal band. Fringe fuscous and white. Secondaries silky white, a very faint discal mark; a slight pale crimson tint at apex; fringe white with fuscous at nervules, which latter colour does not extend to the terminations thereof. By far the largest, and with the exception of *Regia* the most beautiful of that group of which *Rivulosa* is the type.

EUDRYAS GLOVERII, Grote (*Eusciirrhopterus G.*). *Larva.* Length 1¼ inches. Same form as *Grata*. Ground colour pale olive green.* Head and legs red, former with some small black spots. First segment red above and also with a number of small black spots. On the sides of all save the first and last segments is a transverse rather narrow velvet black bar which extends nearly to the middle of the back near which it is widest, these bands do not connect dorsally, but the space between them in each segment except the the first and two last is supplied with a short red transverse band; on the next to last segment the black bands are only on the sides; on the back are two parallel red transverse bands, on the first of which are eight and on the second four small black dots; the anal segment has also some round black dots. On each segment, especially dorsally, are a number of very fine dotted black lines. From head to anus extends a somewhat broad red band. Beneath on each of the fourth, fifth, tenth and eleventh segments is a transverse line formed by an almost connected row of small brown spots.

CATOCALA ULALUME, n. sp. Size and shape of *C. Robinsonii*, Grote, and in position it might stand between that species and *C. Desperata*, Guen. The lines run much as in the latter, but are not so plainly distinguishable owing to the whole wing being heavily dusted with black points, thus obscuring the pale ground colour very much more than in *Desperata*; in the latter there is a suffusion of brown between the transverse lines and the submarginal line, as well as no other parts of the wing; nothing of this is noticeable on the present species; in many respects it resembles *Lacrymosa*, Guen., but is not as dark as that species nor as large; under surface nearly as in *C. Desperata*, but with more tendency to suffusion in the latter. Perhaps the best idea I can convey of *Ulalume* is by saying that were the heavy dark brown shadings which accompany the transverse lines and are on other parts of the primaries of *Desperata* away, and the whole surface peppered with black atoms, it would make a fair counterpart of the insect I am now hopelessly attempting to describe in some such way that the reader may be able to identify it.

Mr. Boll also took PRILAMPELUS LINNEI, G.-R.,† near San Antonio; I formerly received the same species from Mr. Doll, who captured it in east Florida.

September, 1877.

ON SOME LEPIDOPTERA FROM THE REGIONS WEST OF HUDSON'S BAY, BETWEEN THE LATTER AND LAKE ATHABASCA.

To Mr. Waldemar Geffcken of Stuttgart, Germany, I am indebted for a large number of Lepidoptera from those regions west of the Hudson's Bay, known as New North and New South Wales, mostly from the latter. This tract of country lies between 53° and 63° N. L., and in common with most parts of British Columbia is a trackless wilderness, traversed only by the native Indians or hunters and those in the interest of the fur trade, and it was only after several years of ceaseless efforts, accompanied by repeated disappointments which would have thoroughly disheartened any one save a true lover of nature, that Mr. Geffcken at length succeeded in securing connections that enabled him to receive from time to time large numbers of examples, though unfortunately not always in the best condition, owing to the lack of proficiency of the collectors employed, who were mostly Indian boys and girls. These species, as will be seen, are in great measure the same as or forms of those found in N. W. Labrador, though some indigenous to the latter locality, such as *Oeneis Nastes* and *Arg. Polaris*, were not among the collections at various times received.

PAPILIO TURNUS, L. ♂♀. Examples small, somewhat more heavily marked with black than the United States and Canada examples and agreeing nearly with the description of those from the Island of Anticosti (south of Labrador) near top of page 69 of this work. Not uncommon.

PIERIS (*Napi*) var. FRIGIDA, Scud. ♂ does not differ from those found in south west Labrador; i. e., with upper side immaculate white, and under side of secondaries and apices of primaries yellowish, with veins of secondaries accompanied with brown. Only a few examples received.

COLIAS EURYTHEME, Bdl. ♂. One example not differing in size or colour from those found in the United States.

* This description is from an inflated example in which the colour may not have been as vivid as during life.
† Proc. Ent. Soc., Phil. V, p. 182, t. III, (1865).

COLIAS (*Pelidne*) var. CHRISTINA, W. H. Edwds. Of this a large number of examples, ♂ ♀, were received, mostly of extraordinary great size, the largest (♂) being 2¼ inches in expanse, and the smallest (♂) 1½ inches, the average size is 2 inches. That of the typical N. Labrador Pelidne is about 1½ inches. In shade of colour the majority of examples are same as the N. Labrador ones, the ♂ lemon yellow, the ♀ greenish white; but several of the males which were taken near Lake Athabasca have the upper surface suffused more or less with orange similar to those mentioned and figured in W. H. Edwards' Butt. N. Am., the originals of which I have also seen; with the exception of this orange suffusion they differ not a particle from the citron coloured ones. About one-fourth of the females are lemon coloured; these yellow females show no trace of the blackish margin on upper surface of wings, this form is figured also in Edwards' work. On the under surface the various examples exhibit every degree of depth in the greenish colour of hind wings, some being quite pale yellowish green, others as dark as the darkest of the N. Labrador examples; one male has the row of submarginal points and two females have the little reddish brown mark on costa. Mr. W. H. Edwards in describing *Christina* mentions that "in three specimens out of four there were no traces of the submarginal points." His males were all of the orange variety; the greater number of those received by me were of the yellow form; Mr. Edwards' examples were from Slave River, farther west than mine; he had obtained thence however no white female or lemon coloured male. All my males from near Hudson's Bay were yellow, most of the females white; those that I received from farther west (for they were obtained in various localities from Hudson's Bay to near Lake Athabasca) had both white and yellow females and orange as well as yellow males; I know of only one orange male ever having been taken in N. Labrador, which example was received from there by Mr. Moechler and by him described in Wien. Ent. Mon. IV, p. 354, (1860). No yellow female I believe has yet been found in N. Labrador, but on the southern border along the St. Lawrence River and in Canada and along the shores of Lake Superior the females are always yellow. This more southern form is about the same size as those from more northern Labrador.

From all these facts I would deduce the conclusion that the further west we trace this species the more will we find the orange colour to prevail in the males and the yellow in the females, and that the white females, if they occur at all in South Labrador or west of Lake Athabasca, will be the rare exception. Remarkable as is the difference in colour in the various examples, it is only a matter of usual occurrence with the Coliades, for a more difficult group to define or in which to designate the limits of a species is not to be found, and a far more wonderful instance of difference in colour of males is found in C. *Labonotica* and C. *Sagartia*, both forms of the same insect, in which the male of the first is red and that of the second greenish blue. I am also informed that intermediate forms between the two occur in which the red and blue colours are intermixed. Mr. Edwards has figured a male example of *Philodice* that is orange, I also possess a ♂ of the same colour. C. *Helichta* is also an orange form of C. *Eros* (or perhaps as has been suggested is a hybrid between *Eros* and *Eilora*). Of one thing I am most certain, that the Americans have made far too many species by giving to each local varieties a different appellation and with a view to trying in so doing something towards solving the riddle I have given as much attention as possible to this beautiful and most interesting genus, having with a few exceptions obtained all the known species and varieties. Whilst writing the above I have before me twenty-seven ♂ ♀ *Pelidne* from N. W. Labrador, forty from British Columbia from a region extending from Hudson's Bay to Lake Athabasca, five from South Labrador on the Gulf of St. Lawrence, and seven from Colorado. I have examined the types of *Christina*, *Labradoreansis*, *Scudderii*, *Interior* and *Laurentina*, and can only come to the conclusion that they are but three forms of one species of which I here give a short diagnosis:—

COLIAS PELIDNE, Bdl., Icones, t. 8, (1832); Sp. Gen. I, p. 644, (1836); Dup., Suppl., I, t. 15, (1832); Bdl.-Lec., Lep. Am. Sept., p. 66, t. 21, (1833); Herr.-Sch., Schmett. Eur., t. 7, f. 35, 36, t. 8, f. 43, 44, (1843); Frever, Neue. Beit., VI, t. 511, (1831-1858); Men., Cat. Mus. Petr. Lep. I, p. 84, (1855); Moesh., Wien. Monat., IV, p. 349, (1860); Morris, Syn., p. 30, (1862); Kirby, Cat., p. 493, (1871); W. H. Edwds., Butt. N. Am., II, t. I. Col., (1874).

Col. *Antiquete*, Styr., Cat., p. 5, (1871).

Col. *Labradorensis*, Scud., Proc. Bost. Soc. Nat. Hist., p. 107, (1862); Kirby, Cat., p. 495, (1871).

This is the typical N. W. Labrador form of small size and with ♀ always white, and the ♂ yellow, with the single orange coloured exception previously alluded to.

Col. *Scudderii*, Reak., Proc. Ent. Soc. Phil., IV, p. 217, (1865); Kirby, Cat., p. 495, (1871); W. H. Edwds., Butt. N. Am., I, t. VIII, Col., (1872); Mead, Wheeler's Rep. V, p. 749, (1875).

Occurs in Colorado but differs in nothing of any importance from the Labrador examples, except that the ♀ is occasionally, though not often, yellow like the ♂.

var. a. INTERIOR, Scud., Proc. Bost. Soc. Nat. Hist., IX, p. 108, (1862); Kirby, Cat., p. 495, (1871).

Col. *Pelidne* var. *Streck.*, Lep., Rhop.-Het., p. 69, (1873).

Col. *Philodice* var. *Laurentina*, Scud., Proc. Bost. Soc. Nat. Hist., p. 4, (Oct., 1875).

A form found in S. Labrador and in the Lake Superior region, in which the ♀ is in the majority of instances yellow like the ♂; this bears the same relation to the N. Labrador form as does the C. *Werdandi*, H-S. (see Zett.), to the typical C. *Paleno*.

var. b. CHRISTINA, W. H. Edwds., Proc. Ent. Soc. Phil., II, p. 79, (1863); Butt. N. Am., I, t. II, Col., (1868).

This is the form of great size found west of Hudson's Bay in which the male is sometimes orange and sometimes yellow, and the females both yellow and white.

COLIAS PALÆNO, L. About twenty-five examples taken, of which seventeen, 9 ♂, 8 ♀, are before me. They are all of smaller size than the average of those occurring in Europe, the smallest (♂) expanding 1¼ inches and the largest (♀) a trifle over 1¾ inches; otherwise the males differ only in the yellow colour which is a little less intense; the females also agree with the trans-Atlantic examples; two are the yellow form known as *Werdandi*, H-S., all the others are white; the black marginal bands in both sexes present the same differences of width and outline found in their European congeners, some of the females having this band immaculate and in others enclosing spots of the white or yellow ground colour.

W. H. Edwards has figured both sexes of this species in his Butt. N. Am. under the name of C. *Helena*, which he subsequently changed to *Chippewa*. So close is his female figure to one of the examples before me than it seems almost as if the latter had served as the original of it. I here append the synonymy of this species:

PALÆNO, LINN., (Pap. P.), Faun. Suec., p. 272, (1761); Syst. Nat., I, 2, p. 764, (1767); Fabr., Syst. Ent., p. 476, (1775); Ent. Syst., III, p. 207, (1793); Ochs., Schmett., I, 2, 184, (1808); (*Colias P.*) Godt., Enc. Meth., IX, p. 101, (1819); Bull. Sp. Gen., I, p. 645, (1836); Styr., Cat., p. 5, (1871); Kirby, Cat., p. 493, (1871).

Pap. *Europomene*, Esp., Schmett., I, t. 42, (1778); Hub., Eur. Schmett., 434, 435, (1793-1827).

Pap. *Philomene*, Hub., l. c., 602, 663, 740, 741; (Ocl. P.) Dup. Lep., Suppl., I, t. 47, (1832).

Col. *Paleno* var. *Lapponica*, Styr., Cat., p. 5, (1871).

Col. *Werdandi*, H.-S., Schmett. Eur., f. 403, 404, ♀, (1848). Yellow ♀ form.

Col. *Helena*, W. H. Edwds., Proc. Ent. Soc. Phila., II, p. 80, (1863); Butt. N. Am., I, t. I, Col., (1868).

Col. *Chippewa*, W. H. Edwds., l. c., last page Vol. I; Kirby, Cat., p. 495, (1871).

ARGYNNIS CHARICLEA, Schn.
" FREIJA, Thnb.
" (*Aphirape*), var. TRICLARIS, Hub.
" (*Frigga*), var. SAGA, Kaden.

All four species received in a number of examples, but none present any particular points of difference from those found in Labrador.

VANESSA ANTIOPA, L. All examples of small size.

VANESSA (*Antiopa*) ab. HYGIEA, Hdr. (*V. Lintnerii*, Fitch). One example, the black submarginal band with its enclosed row of blue spots entirely wanting; the yellow marginal band nearly twice the width that it is in the normal form of *Antiopa*.

VANESSA MILBERTI, Godt. The same in all respects as in other localities. Common.

VANESSA (*Zephyrus*) var. GRACILIS, G.-R. Only one example, which differs from those from the White Mts. in the basal half of under side, which is without the reddish suffusion; the white mesial band is not so bright, being more covered with fine abbreviated dark lines.

LIMENITIS ARTHEMIS, Drn. Of small size, expanding not quite 2¼ inches. The white bands, especially on primaries, broader than usual. On under side the ground colour very strongly suffused with reddish though not much more so than in some few extreme varieties I have seen from N. York and elsewhere; no indications of greenish at base of secondaries; all the space between the white band on secondaries, and greenish marginal lunules, is brownish red, not set in red lunules bordered above and below with black as is commonly the case. Several examples were received.

EREBIA DISCOIDALIS, Kirby. Of this beautiful species, so rare in collections, over a hundred examples were received. It is a constant species exhibiting scarce any variation.

CHIONOBAS JUTTA, Hub. Several examples, ground colour of both surfaces darker, giving them a more opaque appearance than is exhibited in those from North Europe.

ALYPIA MACCULLOCHII, Kirby. Two examples.

MACROGLOSSA FLAVOFASCIATA, Barnston. One ♀ example.

ARCTIA PARTHENOS, Harris (*A. Borealis*. Mosch.). One ♀ example in very poor condition.

☞ Of the following species I am anxious to obtain examples, either by exchange or purchase; any Naturalists having duplicates will confer a great favor by communicating with

HERMAN STRECKER,
Box 111 Reading P. O.,
Berks Co., Pennsylvania, U. S. of N. America.

Teinopalpus Imperialis, Hope. ♀
Ornithoptera Urvilliana, Guer.
Ornithoptera Magellanus, Feld. ♀
Ornithoptera Criton, Feld.
Ornithoptera Arruanus, Feld.
Ornithoptera Hippolytus, Cram.
Ornithoptera Lydius, Feld.
Ornithoptera Helena, Linn. ♀
Ornithoptera Croesus, Wall. ♂
Ornithoptera Brookiana, Wall.
Papilio Antimachus, Dru.
Papilio Evan, Doubl.
Papilio Pericles, Wall.
Papilio Blumei, Boisd.
Papilio Macedon, Wall.
Papilio Philippus, Wall.
Papilio Phorbanta, Linn.
Papilio Homerus, Fabr.
Papilio Garamas, Hub.
Papilio Caignanabus, Poey.
Papilio Ascanius, Cram.
Papilio Wallacei, Hew.
Papilio Slateri, Hew.
Papilio Dionysos, Dbldy.
Papilio Gundlachianus, Feld.
Papilio Elephenor, Dbldy.
Papilio Disparilis, Herr–Sch.
Papilio Aristagoras, Feld.
Papilio Theroclamus, Feld.
Papilio Vollenhovii, Feld.
Papilio Cornelius, Feld.
Papilio Bachus, Feld.
Papilio Araspes, Feld.
Papilio Adrastus, Feld.
Papilio Tydeus, Feld. ♀
Papilio Adamantius, Feld.
Papilio Telegonus, Feld. ♀
Papilio Annae, Feld.
Papilio Alcumenor, Feld.
Papilio Ambrax, Bdl.
Papilio Godeffroyi, Semp.
Papilio Ormenus, Guer.
Papilio Pandion, Wall.
Papilio Amphitrion, Cram.
Papilio Euchenor, Guer.
Papilio Amphiaraeus, Feld.
Papilio Pilumnus, Bdl.
Euryades Corethrus, Bdl.
Euryades Duponchelii, Luc. ♀
Thais Honoratii, Bdl. ♀
Bhutanitis Lidderdalii, Atk.
Parnassius Stoliczkanus, Feld.
Parnassius Apollonius, Evers.
Parnassius Actius, Evers.
Parnassius Delphius, Evers.

Parnassius Acco, Gray.
Parnassius Simo, Gray.
Parnassius Charltonius, Grav.
Parnassius Jacquemontii, Bdl.
Parnassius Tenedius, Evers.
Parnassius Eversmanni, Men.
Parnassius Glacialis, Butl. ♀
Parnassius Citrinarius, Motsch.
Mesapia Peloria, Hew.
Euterpe Uricoechae, Feld.
Euterpe Tamyris, Feld.
Pieris Celestina, Boisd.
Pieris Blanca, Feld.
Pieris Lorquinii, Feld.
Colias Theraspis, Feld.
Colias Ladakensis, Feld.
Colias Viluiensis, Men.
Colias Pontoni, Wallengr.
Euploea Cuvieri, Feld.
Euploea Westwoodii, Feld.
Euploea Hopfferi, Feld.
Euploea Assimulata, Feld.
Euploea Eurypon, Hew.
Ithomia Susiana, Feld.
Queides Helieroniodes, Feld.
Cethosia Leschenaulti, Godt.
Cethosia Cydalina, Feld.
Cethosia Myrina, Feld.
Cirrochroa Semiramis, Feld.
Cirrochroa Thule, Feld.
Cirrochroa Regina, Feld.
Argynnis Jerdoni, Lang.
Argynnis Dexamene, Boisd.
Argynnis Jainadeva, Moore.
Argynnis Ruslana, Motsch.
Argynnis Aruna, Moore.
Acrea Perenna, Dbldy.
Eresia Castilla, Feld.
Euchinia Megalonice, Feld.
Pyrameis Gonerilla, Fabr.
Pyrameis Abyssinien, Feld.
Pyrameis Dejeanii, Godt.
Pyrameis Tameamea, Esch.
Vanessa v. Elymi, Rbr
Limenitis Lymire, Hew.
Athyma Jocaste, Feld.
Athyma Urvasi, Feld.
Parthenos Tigrina, Voll.
Diadema Boisduvalii, Dbldy.
Diadema Tydea, Feld.
Diadema Polymena, Feld.
Euripus Clytia, Feld.
Paudora Chalcothea, Bates.
Any species of Callithea.
Romaleosoma Sophron, Dbldy.

Romaleosoma Pratinos, Dbldy.
Romaleosoma Arcadius, Fabr.
Any species of Agrias.
Charaxes Epijasius Reiche.
Charaxes Kadenii, Feld.
Charaxes Jupiter, But.
Charaxes Etheocles, Cram.
Charaxes Achaemenes, Feld.
Charaxes Hausalii, Feld.
Charaxes Brennus, Feld.
Nymphalis Calydonia, Hew.
Dynastor Napoleon, Doubl., Hew.
Penetes Pamphanis, Dbl., Hew.
Pavonia Aorsa, West.
Opsiphanes Boisduvalii, Dbldy.
Dasyopthalmia Rusina, Godt.
Morpho Phanodemus, Hew.
Morpho Cypris, Bdl. ♀
Zeuxidia Aurelias, Cram.
Zeuxidia Semperi, Feld.
Zeuxidia Horsfieldii, Feld.
Zeuxidia Wallacei, Feld.
Eurytela Castelnaui, Feld.
Melanitis Egialina, Feld.
Dyctis Biocnlatis, Guerin. et var.
Necyria Lindigii, Feld.
Necyria Fulminatrix, Feld.
Antirrhaea Lindigii, Feld.
Antirrhaea Geryon, Feld.
Daedalma Dorinda, Feld.
Tamyris Pandalma, Feld. (Hesp.)
Ismene Subcaudata, Feld. (Hesp.)
Nyctalemon Metaurus.
Nyctalemon Cydnus, Feld.
Castnia Rutila, Feld.
Castnia Zagraea, Feld.
Castnia Miraica, Feld.
Castnia Unifasciata, Feld.
Desmopoda Bombiformis, Feld.
Acge Venata. Feld.
Microlophia Sculpa, Feld.
Sphinx Cluentius.
Chaerocampa Hystrix, Feld.
Euryteryx Molucca, Feld.
Philampelus Dolichoides, Feld.
Philampelus Orientalis, Feld.
Ambulyx Hypostieta, Feld.
Ambulyx Tigrina, Feld.
Ambulyx Rostralis, Feld.
Ambulyx Eurysthenes, Feld.
Ambulyx Substrigilis, West.
Smerinthus Amboinicus, Feld.
Smerinthus Dumolinii, Guer.
Smerinthus Heuglini, Feld.
Smerinthus Modesta, Fab. (nec Har.)

[OVER

Smerinthus Tartarinovii, Brem.
Smerinthus Dentatus, Cram.
Smerinthus Meander, Bdl.
Smerinthus Panopus, Cram. ♂
Smerinthus Argus, Men.
Smerinthus Kindermanni, Ld.
Smerinthus Maackiae, Brem.
Smerinthus Tremulae, Tr.
Smerinthus Dissimilis, Brem.
Smerinthus Gaschkevitchii, Brem.
Smerinthus Sperchius, Men.
Hepialus Giganteus, H.-S.
Charagia Fischeri, Feld.
Charagia Argyrographa, Feld.
Pielus Hydrographus, Feld.
Pielus Maculosus.
Gynautocera Virescens, Feld.
Opsirhina Crinudes, Feld.

Eochroa Trimenii, Feld.
Thyella Zambesia, Feld.
Hyperchiria Titania, Feld.
Hyperchiria Caudatula, Feld.
Attacus Zacateca, Westw.
Bathyphlebia Aglia, Feld.
Polythysana Apollina, Feld.
Saturnia Stoliczkana, Feld.
Saturnia Semioculata, Feld.
Holocera Smilax, Westw.
Saturnia Epimethea, Dru.
Saturnia Larissa, West. ♀
Saturnia Argus, Drury. ♀
Saturnia (Actias?) Artemis, Brem.
Saturnia Atlantica, Luc.
Actias Idae, Feld.
Bunaea Deroyllei, Thom.
Bunaea Phaedusa, Dru.

Citheronia Phoronea.
Eacles Kadenii, Herr-Sch.
Aricia Auster, Feld.
Aricia Batesii, Feld.
Eacles Kadenii, H.-S.
Eudaemonia Semiramis, Cram.
Eudaemonia Derceto, Mssn.
Eudaemonia Jehovah, Streck.
Ormiscodes Trisignata, Feld.
Ormiscodes Ramigera, Feld.
Any species of Phyllodes.
Any species of Asiatic Catocalae.
Erateina Oriolata, Feld.-Rog.
Erateina Goniaris, Feld.-Rog.
Erateina Pohliata, Feld.-Rog.
Erateina Thyrisiata, Febl.-Rog.
Erateina Palomira, Feld.-Rog.
Erateina Muisenta, Feld.-Rog.

These are a few of the very many of the rarer species that I am eager to procure; of course there are numberless others from all parts of the world, equally desirable and coveted by me.

Lepidoptera, native and exotic, either on hand or can be obtained for clients at short notice.
COLEOPTERA and other Insects occasionally on hand and can always be obtained if ordered—particulars given on application by letter.
I will also sell Insects on commission for persons having such to dispose of.
I am always glad to exchange for any species of Lepidoptera not in my collection, or to obtain such by purchase if exchanging be not desirable.

HERMAN STRECKER,

Box 111 Reading P. O., Pa.

W. V. ANDREWS,

Entomologist, &c.,

187 State Street, Brooklyn, New York.

☞ Purchasing Agent for Books and Apparatus in connection with Natural History. Also, Cork Pins, &c. Eggs of the different varieties of Silk Worms, to order. Lepidoptera and Coleoptera for sale or exchange. Agent for WALLACE'S SILK REELER, and for KIRBY'S SYNONYMIC CATALOGUE OF DIURNAL LEPIDOPTERA.

No. 15.   Issued Quarterly, at 50 Cents per Part in U. S.
In Europe, 2 Shillings.

LEPIDOPTERA,

RHOPALOCERES AND HETEROCERES.

INDIGENOUS AND EXOTIC;

WITH

Descriptions and Colored Illustrations,

BY

HERMAN STRECKER.

Reading, Pa., 1877.

Reading, Pa.:
Owen's Steam Book and Job Printing Office, 515 Court Street,
1877.

MELITÆA ALMA. Nov. Sp.

(PLATE XV, FIG. 1 ♂.)

MALE. Expands 1½ inches.

Same form as *M. Leanira*, Bdl. Head and body above black, beneath whitish yellow, legs fulvous, antennæ black narrowly ringed with white. Upper surface of all wings bright fulvous of nearly the same shade as *M. Whitneyi*, Behr; all nervures and nervules black; fringe white alternated with black at terminations of veins. Primaries have a black border to exterior margin within which are seven bright fulvous spots, the apical one much the smallest and the third from inner angle by far the largest; these fulvous spots occupy by far the greater part of the black border; this border is succeeded immediately on its inner margin by a row of almost confluent pale yellow spots, the two at costa being the largest; interior to this across the middle of the wing is another row of pale yellow spots joined inwardly by a rather broad irregular black band broken at the disco-cellular nervule by a fulvous patch; the colour between this last black band and base of wing is pale yellow; within the discoidal cell a broad transverse black band; costa and base of wing blackish. Secondaries margined exteriorly by a black line; basal third of wing black with two pale yellow marks in discoidal cell; this black basal part is succeeded outwardly by a fair band of pale yellow, about midway between which and the exterior margin is a transverse row of almost confluent pale yellow spots; costa blackish.

Under surface; primaries pale fulvous, exterior black marginal border enclosing pale spots as above, but these latter are pale yellow here instead of fulvous except the second and third from inner angle which are faintly tinged with the latter colour; all other yellow spots of upper surface faintly reproduced. Secondaries whitish yellow, veins black, a black line edges the exterior margin; a black mesial band enclosing six whitish yellow spots; a small black mark or streak in upper part of discoidal cell and another at costa. The under surface of this insect resembles very closely that of *Leanira* which is the nearest allied species.

Two examples, one from Arizona, one from S. Utah.

This is another of those anomalous species from the wonder-producing salt regions. A glance at the under side would lead any one to pronounce it *Leanira*, but the difference on the upper side is astonishing; the black ground colour in that species being here replaced with bright fulvous; nevertheless, despite this great dissimilarity of colour, I am inclined to think it is something of the white peacock business, and that this may possibly be, after all, another of those pallid aberrant sub-species in which Arizona and Utah seem to be so rich.

ÆGIALE COFAQUI. Strecк.

Proc. Acad. Nat. Sc. Phil., p. 148, (1876).

(PLATE XV, FIG. 2 ♀.)

Since the description of this species appeared in the Proc. Phil. Acad. I have had the opportunity of comparing my type (which was taken in Georgia) with two examples, both females, collected in May, 1876, by the expedition under Lieut. E. H. Ruffner during the surveys and explorations of the region of the head waters of the Red River of Texas. They were taken in the *Llano Estacado* or Staked Plains. The only difference of any moment between these Texan examples and the type is in the great size of the first which expand 3¼ inches.

The differences between this and *Æg. Yucca*, Bdl.-Lec., I have fully demonstrated in the paper above alluded to and will merely repeat here that the most noticeable are the entirely different shape of the wings, the greater profusion of yellow markings on upper surface and the many white spots of under surface of secondaries.

This insect as well as *Yucca* undoubtedly belongs in Felder's genus *Ægiale*, of which I consider Scudder's *Megathymus* but a synonym.

MACROGLOSSA ULALUME. Nov. Sp.

(PLATE XV, FIG. 3 ♂.)

MALE. Expands 1¾ inches.

Head above sulphur yellow, below black, antennæ black. Thorax on back black mixed sparingly with yellow hairs, patagiæ sulphur yellow, collar intense velvety black; beneath black; legs black. Abdomen above velvety black with sulphur yellow side tufts to the two last segments, anal brush black above, yellow beneath; under side of abdomen black.

Upper surface all wings blackish, darkest on basal half and at abdominal margin of secondaries; a common broad semi-diaphanous band a shade or so paler than ground colour crosses both wings, this band on the secondaries does not extend to inner margin and shows towards its inner extremity a few scattered scarce noticeable yellow scales.

Under surface as above, dark colour more dull; some loose orange hairs on basal part of primaries, and the inner termination of the mesial band of secondaries is slightly clothed with sulphur coloured scales.

One ♂ from Oregon.

This beautiful species is near to *M. Flavofasciata*, Harnston, but differs notably in the black collar and thorax, in the absence of the bright yellow mesial band of secondaries, as well as in its greater size.

SPHINX VASHTI. Nov. Sp.

(PLATE XV, FIG. 4 ♂.)

MALE. Expands 3 inches.

Head and thorax above whitish grey, patagiæ edged with a velvety black line; abdomen darker grey with heavy black dorsal line and white and black demi-bands on the sides. Beneath grey with a conspicuous black ventral line on abdomen; legs brown.

Upper surface; primaries dark grey; a broad whitish or whitish grey dash extends from base, half the length of the wing, along the costa, not reaching quite to the edge of the latter; at and within the inner edge of this whitish band the wing is strongly suffused with blackish; a black apical line, also four other abbreviated black lines in the cells, the last of which is almost lost in the darkness interior to the great pale basal band; exteriorly a whitish submarginal band, edged inwardly by a black line broad near the inner angle but decreasing to a point not far from apex.

Secondaries whitish grey with broad black mesial and submarginal bands. Fringes of all wings dark grey except near anal angle and on abdominal margin of secondaries where they are white.

Under surface primaries smoky grey, a little paler at exterior margin; an inconspicuous black apical line. Secondaries whitish grey obscured on basal half of costa with darker grey, mesial and submarginal bands well defined but not as dark as above.

One ♂ from Arizona in coll. Neumoegen.

Belongs to the group of *S. Chersis*, Hub, and occupies a position between that species and *S. Perelegans*,* Hy. Edwds., from California, which latter is easily distinguished from the present species by its coal-black occiput and thorax which serve to connect it with *Drupiferarum*, Ab.-S., and *Gordius*, Cram. The relative position of the N. Am. Sphingidæ comprising this group would be somewhat in this wise:

Sphinx (*Lethia*) Chersis, Hub.
" Vashti.
" Perelegans, Hy. Edwds.
" Drupiferarum, Ab.-S.
" Gordius, Cram.
" Lucitiosa, Clemens.

HEPIALUS SANGARIS. Nov. Sp.

(PLATE XV, FIG. 5.)

MALE. Expands 1 to 1¼ inches.

Head and thorax salmon coloured; it is impossible to determine the colour of the abdomen owing to its being badly greased.

Upper surface; primaries salmon red; brighter red on costa; two transverse white bands edged with red lead colour do not reach to the costa; a small white half obsolete mark towards base. Secondaries smoky fringed with reddish. Under surface; primaries reddish grey with edge of costa red, secondaries much as above.

Two males from Arizona.

GLOVERIA ARIZONENSIS. Packard.

Annual Report Peabody Acad. Sc., p. 90, (1871).

(PLATE XV, FIG. 6 ♀.)

FEMALE. Expands 3⅞ inches.

*Sph. Perelegans, Hy. Edwds., described in Proc. Cal. Acad. Sc., (July, 1873).

GLOVERIA ARIZONENSIS.

Pale grey or ashen; primaries with a small round white discal spot beyond which is a brown double transverse line extending from inner margin to the third subcostal nervule; between this line and the exterior margin is a transverse zigzag line, the points of which between the veins are very acute; between this line and the exterior margin as well as at base and along the costa the colour is darker than in the median space. Secondaries without marks, a little darker towards the exterior margin.

Under surface of all wings grey without marks of any kind.

The male is not yet known.

One example from S. W. Arizona.

The genus Gloveria, erected by Dr. Packard for this insect, is as its author stated closely allied to Laslocampa and the species to L. Otus, Dru., to which it bears a considerable resemblance though it is by no means as ponderous or heavily furred as that species. Never was honour more worthily bestowed than in the instance of the dedication of this genus to the most hard-working, overworked, indefatigable entomologist in all America, Prof. Townend Glover, the extent of whose labors in economic entomology are yet to be estimated at their true value.

COLORADIA PANDORA. Blake.

Proc. Ent. Soc. Phil. II, p. 279, t. VII, ♀, (1863).

(PLATE XV, FIG. 7 ♂.)

Male. Expands 3¼ inches.

Antennae yellow and pectinated as in *Pseudohazis* and *Hemileuca*; head and thorax dark brown with scattered white hairs; abdomen above heavily clothed with dark brown and white hairs, beneath brown, the segments widely fringed with white; legs dark brown. Wings inclined to semi-transparency and with the exception that the primaries are a little more pointed apically are nearly the same shape as in the female. Primaries dark brown with broad irregular transverse sub-basal and median bands; between the latter and the exterior margin is another broader transverse band or shade edged outwardly with white scales; the spaces between all bands more or less covered with scattered white scales; a round black discal spot. Secondaries very pale rose colour more strongly tinged at base and abdominal margin; a round black discal spot; a narrow brown median band; a broad brownish border to exterior margin.

Under surface pale rose colour, darkest on primaries and at abdominal margin of secondaries. All wings broadly bordered with brown on exterior margins, also with a narrow brown median band and black discal spots.

One ♂ from Oregon in Mus. Streckeri.

The female type, and only example of that sex yet known, is in the museum of the Am. Ent. Soc. The markings are not near as dark or well defined as in the male, and scarcely any indication of pink or rose colour is on the upper side of secondaries. It was taken at Pike's Peak, Colorado.

I here append Mr. Blake's original description:—

"*Female.*—Brownish-grey. Head not prolonged, palpi extending rather beyond the head. Antennae bright buceus, biserrate, a little longer than the thorax. Thorax densely villose. Abdomen above fuliginous, sides mixed with griseous, apex tufted, extending a little beyond the hind wings. Wings semi-transparent. Fore wings with two indistinct, oblique, somewhat undulating, fuliginous bands, the exterior one paler than the other, the space between the bands covered somewhat sparsely with distinct white scales, a small black spot on the discal nervure. Hind wings with an indistinct cloudy band, broader at the interior margin, gradually tapering to the exterior. A pale fuliginous spot on the disc. Base of the wings clothed with pale pinkish hairs, cilia whitish at the extremity of the veins. Under side brownish-grey, tinged with pink; the discal spots more distinct than on the upper side. Length of the body 15 lines. Expanse of the wings 38 lines."

PSEUDOHAZIS HERA. Harris.

Saturnia Hera, Harris, Rep. Ins. Mass., p. 286, (1841); *Morris*, Syn., p. 221, (1862); (*Hemileuca H.*) *Packard*, Proc. Ent. Soc. Phil., III, p. 383, (1864); (*Pseudohazis H.*) *G.-R.*, Ann. Lyc. Nat. Hist. N. Y. VIII, p. 377, (1866).

Hemileuca Pica, Walker, Cat. Lep. B. M. VI, p. 1318, (1855); (*Saturnia P.*) *Morris*, Syn. p. 222, (1862); (*Hemileuca P.*) *Packard*, Proc. Ent. Soc. Phil., III, p. 383, (1864).

Var. Eglanterina, Bdl., (*Saturnia E.*) Ann. Soc. Ent. Fr., 2me Ser. X, p. 323, (1852); (*Hemileuca E.*) *Wlk.*, Cat. Lep. B. M. VI, p. 1318, (1855); (*Telea E.*) *H.-S.*, Lep. Exot., p. 445, (1855); (*Saturnia E.*) *Morris*, Syn., p. 222, (1862); (*Hemileuca E.*) *Packard*, Proc. Ent. Soc. Phil., III, p. 383, (1864); (*Pseudohazis E.*) *G.-R.*, Ann. Lyc. Nat. Hist. N. Y. VIII, p. 377, (1866).

Var. Nuttalli, Streck., Lep., Rhop.-Het., p. 107, (1875).

Var. Arizonensis, nobis.

*PLATE XV, FIG. 8, PSEUD. EGLANTERINA, Bdl. ♂, California.
          FIG. 9, PSEUD. EGLANTERINA. ♂ aberration, California.
          FIG. 10, PSEUD. HERA, Harris, (*Pies, Wlk.*) ♂, Utah.
          FIG. 11, PSEUD. HERA, yellow var. ♂, Colorado.
          FIG. 12, PSEUD. HERA. ♂ black aberration, "Rocky Mts."
          FIG. 13, PSEUD. NUTTALLI, Streck. ♂ Rocky Mts., head of Snake River.
          FIG. 14, PSEUD. NUTTALLI, ♀ Rocky Mts., head of Snake River.

The ♀ of *P. Hera*, the earliest described of the above forms, was figured by Audubon on plate 359 in Vol. IV of his great work on the Birds of N. Am. and on plate 53, Vol. I, of his later smaller edition, but no name or word regarding the insect appeared in the text. Dr. Harris afterwards described and named the species from the example that had furnished Audubon with the original of his figure, which was in the possession of Mr. Ed. Doubleday of Epping, Esq. This, as well as other examples, were taken by the ornithologist Nuttall in the Rocky Mts. in 1836. Audubon's figure is apparently a female to judge from the antennæ, though Harris describes it as a male, and states that the other figure " is probably the female of the preceding, apparently differs from it only in being of a deep Indian-yellow colour and in having the crescent in the saddle of the kidney shaped spots very distinct, whereas in the male it is almost obsolete." This latter figure however is more likely the female of one of the other forms, *Eglanterina* or *Nuttalli* probably, as I have seen and examined a number of *P. Hera* from Utah in which the females as well as the males have the wings either quite white or else white with a very faint yellowish tint or cast. This white form appears to be indigenous to the salt regions of Utah and nowhere else. I have only figured the male, but if those of my readers, who have not easy access to Audubon's work, will glance at my figure (14) of *P. Nuttalli*, ♀, and imagine the ground colour of all wings white and the abdomen ringed with black they will have a very correct idea of the female of the form or var. *Hera*, Harr.

Both sexes of the Colorado variety of *P. Hera* have all the wings yellow, the primaries not however as deep in colour as the secondaries and body; the male and female present scarcely any difference in the markings or outline of wings. Fig. 11 on plate XV represents the ♂ of this form; fig. 12 on same plate depicts a melanotic aberration of *P. Hera*, the original of which, taken by Mr. Nuttall in the "Rocky Mts." in 1836, is now in the coll. of Mr. Titian R. Peale, to whose goodness I am indebted for the privilege of figuring it, as well as Nos. 13, 14, which illustrate both sexes of *P. Nuttalli* described on page 107 of this work. At the time I designated this latter as a distinct species I considered the total absence of the black bands on the abdomen as entitling it to have some claims as such, but lately having examined a number of both sexes of an intermediate form received by Mr. Neumoegen from Arizona I am convinced that *P. Nuttalli* is but an extreme variation after all. Both sexes of these Arizona examples just alluded to (which came into my hands too late to introduce on plate XV) resemble closely in outline of wing, color and markings, *P. Nuttalli* ♀ (fig. 14, plate XV) with the exception that the abdomen over half the length from the thorax is banded with black, the two bands nearest the thorax being broader and thence out becoming narrower until but few traces are noticeable on the terminal segments; the black marks on wings are but very little heavier on the male than in the female. This form I would propose to designate as variety *Arizonensis*; it seems to be intermediate between the Colorado form of *P. Hera* (fig. 11) and *P. Nuttalli*.

The best known and by far the commonest is the Californian form *Eglanterina* in which the upper surface of primaries is more or less suffused with pinkish; it is very variable in the black markings; in some instances being almost as heavily blacked as the variety of *P. Hera* (fig. 12), in others it is scarcely more so than in *P. Nuttalli* ♀ (fig. 14); nor is this disunion of the black confined to the females only as I have males with as little black on as any female I have yet seen, and even less. An extreme case in point is the male aberration (fig. 9) in which the black marks are almost totally obliterated on both surfaces. I have three of this type, all five unblemished examples, but in neither of the remaining two are the dark bands and spots so completely obscured in the one figured. The Californian examples are not even constant in outline of wing, some being narrow winged like form *Hera*, others broad as in fig. 8, pl. XV; in fact this is fairly demonstrated by comparing the outline of fig. 8 with that of its aberration fig. 9, which presents an entirely different shape of wing. Though the upper surface of primaries is more generally flesh coloured or pinkish, this is not always the case, as I have seen and possess examples of both sexes in which the primaries are the same yellow colour as the secondaries, and others in which part are yellow and part flesh coloured; in fact the number of variations and sub-variations of this and the other form is truly wonderful; I could easily have filled a plate with these had it been worth the while, but I trust I have figured enough to illustrate the fact that all are but forms, or sports, or variations of one species.

The two examples in my collection taken by Nuttall in 1836 in the "Rocky Mts." are the ordinary form of *Eglanterina*; in the same expedition Nuttall also took the three insects figs. 12, 13 and 14 on plate XV, as well as the originals of Audubon's figures, one of which furnished the type for Harris' *Hera*.

HYPERCHIRIA VARIA. Walker.

Cat. Lep. B. M. VI, p. 1278, (1855).

(PLATE XV, FIG. 15 ♂ aberr., 16 Hermaphrodite.)

This common but none the less beautiful species is subject to many and most startling variations, two of which I have figured on plate XV, of which, as well as of several others, I will proceed to make further note. Fig. 15, a ♂, has on upper surface all the usual brownish marks of primaries, and the red abdominal margin and submarginal line of secondaries replaced by white or very pale yellowish. Beneath the primaries are yellow with no other colour or mark save a black discal spot papilled with white; the secondaries are yellow from where the transverse line usually is to outer margin, interior to this yellow part the wing is yellowish white. Two examples of this aberrant were raised from a large brood, the remaining members of which were all of the ordinary form.

Fig. 16 represents one of those incomprehensible freaks, a partial hermaphrodite. The left antenna is male, the right one female; the thorax above is yellow like the male, with several isolated patches of the reddish female colour; beneath the thorax in front is red, rest yellow; legs reddish; abdomen above and below yellow and in all appearance is that of a male. The left primary is male except a small patch on interior margin, not far from the inner angle, and a not broad mark extending along inner margin from the aforesaid patch inwards to the transverse sub-basal line; the right hand primary is female excepting the parts along inner margin which on the left wing are female are here male, also at the inner angle is an irregular triangular patch of the yellow male colour. Secondaries on upper surface are both alike and appear to be, from the produced abdominal angle, male. Under surface all wings yellow and in all other aspects like the normal males with the single exception that the costa of the right hand secondary is bordered

* On this accompanying plate (XV) figs. 8 and 9 are marked by mistake as *Hera* instead of *Eglanterina*, and figs. 10, 11 and 10 as *Pies* instead of *Hera*, so that they should read at bottom of plate correctly thus: 8, *Pseudohazis Eglanterina*, Bdl., ♂; 9, *P. Eglanterina*, ♂ aberr.; 10, *P. Hera*, Harris, ♂; 11, *P. Hera*, ♂ var.; 12, *P. Hera*, ♂ aberr.

whole length by a broad nearly even band or margin of reddish brown, the same colour as on the under side of the usual female form.

Although there was some slight disarrangement in the general sexual make-up of this individual still it had sense enough to assert its manhood, though by so doing it sacrificed both liberty and life: it flew into an open window attracted by a captive virgin female which had that day emerged from the chrysalis.

For both the above remarkable insects, as well as numberless other kindnesses extending through long years, am I indebted to my old friend Herman Sachs who bred the first and captured the second at his residence in Hoboken, New Jersey, some years ago; and as I now gaze at them many and many a pleasant recollection arises of the days of "auld lang syne."

From the state of Maine I received a male example in which the median or second submarginal line is crimson like the outer one instead of black as in all other examples I have ever seen.

Another male, from Maryland, has the ocellus of secondaries entirely black without any shining blue, or white discal mark.

Two females have the lower edge of the ocellus of secondaries resting on the black transverse line.

One female has upper surface of primaries purplish grey, transverse lines and discal marks white. Ocellus of secondaries of immense size filling nearly the whole space interior to the black line; under surface of this example is greyish yellow. It is from Ohio.

A female in Mus. Comp. Zool., Cambridge, is of small size and has the ocellus of secondaries suffused and irregularly spread over a great portion of the wing.

In collection of Mr. J. Meyer of Brooklyn is a female of very large size in which the upper side primaries are ornamented with rays of darker colour which start from the base and diverge outwards wider and wider until their points reach nearly to the exterior margin.

In Mr. Neumoegen's coll., also in my own, are female examples with the upper side of primaries pale grey or ashen; one of these was taken near Morristown, N. Jersey, the other I bred from larva found here.

HYPERCHIRIA LILITH. Nov. Sp.

(PLATE XV, FIG. 17 ♀.)

FEMALE. Expands 2⅞ inches.

Head, thorax and legs dark Indian red; abdomen same colour with the exception that the segments above are edged with a somewhat yellowish hue.

Upper surface; primaries dark reddish brown more inclined to red at the base, and somewhat paler and tinged with grey at outer margin; discal mark scarcely discernible; a darker inconspicuous transverse median line or rather shade. Secondaries brownish not as dark as the primaries, broadly bordered at abdominal margin with dull crimson; a large central ocellus formed by a distinct black ring enclosing shining blue or steel colour and with a small white discal mark, the black and blue do not merge into each other as in *H. Varia* but the black ring is not wide and is clear and distinct on its inner edge as on the outer; outside the ocellus is a broad black line, between this latter and the exterior margin is another broader line of reddish brown; the exterior margin is also bordered with the same colour.

Under surface dark Indian red shaded towards exterior margins with brownish; transverse lines as in *H. Varia*; on primaries a very large black oval discal spot with small white round spot in centre, on secondaries a small white discal spot.

Hab. Georgia.

I have only had the opportunity of examining the females, of which there were eight or nine, all bred at one time; the male, of which there were only a few examples, was described to me as being much like the female, but the primaries darker or more greyish and the secondaries paler. Of the eight or nine females bred I have examined six, and all are remarkably alike, presenting scarcely any difference from each other in size, shape or colour. The wings are broader and shorter than in *H. Varia*, which is the nearest allied species. My friend did not take any particular note of the larva more than that they were of the *Varia* type and that he found them feeding on some small weed which soon gave out, he then fed them on wild cherry which they ate readily until they were ready to change into the chrysalis state. This could scarce be a local form of *H. Varia* as the latter species I received from the same locality and bred at some time in large numbers of both sexes which were in all respects the same as those found in Penna., N. York and other more northern localities. I have every hope that I will in a future plate be able to depict the male of this beautiful insect as my informant is confident that he has in chrysalis state another brood of it.

I have named this species after a lady of considerable celebrity in the olden time long ago, to wit: No less a personage than Adam's first wife Lilith, the mother of the giants, who was eventually turned into a demon, as has been not unfrequently the case with members of her sex in subsequent times, not long ago.

ON THE NORTH AMERICAN SPHINGIDÆ IN MR. A. G. BUTLER'S REVISION OF THAT FAMILY PUBLISHED IN TRANS. ZOOL. SOC., LOND., VOL. IX., PART 10, (1877).

Among the various works of interest to the Entomologist that have of late years appeared, two come in for considerable attention: the "Sphingidæ, Sesiides, Castniidæ," by Dr. Boisduval (1874), and the "Revision of the Sphingidæ," by Mr. Butler (1877); as the latter is later and in a measure a revision of the former, I will more particularly direct my attention and remarks to its contents, contenting myself for the present by expressing my delight at the wonderful correctness of the drawing and beauty of colouring of the figures in the work of Dr. Boisduval.

Page 517, No. "2 *Lepisesia victoria*, Grote Bull. Buff. Soc. Nat. Sci., p. 147, (1874). British Columbia." Was described from a faded example of *Pterogon Clarkiæ*, Edl. In appendix p. 634 Butler makes the correction in a measure, "Said to be identical with *Pterogon darkiæ* of Boisduval; see Bull. Buff. Soc., ii, p. 225."

On p. 112 of this work I have slightly alluded to the above.

ON SOME N. AMERICAN SPHINGIDÆ IN A. G. BUTLER'S REVISION.

Page 518, No. "2, HEMARIS FUMOSA. *Macroglossa fumosa*, Strecker, Lep. Rhop. et Het., p. 93, 1874. Albany. Allied to *H. diffinis*; Grote believes it to be *H. tenuis*, in which the scales on the pellucid area of the wings are still adherent." Grote is right in his belief.

Page 519, No. 6, "*Sesia thetis*, Grote and Robinson, Trans. Am. Ent. Soc., Vol. I, p. 3, pl. 6, fig. 36, (Jan., 1868)." Should be p. 325, pl. 6, etc., not p. "3."

Page 521, No. "16, HEMARIS AXILLARIS. *Sesia axillaris*, Grote and Robinson, Trans. Am. Ent. Soc., ii, p. 180, (1868). HEMARIS AXILLARIS, Grote, Bull. Buff. Soc. Nat. Sci., p. 6, pl. I, fig. 9, (1873). *Sesia gratci*, Butler, Ann. & Mag. Nat. Hist., Ser. 4, Vol. XIV, p. 365, (1874). Texas, (*Belfrage*)."
No. "17, HEMARIS MARGINALIS. Grote, Bull. Buff. Soc. Nat. Sci., p. 6, pl. 1, fig. 10, (1873). Michigan. (*Strecker.*)"
These two are unquestionably the same species: the type of *Marginalis* has the dentations on inner edge of margin of primaries not as deeply cut as in the type of *Axillaris*, but in a number of examples all the gradations between the two extremes can be found and in one example which I possess the teeth are prolonged inwardly even more than in Grote's figure in Buff. Bull.

P. 522 No. "20 HEMARIS BUFFALOENSIS. *Hæmorrhagia buffaloensis*, Grote and Robinson, Ann. Lyc. Nat. Hist. New York, Vol. viii, p. 437, pl. 16, figs. 18, 19, (1867). Buffalo. Very closely allied to, if not identical with *H. rufocaudis* of Walker (? Kirby); the body, however, seems greener in colouring, and the cell of primaries less open."
This is, as the latter part of the above quotation would lead us to infer, indentical with *Rufocaudis* of which *Sesia Uniformis*, G.-R., is also a synonym.
On the same principle that certain individuals of this *Rufocaudis* were erected into the species *Buffaloensis*, all those found in Reading would be designated as *Reafingensis*, those from Kutztown as *Kutztownensis*, those from Folly-hill as *Folly-hillensis*, and so on.
The synonymy of this species is:
MACROGLOSSA RUFICAUDIS, KIRBY, (*Sesia R.*), Faun. Bor. Am. IV, 303, (1837). *Walker*, C. B. M. VIII, 82, (1856). *Morris*, Cat. Lep. N. Am. 17, (1860). Syn. Lep., 149, (1862). *Cooper*, Can. Ent. IV, 205, (1872).
Hæmorrhagia Rufocaudis, Grote & Robinson, Proc. Ent. Soc. Phil. V, 149 & 175, (1865).
Hemaris Rufocaudis, Butler, Trans. Zool. Soc. Lond. IX, 521, (1877).
Hæmorrhagia Buffaloensis, Grote & Robinson, Ann. Lyc. Nat. Hist. N. Y. VIII, 437, t. 16, figs. 18, 19, (1867), List Lep. N. Am., 3, (1868). Grote. Bull. Buff. I. 18, (1873), II, 224, (1875).
Hemaris Buffaloensis, Butler, Trans. Zool. Soc. Lond. IX, 522, (1877).
Sesia Uniformis, Grote & Robinson, Trans. Am. Ent. Soc. II, 184, (1868). *Lintner*, 23d Rep. N. Y. State Cab. Nat. Hist., 172,(1872).
Hæmorrhagia Uniformis, Grote & Robinson, List Lep. N. Am., 3, (1868). *Grote*, Bull. Buff. I. 18, (1873), II, 224, (1875).
I would further refer the student to page 109 of this work where I have dwelt at some length on this species.

No. "21 HEMARIS FUSCICAUDIS. *Sesia fuscicaudis* Walker, Lep. Het. viii, p. 83, No. 6 (1856). *Hæmorrhagia fusicaudis*, Grote & Robinson, Proc. Ent. Soc. Phil. vol. v, p. 174 (1865).
Georgia (*Abbot*). Type, B. M."
This is the southern form of *Thysbe*, from which it differs in nothing except its greater size; the absence of the greenish colour on the sides of the two last segments of abdomen is not specific as I have taken as many of *Thysbe* in Pennsylvania destitute of this greenish colour on abdomen as I have with it, and one example in my cabinet has all the segments of the abdomen dark red and only the thorax green; otherwise it is the same as the ordinary *Thysbe*.

On p. 519 No. "3 Hemaris palpalis Grote" from British Columbia.
No. 7 "Hemaris matuthetis" Butler from Texas.
Of "*Hemaris rubens*, H. Edwds." from Oregon and "*H. cynoglossum* H. Edwds." from California and Vancouver's Island. Are all unknown to me save through the author's descriptions.

Page 529, No. "43, MACROGLOSSA ERATO, Bdl., Lep. Cal. in Ann. Soc. Ent. Belge., xii, p. 65, no. 67 (1868)."
Page 536, "*Eupraserpinus phæton*, Grote and Robinson, Proc. Ent. Soc. Phil. vol. v, page 178, (1865). California (*Weidemeyer*)."
Further on page 636, "*Eupraserpinus phæton* of Grote is said to be identical with *Macroglossa erato* of Boisduval; *see* H. Edwards in Proc. Cal. Acad. Sc. 1875, p. 3."
On page 113 (foot note) and page 124, I have explained fully in regard to the confusion of names in this species which is fig. 1 on plate XIV.

Page 560, Deilephila Galii and D. Chamænerii are cited as separate species, but the author adds: "according to Strecker (Can. Ent. IV, p. 200), *D. chamænerii* is *= D. galii*," and I must here repeat that they undoubtedly are but one species, the only difference between them being in the name.

Page 560 No. "8 *Deilephila intermedia*, Kirby, Fauna Amer.-Bor. vol. iv, p. 302 (1837). "Canada" (*Kirby*)."
This also I believe to be nothing more than *Galii* (*Chamænerii*, Harr.).

Page 574 No. "2, *Philampelus linari* Grote & Robinson Proc. Ent. Soc. Phil. Vol. v, pp. 157, 179, 182, pl. 3, fig. 3, (1865). *Sphinx vitis*, Cramer, Pap. Exot. vol. iii, pl. 268, fig. E (1782).
Dupo vitis, Hübner, Verz. bek. Schmett. p. 137, no. 1496 (1816).
Philampelus vitis, Walker, Lep. Het. viii, p. 176, no. 4 (1856).
Philampelus fasciatus, Grote, nono Cub. Sphi., Proc. Ent. Soc. Phil. Vol. v, pp. 59, 84, (1865).
Mexico (*Herwig*); Haïti (*Ouning of Tweedie*); ——? (*Strauss*)."
I have an example of this species taken by Boll near San Antonio, S. W. Texas; Mr. J. Doll also took it in Florida. Some examples which I received from Surinam and the upper Amazon are much larger than the West Indian ones or those from Florida and Texas, expanding 4½ to 4¾ inches; these S. Am. examples are also of a general darker hue, the dorsal stripe of abdomen being not particularly noticeable; the greenish of upper side of secondaries more inclined to grey, the rose-coloured inner margin darker, the pale lines and bands of primaries clouded or shaded with brown, and the veins are accompanied with white to the extreme edge of the exterior margin, whilst in the Cuban and U. S. examples they extend only to the grey border of exterior margin. They remind one forcibly of Ménétriés' figure of *P. Strenua* (Cat. Mus. Petrop. Lep. II, f. 12, 1857) and it is only by actual comparison that the mind can be disabused of the idea that they are identical.

The most striking points of difference is in the absence in *P. Strensu* of the broad pale band that crosses the upper surface of primaries lengthwise from base to the great pale medial band in *P. Linnei*, also in the absence of the paler border of exterior margin.

No. "3, *Philampelus hornbeckiana*, Harris, Cat. N.-Am. Sph. Sill. Journ. p. 299, (1839). "St. Thomas, West Indies." *Harris*. Apparently allied to the preceding."

It is quite likely that this and *P. Linnei* are the same species but as Dr. Harris' type is not to be found and his description not fully agreeing with *P. Linnei* it will have to remain one of those plagues to Lepidopterists, a description without a type, unless perchance time or some accident solves the riddle.

Page 575, No. "9, PHILAMPELUS PANDORUS. *Daphnis pandorus*, Hubner."

The author should have added to his synonyms of this species *Philampelus Satellitia*, Harris, instead of citing the latter as *Sphinx Satellitia*, Linn.

Page 578, No. "10, PHILAMPELUS LABRUSCÆ. *Sphinx labruscæ*, Linnæus, Mus. Lud. Ulr. p. 352, (1764)."

This species has to my knowledge twice been taken in the United States—once in New Jersey and once in Florida.

Genus 18, No. "1, PACHYLIA FICUS. *Sphinx ficus*, Linnæus, Mus. Lud. Ulr. p. 352, (1764)."

Has been captured in S. W. Texas.

With the Smerinthus Mr. Butler has taken the same if not more liberty than did Grote, making out of every group a separate genus, though sometimes the species even in these limited genera are not happily grouped. I cannot possibly see why *Smerinthus Querens*, W. V., should be associated in the genus *Mimas* with *S. Tilia*, L. and *S. Decolor*, Wlk., neither of which does it in any way closely resemble, whilst such species as *S. Dyras*, *S. Geschkeritschii*, *S. Albiornus*, etc., which it closely resembles, are made to constitute the genus *Triptogen*, and I fear it is tript and tript again all through in these Hubnerian–Grotetian–Butlerian coitus-generic arrangements which seem to be the only exceptionable points of any moment in the work I am now examining. In the aforesaid genus *Triptogen* is placed our *S. Modesta*, of which the author says: "this is unquestionably the proper place for this species," to which no particular objection can be made as it is as near to the *Dyras* group or nearer than to any other, but why, I would again ask, is *S. Querens*, which resembles *Dyras* and allies much more than does *Modesta*, removed so far away, with four genera intervening?

Page 590. Is described under the name of *Cressonia Robinsonii*, what is supposed to be a new species allied to *S. Juglandis*, Ab.-S. The author says: "We have a pair of what seems to be a second species; it is of a greyer tint and half as large again, the transverse lines wider apart, and the primaries with central band not darkened on the inner margin;" and further suggests "it is quite possible that the above may be a large form of *C. Juglandis*; but it differs noticeably from our six examples of that species."

I do not know of anything agreeing with the above description in any American collection. Is Mr. Butler quite sure that "New York" is the true locality of this type?

No. "3, CRESSONIA PALLENS. ♀ *Smerinthus pallens*, Strecker, Lep. Rhop. & Het. pt. 7, p. 54, pl. vii, fig. 14, (1873), Texas."

To which is appended the following foot note: "Mr. Grote is confident that this is only a variety of *C. juglandis*. It looks quite distinct."

Mr. Butler's only ground for stating that "it looks quite distinct" is from examination of my figure, he being in England and the type having never left my cabinet. But how Grote came to be so confident as to assert the species was only *Juglandis* it being impossible for him ever to have seen the type as none but gentlemen enter my house.

Mr. Butler says on page 590, "I find that dissimilarity in the outline of wings is almost always accompanied by modification of the discocellular nervules, which would be sufficient in the eyes of any Lepidopterist to warrant generic separation," and on same page commences his genus *Paonias*, comprised of two species, *Excaecatus* and *Myops*, showing about as much dissimilarity in the outline of wings as can probably be found between any of the species among all the Smerinthus.

Page 591, *Asylus*, which is closer to *Myops* than any other species, is placed in another genus, the *Calasymbolus* of Grote. In regard to my figure the author says, "Strecker's figure of this species has the two opposite primaries rather different in outline; but judging from Drury's figure, I have little doubt as to its genus." As regards this difference of outline he is correct; so was I in my drawing, for on examining the example from which I drew the figure I find the same difference in outline exists as in the figure which I faithfully copied.

In this same genus *Calasymbolus* along with *Asylus* are placed *Geminatus*, *Ceryii*, *Oxeus* and *Kindermanni*, which four species bear no particular resemblance to *Asylus* in outline of wing, colour, or anything else except in the common fact that all have an ocellus on hind wings. This extension of *Calasymbolus* was too much for even Grote who in Can. Ent. IX, p. 132, says: "I am not now prepared to accept the extension of *Calasymbolus*;" but to make amends Mr. Butler here makes a new genus which he calls *Euomerinthus* for the reception of *Geminatus*, in order that he can say *Euomerinthus Geminatus*, Grote, instead of *Sm. Geminatus*, Say.

In his arrangement of species Mr. Butler has No. 2 *Geminatus*, No. 3 *Cerisii*, and No. 4 *Oxeus*. Why *Ceryii* was placed between *Geminatus* and *Oxeus* I cannot imagine, as *Oxeus* is so close to *Geminatus* that were it not for the difference in the first principal transverse line or shade on primaries, which is strongly angulated in the latter, they might be considered identical.

The variety of *Geminatus* figured in Drury and there named *Jamaicensis*, Mr. Butler has cited erroneously as a synonym of *S. Myops*.

Page 605, No. "3, DILOPHONTA MERIANÆ. *Erinnyis meriana*, Grote, Proc. Ent. Soc. Phil. v, pp. 75 and 168, pl. 2, fig. 2, (1865)."
"Tropical insular and continental districts." (*Grote*.)

I have received examples of the above bred from larvæ found in S. W. Texas near San Antonio and New Braunfels.

Page 613, No. "2, *Sphinx leucophæata*, Clemens, Journ. Acad. Nat. Sci. Phil. 1859, p. 168. *Sphinx lugens* (part.), Walker, Lep. Het. viii, p. 219, No. 11, (1856). Oaxaca, Mexico. (*Bartung*)."

S. Leucophæata is unknown to American Lepidopterists further than by Clemen's description. For my part I have little doubt but that it is a synonym of *S. Lugens*, Wlk., although of this latter Mr. Butler says, "although coming from the same locality as the preceding, and very like it in its general character, I believe this species to be quite distinct. It is altogether shorter, broader and darker, and has the pale bars of secondaries much narrower and whiter."

By whom were the examples in the British Museum, cited by Mr. Butler as *Leucophæata*, identified?

142   ON SOME N. AMERICAN SPHINGIDÆ IN A. G. BUTLER'S REVISION.

No. "8 *Sphinx occidophos*, H. Edwards, Proc. Calif. Acad. Sci. v. p. 109 (1874). California."
The above is a synonym of *Chersis*, Hub. The Californian and Oregon examples are not as large as those from the Atlantic State.

Page 619. "12 SPHINX ? LANCEOLATA. *Sphinx lanceolata*, Felder, Reise der Nov., Lep. iv. tab. lxxviii. fig. 3 (Nov., 1874). Guatemala and Mexico.
Seems allied to *S. chersis*, but may possibly belong to the genus *Pseudosphinx*; without seeing the insect it is impossible to decide."
I have an example in my collection from Panama which agrees exactly with Felder's splendid figure; it is close to *S. Chersis*.

On page 621 No. "3 LINTNERIA EREMITOIDES. *Sphinx eremitoides*, Strecker, Lep. Rhop. and Het. p. 93 (1874)."
This is a synonym of *S. Lugens*. Mr. Butler says, "Mr. Grote thinks it probable that *S. eremitoides* is = *S. Lugens* of Walker; but, judging from Mr. Grote's previous paper on the Sphingidæ! I am doubtful whether he knows the *S. lugens* of Walker. It is certain that Clemens did not; for he separated it by a wide interval from his *S. leucophæata*."
Lugens was unknown to American entomologists until after I redescribed it on p. 93 under the name of *Eremitoides*; it was Grote's ignorance of the species that led me into the error, as in the collection of Lepidoptera made by Grote and the late Coleman Robinson was an example of *Sphinx Justiciæ*, Bdl., erroneously labeled *S. Lugens*, Wlk., and inasmuch as Grote and his collaborator made their identifications of the Walkerian species by comparison with the types in the British Museum, during a visit to England, I had not the remotest idea that they would blunder on so large and conspicuous an insect as *Lugens*.

Mr. Butler makes a new genus which he calls *Lintneria* for the reception of *Sphinx Eremitus*, Hb. and *Sph. Perelegans*, Hy. Edwds., though the latter he prefixes with a ?. *Perelegans* is nearer in general appearance to *Chersis* Hub., and also to *Drupiferarum*, Ab.-S., than to any others, as I have shown on page 106.

"*Ceratomia lugens*, Grote," is closely allied to *Daremma Undulosa*, Wlk., and not at all to *Ceratomia Amyntor*, Hub., from which it differs remarkably in both the larva and winged state, as will be seen by referring to page 127 of this work where I have treated on this species at length.

In appendix 1, p. 629:
———— Canadensis, Boisd.* p. 93, No. 29, = ?, *Sphinx leucophæata*."
The species figured on plate XIII and described on page 115 under the name of *Sph. Plota* may be the same as *S. Canadensis*; if such should prove to be the case the name given by me must fall. Dr. Boisduval's having priority by several years.

Sphinx strobi, Boisd. figured pl. 5 fig. 3.
———— *cupressi*, Boisd. p. 102, p. 41, pl. 2, figs. 3–5."
Neither of these are in any American collection as far as I am aware of, nor are they known here save through the figures of Boisduval. They appear to me to belong to the *Pinastri* group. The habitat of *S. Cupressi* is given as Georgia.

"*Sphinx catalpæ*, Boisd. p. 163, no. 42, pl. 2. figs. 1, 2"
Prof. C. V. Riley has found the larva of this species on Catalpas, but so far has not been successful in securing the imago. Boisduval states, on the authority of Abbot, that the larva was found on *Catalpa Cordifolia* in Georgia.
He also says his description was drawn up from a good figure by Abbot and the notes of Leconte; he had failed to receive the species owing to the death of Abbot.
The insect I believe is unrepresented in American collections; the figure of the imagine on Boisduval's plate looks a good deal like something between *S. Undulosa* and *S. Hageni*.

Page 634, "*Macroglossa æthra*, Strecker, Lep. Rhop. and Het. i, p. 107, (1875); pl. xiii, fig. 2, (1876)."
In my description of this species on page 107 I have stated that the type was from Montreal, Canada, which it seems Mr. Butler overlooked. I have since received other examples from the same place.

Page 635, "HEMARIS RUFICAUDIS, (synonym). *Macroglossa rufecaudis*, Strecker, Lep. Rhop. and Het. i, pl. xiii, fig. 1, (1876)."
"Synonym" of what?

Page 636, "DEIDAMIA INSCRIPTA. *Pterogon inscriptum*, Strecker, Lep. Rhop. and Het. pl. xiii, fig. 8, (1876)."
This belongs in the same genus with the Russian *Gorgoniades*, Hub., and wherever the one is placed the other likewise belongs.

Page 637, "Genus ELIBIA, Walker. ELIBIA VERSICOLOR. *Darapsa versicolor*, Strecker, Lep. Rhop. and Het. i, pl. xiii, fig. 9, (1876). It is evident from Strecker's figure, that this species has been erroneously referred to the allied genus *Otus*."
What in all the earth could move Butler to place this species in the genus *Elibia* is beyond all comprehension. In *Elibia* are but two species, both from India, *Elibia dolichus*, West., and *Dolichoides*, Feld.; with the first only am I acquainted in nature and the only point in common between it and *Versicolor* is the pale dorsal line, which decoration is also shared in by ground squirrels (*Tamias*) and garter snakes, and the most rabid genus-fabricator would scarce on that account place these animals in one genus.
Mr. J. Meyer of Brooklyn, N. Y., who was the first person that bred *Versicolor* from the larva, informed me that the latter are, with the exception of being a little larger, almost precisely like those of *D. Myron*, Cram. And there cannot be the least doubt but that *Versicolor* belongs to the same genus with *Chœrilus* and *Myron*.
Darapsa was used to replace *Otus* on account of the latter being preoccupied in ornithology.

Page 638. The author says of *Chærocampa Procne* (fig. 10, plate XIII of this work) from California: "It is much more probable that this is an Asiatic species allied to *C. lucasii*."
The example from which my figure was drawn agrees in all particulars exactly with Clemens's description in Jnl. Acad. Nat. Sc. Phil., 1859. I obtained it along with the collection of Rev. Dr. J. G. Morris some twenty years since; it had a small slip on the pin with the locality "S. California" written thereon, and I candidly confess that I still think this locality the correct one; there is no

---

* Lep. Het. 1, Sph., Sesiides, Castnides, Sphex a Buffon.

ON SOME N. AMERICAN SPHINGIDÆ IN A. G. BUTLER'S REVISION.

reason why it should not be; even if the insect is allied to an Asiatic species, is it more wonderful to find a species of *Chærocampa* on the Pacific coast allied to an Asiatic one than to find such closely allied things as *Smerinthus Cerisy* and *S. Geminatus*, the first in Asia, the latter in the Atlantic United States, or *Parnassius Intermedius* and *P. Smintheus*, which I believe are identical, the former in the Altai Mts. and the latter in the Mts. of Colorado?

Page 642, "HYLOICUS SANIPTRI. *Sphinx Saniptri*, Strecker, Lep. Rhop. and Het. i, pl. xiii, fig. 18 (1876)."

I am now convinced this is identical with *Sph. Pinastri*. My principal grounds, apart from its being found in the United States, which is of small moment, was the absence of the broad dark transverse shade of primaries, but I have since received examples from Germany which are also destitute of this band or shade.

Attached to this monograph are five coloured plates representing various species, mostly new, of Sphingidæ, and also a number of larvæ.

I cannot say I am enamoured with the frightful number of genera adopted, which is the one objectionable feature to this otherwise excellent work, but it appears Mr. Butler has equal want of affection for the paucity of genera accepted by myself, for he alludes pleasantly on p. 621 to "Mr. Strecker's incomprehensible affection for unmanageably extensive genera."

But in truth it is greatly to be deplored that the plan (insanity offspring of Grote's vanity) of dividing and subdividing so natural a genus as Smerinthus should be here adopted; but I have treated fully this subject on pp. 52, 55, as well as elsewhere in this volume. What better proof of the compactness of a genus is required than the knowledge that two of its most dissimilar-looking species will copulate and produce hybrids as in the case of *S. Populi* and *S. Ocellata*.

In speaking on p. 613 of *Dilutia Brontes*, Dru., (*Sphinx Cubensis*, Grote, is a synonym,) the author expresses himself in the following language which certainly will meet the sincere approval of all true lovers of science. He there says: "I cannot but regret that Mr. Grote has thought it necessary to add to the synonymy by proposing names for species before they were required. It is true that he might otherwise have been superseded; but as a fact it does not matter who names a species, so long as the name given be euphonious, whilst on the other hand a cumbrous synonymy is a great evil."

In the Can. Ent. IX, p. 130–133, Grote gave what he calls a "Notice of Mr. Butler's Revision of the Sphingidæ," though as usual it is a dissertation on himself, in which the first seven lines are devoted to praising Mr. Butler, being prefatory to the remaining seventy odd which are mainly devoted to the highly gratifying and instructive purpose of praising himself.

November, 1877.

www.ingramcontent.com/pod-product-compliance
Lightning Source LLC
Chambersburg PA
CBHW021823230426
43669CB00008B/852